Springer Tracts in Modern Physics
Volume 239

Springer Tracts in Modern Physics

Springer Tracts in Modern Physics provides comprehensive and critical reviews of topics of current interest in physics. The following fields are emphasized: elementary particle physics, solid-state physics, complex systems, and fundamental astrophysics.
Suitable reviews of other fields can also be accepted. The editors encourage prospective authors to correspond with them in advance of submitting an article. For reviews of topics belonging to the above mentioned fields, they should address the responsible editor, otherwise the managing editor.
See also springer.com

Alexander Petrovich Potylitsyn
Mikhail Ivanovich Ryazanov
Mikhail Nikolaevich Strikhanov
Alexey Alexandrovich Tishchenko

Diffraction Radiation from Relativistic Particles

Springer

Prof. Alexander Petrovich Potylitsyn
Tomsk Polytechnic University
Lenin Ave. 30
634050 Tomsk
Russia
pap@interact.phtd.tpu.ru

Prof. Mikhail Ivanovich Ryazanov
National Research Nuclear University
MEPhI
Kashirskoe Sh. 31
115409 Moscow
Russia
ryazanov-m@mail.ru

Prof. Mikhail Nikolaevich
Strikhanov
National Research Nuclear University
MEPhI
Kashirskoe 31
115409 Moscow
Russia
MNStrikhanov@mephi.ru

Prof. Alexey Alexandrovich
Tishchenko
National Research Nuclear University
MEPhI
Kashirskoe Sh. 31
115409 Moscow
Russia
tishchenko@mephi.ru

A.P. Potylitsyn et al., *Diffraction Radiation from Relativistic Particles*, STMP 239 (Springer, Berlin Heidelberg 2010), DOI 10.1007/978-3-642-12513-3

ISSN 0081-3869 e-ISSN 1615-0430
ISBN 978-3-642-26546-4 ISBN 978-3-642-12513-3 (eBook)
DOI 10.1007/978-3-642-12513-3
Springer Heidelberg Dordrecht London New York

Cover design: Integra Software Services Pvt. Ltd., Pondicherry

Printed on acid-free paper

Springer is part of Springer Science+Business Media (www.springer.com)

Foreword

This monograph presents the results of theoretical and experimental investigations of diffraction radiation accompanying the motion of charged particles near a conducting surface. The foundations of the theory of diffraction radiation are systematically presented for the first time. Particular attention is focused on radiation from periodic structures (Smith—Purcell radiation). Although some results refer to the nonrelativistic and moderately relativistic cases, the major part of the monograph is devoted to radiation generated by relativistic and ultrarelativistic charged particles and from bunches of these particles. The experimental results on diffraction radiation are presented in comparison with the theory. The problems of the application of diffraction radiation to non-invasive diagnostics of charged particle beams in modern accelerators are discussed. The monograph is addressed to researches in such fields as electromagnetic radiation and the accelerator physics. It also can be useful for higher year and postgraduate students.

Tomsk-Moscow, October 2009

Alexander Petrovich Potylitsyn
Mikhail Ivanovich Ryazanov
Mikhail Nikolaevich Strikhanov
Alexey Alexandrovich Tishchenko

Preface

Diffraction radiation appearing in the optical range when charged particles move in vacuum along a periodically deformed surface (grating) was observed for the first time in the early 1950s by S.J. Smith and E.M. Purcell [1]. This radiation was theoretically predicted by I.M. Frank in the early 1940s [2]. In the next two decades, this type of radiation was investigated in detail on beams of nonrelativistic electrons in the centimeter wavelength range. At the same time, theoretical methods were developed for calculating the characteristics of diffraction radiation for various configurations of measuring instruments and various parameters of a beam. A new field appeared in microwave electronics [3] and development of this field actively continues to date [4, 5].

At present, it has been shown that the intensity of visible and ultraviolet diffraction radiation generated by relativistic particles can be comparable with the intensity of transition radiation, which is widely used in high energy physics and accelerator physics. In contrast to transition radiation, diffraction radiation is not accompanied by the direct interaction of beam particles with a target and this circumstance opens prospects for non-invasive diagnostics of beams in modern accelerators.

Diffraction radiation can be used to analyze the structure of micron objects for which traditional X–ray methods are ineffective because of the absence of X–ray lenses with the required luminosity.

We point to the potentialities of coherent diffraction radiation generated by a beam of moderately relativistic electrons that are grouped into bunches shorter than 1 mm. In this case, the radiation spectrum covers the terahertz range, which is of considerable interest for applied investigations in physics, chemistry, and biology [6].

Diffraction radiation generated by relativistic particles is presented very briefly in modern monographs. Monographs [3] and [4] are completely devoted to diffraction radiation generated in periodic structures by nonrelativistic electrons. Among other problems, some applications of diffraction radiation generated by both nonrelativistic and relativistic particles were considered in monograph [5], but with emphasis on the specific features of microwave instruments (modulation of a beam in the process of its interaction with a target, comparatively low energies of the beam particles, and nonlinearity of physical phenomena), whereas the problems of diffraction radiation itself and modern experimental results in this field remained beyond the scope of that monograph. In each of more general monographs [7, 8] devoted to radiation generated by fast charged particles in a medium, diffraction radiation is discussed only in one section. At the same time, there are many theoretical and experimental studies, where the application of diffraction radiation to non-invasive diagnostics of electron beams and bunches is justified and the corresponding experimental methods are developed. This circumstance stimulates interest of both theorists and experimentalists in the properties of diffraction radiation.

Successes achieved in the past decade in this field of physics lead to significant progress in the investigation and application of diffraction radiation. In this monograph, we review the current status of theoretical and experimental investigations of diffraction radiation generated by ultrarelativistic particles.

Diffraction radiation is very close in nature to transition radiation. Indeed, both kinds of radiation can be treated as radiation from dynamical polarization currents induced in the target material by the Coulomb field of moving charged particles. However, in contrast to well-studied transition radiation, the situation with diffraction radiation is much more complicated, because the expressions for transition radiation (at least, widely known Ginzburg-Frank formulas) are derived for an infinite planar interface and for the far zone (wave zone or Fraunhofer zone in the optical terminology). However, diffraction radiation always implies a much more complex interface. As known, the strict solution of boundary value problems with complex boundary conditions involves significant mathematical difficulties. A number of physical approximations are usually used in real problems, e.g., in physical optics; they make it possible to obtain the results acceptable for applications (see, e.g., the wonderful monograph [9], where the approaches for solving diffraction problems of current interest in stealth technologies are analyzed). For this reason, many approaches presented in this book are different from each other and are based on some physical approximations. These approaches are obviously of interest for researchers working in this field and related fields; in addition, they are useful for young physicists to develop scientific insight into solving particular physical problems by the methods of classical electrodynamics.

Addressing to theoretical and experimental physicists, we aim to strictly justify the approaches used and briefly presenting recent experimental results. We hope that the monograph helps young researchers to acquire knowledge for active investigations in this field.

The problems concerning the effect of currents induced in the target on the characteristics of the beams are not included because of a limited volume of the monograph. Experimental data indicate that such a simplification is justified when energy loss to radiation is much lower than the kinetic energy of the beam. This simplification allows us, on the one hand, to avoid the inclusion of nonlinear phenomena and, on the other hand, to develop the foundations of non-invasive diagnostics of charged particle beams.

The list of references presents available studies used for writing this monograph.

We are grateful to B.M. Bolotovskii, N.F. Shul'ga, N.N. Nasonov, M. Ikezawa, Y. Shibata, J. Urakawa, G. Kube, K.A. Ispiryan, G.A. Naumenko and P.V. Karataev for numerous stimulating discussions and to L.V. Puzyrevich,

N. A. Potylitsina-Kube, D. V. Karlovets, A. R. Wagner and L. G. Sukhikh for assistance in the preparation of the manuscript. We are indebted to R. Tyapaev, translator of the book, for effective cooperation.

Tomsk-Moscow, October 2009

Alexander Petrovich Potylitsyn
Mikhail Ivanovich Ryazanov
Mikhail Nikolaevich Strikhanov
Alexey Alexandrovich Tishchento

References

1. Smith, S.J., Purcell, E.M.: Visible light from localized surface charges moving across a grating. Phys. Rev. **92**, 1069 (1953).
2. Frank, I.M.: Doppler effect in a refractive medium. Izv. Akad. Nauk USSR. Fizika. **6**, 3 (1942).
3. Shestopalov, V.P.: Diffraction Electronics. Kharkov, Ukraine (1976).
4. Shestopalov, V.P.: The Smith-Purcell effect. Nova Science Publishers, Commack, NY (1998).
5. Tsimring, S.E.: Electron Beams and Microwave Vacuum Electronics. Wiley, Hoboken, NJ (2003).
6. Williams, G.P.: Filling the THz gap-high power sources and applications. Rep. Prog. Phys. **69**, 301 (2006).
7. Ter-Mikaelyan, M.L.: High-Energy Electromagnetic Processes in Condensed Media. Wiley-Interscience, New York, NY (1972).
8. Rullhusen, P., Artru, X., Dhez, P.: Novel Radiation Sources Using Relativistic Electrons. World Scientific, Singapore (1998).
9. Ufimtsev, P.Ya.: Theory of Diffraction Boundary Problems in the Electrodynamics, Moscow, Binom (2007) in Russian.

Contents

Chapter 1
Radiation from Relativistic Particles

1.1 General Properties of Radiation from Relativistic Particles

It is well known that emission is the process of the formation of transverse electromagnetic waves by moving charged particles. Let us consider the emission process occurring when a relativistic charged particle moves according to the law $\mathbf{r} = \mathbf{R}(t)$ in a medium with relative permittivity $\varepsilon(\omega)$. At optical and higher frequencies, the relative permeability of the medium is negligibly different from one and the medium can be treated as nonmagnetic. In order to determine the emitted energy, we first calculate the magnetic field generated due to this motion of the charged particle. Maxwell's equations for the time Fourier transforms of fields and currents in the homogeneous isotropic nonmagnetic medium have the form

$$\begin{aligned} &\operatorname{rot}\mathbf{H}(\mathbf{r},\omega) = \frac{4\pi}{c}\mathbf{j}(\mathbf{r},\omega) - i\frac{\omega}{c}\varepsilon(\omega)\mathbf{E}(\mathbf{r},\omega) \\ &\varepsilon(\omega)\operatorname{div}\mathbf{E}(\mathbf{r},\omega) = 4\pi\rho(\mathbf{r},\omega) \\ &\operatorname{rot}\mathbf{E}(\mathbf{r},\omega) = i\frac{\omega}{c}\mathbf{H}(\mathbf{r},\omega) \\ &\operatorname{div}\mathbf{H}(\mathbf{r},\omega) = 0 \end{aligned} \tag{1.1}$$

where $\mathbf{j}$ and ρ are the current and charge densities, respectively, that correspond to the motion of the charged particle.

Application of the curl operator to the first of Eq. (1.1) immediately yields the following equation for the magnetic field:

$$\left(\Delta + k^2\right)\mathbf{H}(\mathbf{r},\omega) = -\frac{4\pi}{c}\operatorname{rot}\mathbf{j}(\mathbf{r},\omega), \tag{1.2}$$

where

$$k^2 = \left(\frac{\omega}{c}\right)^2\varepsilon(\omega). \tag{1.3}$$

Equation (1.2) is similar to the well-known equation for the vector potential created by a given current and its solution is given by the expression:

A.P. Potylitsyn et al., *Diffraction Radiation from Relativistic Particles*,
STMP 239, 1–28, DOI 10.1007/978-3-642-12513-3_1,

$$\mathbf{H}(\mathbf{r},\omega)=\frac{1}{c}\int d^3r'\frac{\exp\left(ik|\mathbf{r}-\mathbf{r}'|\right)}{|\mathbf{r}-\mathbf{r}'|}\mathrm{rot}'\mathbf{j}\left(\mathbf{r}',\omega\right). \tag{1.4}$$

At large distances, when $|\mathbf{r}|\gg\left|\mathbf{r}'\right|$, this solution can be represented as:

$$\mathbf{H}(\mathbf{r},\omega)=\frac{i}{c}\frac{e^{ikr}}{r}\int d^3r'\left[\mathbf{k}\mathbf{j}\left(\mathbf{r}',\omega\right)\right]\exp\left(-i\mathbf{k}\mathbf{r}'\right). \tag{1.5}$$

Here, $\mathbf{k}=\mathbf{n}k$, where $\mathbf{n}=\frac{\mathbf{r}}{r}$, is the wave vector of the radiation field at the observation point $\mathbf{r}$.

The energy of the field at large distances inside solid angle $d\Omega$ in the direction of vector $\mathbf{n}$ in frequency interval $d\omega$ can be determined as follows. The energy emitted in the solid angle $d\Omega$ in the total time of the motion of charges is given by the expression:

$$\begin{gathered}dW(\mathbf{n})=r^2d\Omega\frac{c}{4\pi}\int dt\left(\mathbf{n},\left[\mathbf{E}(\mathbf{r},t)\,\mathbf{H}(\mathbf{r},t)\right]\right)=\\ r^2d\Omega\frac{c}{2}\int_{-\infty}^{\infty}d\omega\int_{-\infty}^{\infty}d\omega'\left(\mathbf{n},\left[\mathbf{E}(\mathbf{r},\omega)\,\mathbf{H}\left(\mathbf{r},\omega'\right)\right]\right)\int\frac{dt}{2\pi}e^{i\left(\omega+\omega'\right)t}.\end{gathered} \tag{1.6}$$

The integral with respect to time is equal to the Dirac delta function $\delta\left(\omega+\omega'\right)$, which makes it possible to calculate one integral with respect to the frequency. Since the integrand is an even function, the integral over the entire frequency interval can be replaced by the doubled integral over the positive frequencies:

$$dW(\mathbf{n})=cr^2d\Omega\int_0^{\infty}d\omega\left(\mathbf{n},\left[\mathbf{E}(\mathbf{r},\omega)\,\mathbf{H}^*(\mathbf{r},\omega)\right]\right). \tag{1.7}$$

At large distances the Fourier transforms of the fields at frequency ω are related by the same expressions as the fields in the plane wave

$$\mathbf{E}(\mathbf{r},\omega)=\varepsilon^{-1/2}(\omega)\left[\mathbf{H}(\mathbf{r},\omega)\,\mathbf{n}\right], \tag{1.8}$$

because the curvature of the wave front is on the order of $1/r$. As a result, the angular distribution of the emitted energy is represented as

$$dW(\mathbf{n})=cr^2d\Omega\int_0^{\infty}d\omega\,\varepsilon^{-1/2}(\omega)\left|\mathbf{H}(\mathbf{r},\omega)\right|^2. \tag{1.9}$$

The integrand in Eq. (1.9) is positive everywhere; therefore, the spectral–angular distribution of the emitted energy can be written by omitting the integration sign:

$$d^2W(\mathbf{n},\omega)=cr^2d\Omega d\omega\,\varepsilon^{-1/2}(\omega)\left|\mathbf{H}(\mathbf{r},\omega)\right|^2. \tag{1.10}$$

The substitution of expression (1.5) for $\mathbf{H}(\mathbf{r},\omega)$ yields the spectral–angular distribution of the energy of radiation generated in the medium by an arbitrary current density $\mathbf{j}(\mathbf{r},t)$:

$$\frac{d^2W(\mathbf{n},\omega)}{d\Omega\, d\omega} = \frac{\omega^2}{c^3}\sqrt{\varepsilon}\left|\int d^3r' \exp\left(-i\mathbf{k}\mathbf{r}'\right)\left[\mathbf{n}\mathbf{j}\left(\mathbf{r}',\omega\right)\right]\right|^2. \tag{1.11}$$

From this expression, it is easy to obtain the distribution of the energy emitted by charge e whose motion is specified by the law $\mathbf{r} = \mathbf{R}(t)$. In this case,

$$\begin{aligned} \mathbf{j}(\mathbf{r},t) &= e\mathbf{v}(t)\,\delta(\mathbf{r}-\mathbf{R}(t)) \;, \\ \mathbf{j}(\mathbf{r},\omega) &= \frac{e}{2\pi}\int dt\,\mathbf{v}(t)\,\delta(\mathbf{r}-\mathbf{R}(t))\exp(i\omega t)^{\cdot} \end{aligned} \tag{1.12}$$

where $\mathbf{v}(t) = \dfrac{d\mathbf{R}}{dt}$. After the substitution of this expression into Eq. (1.11), the integration with respect to the coordinates is performed with the use of the Dirac delta function and only the integral with respect to time remains.

The spectral–angular distribution of the energy emitted by the charge whose motion is specified by the law $\mathbf{R}(t)$ is obtained in the form:

$$\frac{d^2W(\mathbf{n},\omega)}{d\omega d\Omega} = \frac{e^2\omega^2\sqrt{\varepsilon}}{4\pi^2c^3}\left|\int dt\,[\mathbf{n}\mathbf{v}(t)]\exp\{i\omega t - i\mathbf{k}\mathbf{R}(t)\}\right|^2. \tag{1.13}$$

We emphasize that relation (1.13) for distances much larger than the wavelength and sizes of the emitter (this region is usually called wave (far) zone) is exact for any, from nonrelativistic to ultrarelativistic, velocity of the charged particle. It is only necessary to know the law of the motion of this particle.

In certain cases, it is convenient to represent the emitted energy in another form. To obtain this form, we substitute the following expansion of the current density in the Fourier integral in all the coordinates and time into Eq. (1.11):

$$\mathbf{j}(\mathbf{r},t) = \int d\omega\int d^3q\,\mathbf{j}(\mathbf{q},\omega)\exp(i\mathbf{q}\mathbf{r} - i\omega t)\,. \tag{1.14}$$

In this case, Eq. (1.11) is represented in the form

$$\frac{d^2W(\mathbf{n},\omega)}{d\Omega\, d\omega} = \frac{(2\pi)^6}{c\sqrt{\varepsilon}}\left|\left[\mathbf{k}\mathbf{j}(\mathbf{k},\omega)\right]\right|^2. \tag{1.15}$$

To calculate the radiation spectrum, it is convenient to transform this expression as follows. Representing Eq. (1.15) in the form

$$\frac{d^2W(\mathbf{n},\omega)}{d\Omega\,d\omega}=\frac{(2\pi)^6}{c\sqrt{\varepsilon}}\left(k^2\left|\mathbf{j}(\mathbf{k},\omega)\right|^2-\left|\mathbf{kj}(\mathbf{k},\omega)\right|^2\right), \tag{1.16}$$

and using the continuity equation for the Fourier transforms of the charge density $\mathbf{j}(\mathbf{q},\omega)$ and charge density $\rho(\mathbf{q},\omega)$,

$$\mathbf{qj}(\mathbf{q},\omega)=\omega\rho(\mathbf{q},\omega), \tag{1.17}$$

one can transform Eq. (1.16) to the form

$$\frac{d^2W(\mathbf{n},\omega)}{d\Omega\,d\omega}=\frac{(2\pi)^6}{c\sqrt{\varepsilon}}\left(k^2\left|\mathbf{j}(\mathbf{k},\omega)\right|^2-\omega^2\left|\rho(\mathbf{k},\omega)\right|^2\right). \tag{1.18}$$

For a charge whose motion is described as $\mathbf{r}=\mathbf{R}(t)$, the Fourier transforms in space and time for the current density and charge density are easily calculated as:

$$\begin{aligned}\mathbf{j}(\mathbf{q},\omega)&=e(2\pi)^{-4}\int dt\,\mathbf{v}(t)\exp\{i\omega t-i\mathbf{kR}(t)\}\\ \rho(\mathbf{q},\omega)&=e(2\pi)^{-4}\int dt\exp\{i\omega t-i\mathbf{kR}(t)\}.\end{aligned} \tag{1.19}$$

The substitution of these expressions into Eq. (1.18) yields

$$\begin{aligned}\frac{d^2W(\mathbf{n},\omega)}{d\Omega\,d\omega}&=\frac{e^2\omega^2}{4\pi^2c^3\sqrt{\varepsilon}}\int dt\int dt'\left\{\mathbf{v}(t)\mathbf{v}(t')\varepsilon(\omega)-c^2\right\}\times\\ &\times\exp\left\{i\omega(t-t')-i\mathbf{k}\left(\mathbf{R}(t)-\mathbf{R}(t')\right)\right\}.\end{aligned} \tag{1.20}$$

The advantage of such a representation is that the unit vector $\mathbf{n}=\mathbf{k}/k$ specifying the direction of the solid angle $d\Omega$ enters only into the exponential, whereas Eq. (1.13) contains $\mathbf{n}$ both in the exponential and in the pre-exponential factor. For this reason, the integration of Eq. (1.20) with respect to the angles for the calculation of the radiation spectrum $dW(\omega)$ reduces only to the integration of the exponential with respect to the angles. In view of

$$\int d\Omega\exp(i\mathbf{ng})=4\pi\frac{\sin g}{g}, \tag{1.21}$$

it is easy to integrate Eq. (1.20) with respect to the angles and to obtain the radiation spectrum of the charge whose motion is given by the law $\mathbf{r}=\mathbf{R}(t)$ in the form

$$\begin{aligned}\frac{dW(\omega)}{d\omega}&=\frac{e^2\omega}{\pi c^2\varepsilon(\omega)}\int\int dt\,dt'\left\{\mathbf{v}(t)\mathbf{v}(t')\varepsilon(\omega)-c^2\right\}\frac{\sin\left(k\left|\mathbf{R}(t)-\mathbf{R}(t')\right|\right)}{\left|\mathbf{R}(t)-\mathbf{R}(t')\right|}\times\\ &\times\exp\left\{i\omega(t-t')\right\}.\end{aligned} \tag{1.22}$$

Expressing the exponential in terms of the sine and cosine, one can easily see that the integral with the sine vanishes, because the corresponding integrand changes sign under change of t to t'. It is convenient to reduce the remaining integration with the cosine to the region $t > t'$ and to use the relation

$$2\sin(g)\,\cos(\omega\tau) = \sin(\omega\tau + g) - \sin(\omega\tau - g)\,. \tag{1.23}$$

After that, Eq. (1.22) is represented in the form $\left(t' = t + \tau\right)$

$$\begin{aligned}\frac{dW(\omega)}{d\omega} &= \frac{e^2\omega}{\pi c^2\varepsilon(\omega)}\int_{-\infty}^{\infty} dt \int_0^{\infty} d\tau \left\{c^2 - \mathbf{v}(t)\,\mathbf{v}(t+\tau)\,\varepsilon(\omega)\right\} \times \\ &\times \frac{\sin\{\omega\tau - k\,|\mathbf{R}(t) - \mathbf{R}(t+\tau)|\} - \sin\{\omega\tau + k\,|\mathbf{R}(t) - \mathbf{R}(t+\tau)|\}}{|\mathbf{R}(t) - \mathbf{R}(t')|}\,.\end{aligned} \tag{1.24}$$

The spectral–angular distribution of the emitted energy given by Eq. (1.13) and radiation frequency spectrum (1.24) are obtained for the particle that has charge e and moves according to law $\mathbf{R}(t)$ in the medium with relative permittivity $\varepsilon(\omega)$. The distribution of the emitted energy upon the same motion of the particle in vacuum are obtained from Eqs. (1.13) and (1.24) by changing $\varepsilon(\omega)$ to one. It should be taken into account that relative permittivity $\varepsilon(\omega)$ appears in the result both as an explicit factor and, implicitly, in quantity $k = (\omega/c)\sqrt{\varepsilon(\omega)}$.

The motion of ultrarelativistic particles with energy $E = mc^2\gamma \gg mc^2$ is primarily determined by the external force component transverse to the velocity, whereas the longitudinal component of the external force is smaller by a factor of γ^2 (see, e.g. [1, section 9]). This means that the longitudinal component of the external force and the longitudinal acceleration can be neglected in the first approximation. In this approximation, acceleration is perpendicular to the particle velocity and the absolute value of the particle velocity is conserved. Let us consider the case, where change in the particle velocity is relatively small in the time interval determining the integral with respect to τ in Eq. (1.24).

Since the particle velocity in the approximation under consideration changes only in direction,

$$\mathbf{v}(t+\tau) = \mathbf{v}(t)\left\{1 - \varsigma^2(t)\,\tau^2/2\right\} + \mathbf{e}(t)\,\varsigma(t)\,\tau\,v(t)\,, \tag{1.25}$$

where $\mathbf{e}(t)$ is the unit vector of the force component transverse to the particle velocity, i.e., $\mathbf{v}(\tau)\,\mathbf{e}(t) = 0$. Relatively small change in velocity means that $\varsigma(t)\,\tau \ll 1$. Since the absolute value of the velocity is conserved, the radius vector of the particle can be written in the form

$$\begin{aligned}&\mathbf{R}(t+\tau) = \mathbf{R}(t) + \mathbf{v}(t)\,\tau\left[1 - \varsigma^2(t)\,\tau^2/6\right] + \mathbf{e}(t)\,\varsigma(t)\;v(t)\;\tau^2/2;\\ &|\mathbf{R}(t+\tau) - \mathbf{R}(t)| = \mathbf{v}(t)\,\tau\left\{1 - \varsigma^2(t)\,\tau^2/24\right\}.\end{aligned} \tag{1.26}$$

This makes it possible to transform the spectrum of radiation generated by the ultra-relativistic particle given by Eq. (1.24) to the form

$$\frac{dW(\omega)}{d\omega} = \frac{e^2\omega}{\pi c^2 \varepsilon(\omega)} \int_{-\infty}^{\infty} \frac{dt}{v(t)} \int_0^{\infty} \frac{d\tau}{\tau} \frac{c^2 - v^2(t)\,\varepsilon(\omega)\left[1 - \varsigma^2\tau^2/2\right]}{1 - \varsigma^2\tau^2/24} \times$$
$$\times \left(\sin\left\{(\omega - kv)\,\tau + kv\varsigma^2\tau^3/24\right\} - \sin\left\{(\omega + kv)\,\tau - kv\varsigma^2\tau^3/24\right\}\right) \tag{1.27}$$

The case $\varsigma = 0$ corresponds to uniform rectilinear motion when radiation is absent and Eq. (1.27) vanishes. This is easily shown by passing to new integration variables $u = (\omega - kv)\,\tau$ in the integral with the first sine and $u = (\omega + kv)\,\tau$ in the integral with the second sine; then, these integrals become the same and, hence, their difference vanishes.

1.2 Radiation Formation Length

Let us return to Eq. (1.13) and consider the case, where a charge moves with a constant velocity $\mathbf{v}$ in the time interval between the times $t = -T/2$ and $t = T/2$.This means that the charge is at rest at the points $\mathbf{r} = -\mathbf{v}T/2$ and $\mathbf{r} = \mathbf{v}T/2$ for times $t < -T/2$ and $t > T/2$, respectively. The charge is abruptly accelerated to the velocity $\mathbf{v}$ at the time $t = -T/2$ and stops abruptly at the time $t = T/2$. The applicability of the approximation of abrupt stop and abrupt acceleration will be discussed below. Performing the integration with respect to t in Eq. (1.13) for such a motion of the charge, we arrive at the following expression for the distribution of the emitted energy:

$$\frac{d^2W(\mathbf{n},\omega)}{d\omega d\Omega} = \frac{e^2\omega^2}{4\pi^2 c^3}\varepsilon^{1/2}(\omega)\,[\mathbf{n},\mathbf{v}]^2\,\frac{4\sin^2\{(\omega - \mathbf{kv})\,T/2\}}{(\omega - \mathbf{kv})^2}. \tag{1.28}$$

First, we consider the passage to the limit $T \to \infty$. In this case it is convenient to use the known formula

$$\lim_{T\to\infty} \frac{\sin^2(xT)}{Tx^2} = \pi\delta(x)\,. \tag{1.29}$$

For $T \to \infty$ Eq. (1.28) has the form

$$\frac{d^2W(\mathbf{n},\omega)}{d\omega d\Omega} = \frac{e^2\omega^2}{2\pi c^3}\sqrt{\varepsilon(\omega)}\,[\mathbf{n},\mathbf{v}]^2\,T\delta(\omega - \mathbf{kv})\,, \tag{1.30}$$

which coincides with the energy distribution of Cherenkov radiation.

Note that the terms that do not increase with T are omitted in the passage to the limit $T \to \infty$. For this reason, at $\omega \neq \mathbf{kv}$ when the argument of the Dirac delta function is nonzero (i.e., $c^2 > v^2\varepsilon(\omega)$), more accurate analysis is required. In this case,

$$2\sin^2\{(\omega - \mathbf{kv})\,T/2\} = 1 - \cos\{(\omega - \mathbf{kv})\,T\}, \tag{1.31}$$

where the rapidly oscillating cosine provides zero contribution after integration over a small interval of angles or frequencies in the limit $T \to \infty$. Therefore, at $c^2 > v^2\varepsilon$,

$$d^2W(\mathbf{n},\omega) = 2d\omega d\Omega \frac{e^2\omega^2}{4\pi^2c^3}\sqrt{\varepsilon(\omega)}\frac{[\mathbf{nv}]^2}{(\omega - \mathbf{kv})^2} = 2d^2W_{\mathrm{st}}(\mathbf{n},\omega), \tag{1.32}$$

where $d^2W_{\mathrm{st}}(\mathbf{n},\omega)$ is the distribution of the radiation energy emitted either by the charge that moves with the velocity $\mathbf{v}$ and then stops abruptly or by the charge at rest that is abruptly accelerated to velocity $\mathbf{v}$.

In the opposite limiting case of small T values, i.e., when $(\omega - \mathbf{kv})\,T \ll 1$, it follows from Eq. (1.28) that

$$\frac{d^2W(\mathbf{n},\omega)}{d\omega d\Omega} = \frac{e^2\omega^2}{4\pi^2c^3}\sqrt{\varepsilon(\omega)}\,[\mathbf{nv}]^2\,(2T)^2. \tag{1.33}$$

In this case, the emitted energy is proportional to the motion time squared; hence, the field at the observation point is proportional to the motion time T. This behavior can be explained by the fact that the fields generated in a small section of the particle path arrive at the observation point with almost the same phases and are coherently summed without cancellation. According to Eq. (1.28), the radiation intensity is maximal at $T = (2N + 1)\,\pi/(\omega - \mathbf{kv})$, where N is an integer, and is equal to zero at $T = 2N\pi/(\omega - \mathbf{kv})$. An explanation is as follows: as the length of the particle path vT increases, the difference between the phases of the waves arriving at the observation point increases and, therefore, partial cancellation of different waves also increases. At $T = 2\pi/(\omega - kv)$, the fields from two halves of the path arrive at the observation point in antiphase and the fields are completely cancelled at the observation point. Therefore, the field detected at the observation point is formed in the final section of the path of the charged particle; the length of this section can be estimated as

$$l_c = \frac{2\pi v}{\omega - \mathbf{kv}}. \tag{1.34}$$

This length is usually called the radiation formation length [2] or the coherence length [3]. The coherence length can be defined as the length of the particle path section such that the fields from all its points arrive at the observation point with almost the same phases and are coherently summed. From this point of view, radiation generated by a uniformly moving charge is absent, because the waves emitted from different path sections are completely cancelled at the observation point. Any violation of the wave interference conditions at the observation point gives rise to the appearance of radiation.

In particular, if the charged particle colliding with an atom abruptly changes velocity from $\mathbf{v}$ to $\mathbf{u}$ at the time $t = 0$, then the motion of the charge at $t > 0$ and $t < 0$ is uniform and rectilinear. In all the charge path sections except the section near the turning point, the waves emitted from the neighboring coherence lengths are cancelled. This does not occur only in the path sections that join the turning point and have lengths on the order of $2\pi v/(\omega - \mathbf{kv})$ and $2\pi u/(\omega - \mathbf{ku})$.

It is the path section of the formation of radiation generated by the particle whose velocity changes abruptly. Change in the particle velocity (including the stop of the charge) can be called abrupt if it occurs in time much smaller than the effective radiation formation time $2\pi/(\omega - \mathbf{kv})$. Radiation appearing when the velocity of the charged particle changes due to collision with an atom is usually called bremsstrahlung.

It follows from the Lorentz transformations that electromagnetic waves emitted by an ultrarelativistic source in vacuum are concentrated in the region of small angles ϑ on the order of $\gamma^{-1} = \sqrt{1 - (v/c)^2}$ near the velocity direction. The same estimate is obtained from the condition of the maximum coherence length,

$$\vartheta^2 \sim \gamma^{-2} = 1 - (v/c)^2 . \tag{1.35}$$

As seen from Eq. (1.13), the spectral–angular distribution of radiation generated by the charge moving in a medium is obtained by multiplying the corresponding distribution for the charge undergoing the same motion in vacuum by a factor of $\sqrt{\varepsilon}$ and changing $c \to c/\sqrt{\varepsilon}$. Relative permittivity at frequencies much higher than atomic frequencies is given by the expression

$$\varepsilon(\omega) = 1 - \left(\frac{\omega_p}{\omega}\right)^2; \qquad \omega_p^2 = \frac{4\pi n e^2}{m} \ll \omega^2. \tag{1.36}$$

For high frequencies and ultrarelativistic particles, change $c \to c/\sqrt{\varepsilon}$ transforms estimate (1.35) for the characteristic radiation emission angles to the form

$$(\omega - \mathbf{kv})\omega 1 - \left(1 - 1/(2\gamma^2)\right)\left(1 - \theta^2/2\right)\left(1 - \omega_p^2/(2\omega^2)\right) \tag{1.37}$$

The order of magnitude of the radiation formation length in this frequency range is given by the expression

$$l_c = \frac{2\pi v}{\omega - \mathbf{kv}} = \frac{2\lambda}{\vartheta^2 + \left(\omega_p/\omega\right)^2 + \gamma^{-2}}. \tag{1.38}$$

Hence, for particles with energy in the range

$$mc^2\left(\omega/\omega_p\right) \gg E \gg mc^2, \tag{1.39}$$

the coherence lengths for radiation in the medium and vacuum coincide. This means that the presence of the medium does not affect the emission process in range (1.39).

For sufficiently high particle energy satisfying the inequality

$$E \gg mc^2 \left(\omega/\omega_p\right) \gg mc^2, \tag{1.40}$$

the order of magnitude of the radiation formation length in this frequency range is given by the expression

$$l_c \simeq \frac{2\lambda}{\vartheta^2 + \left(\omega_p/\omega\right)^2}. \tag{1.41}$$

An important difference between radiations from nonrelativistic and ultrarelativistic particles is that $v \ll c$ and $\mathbf{kv} \ll \omega$ for a nonrelativistic particle, so that the radiation formation length $2\pi v/\omega$ is always much smaller than the radiation wavelength $2\pi c/\omega$. In the ultrarelativistic case, the coherence length for radiation in vacuum is always larger than the radiation wavelength and this relation for radiation in the medium is valid for frequencies much higher than atomic frequencies.

For ultrarelativistic particles, the coherence length can be so large that other processes accompanying the motion of the particle through the medium affect the radiation formation process. For example, owing to the multiple scattering of the particle on atoms of the medium, the motion of the particle on the coherence length is not rectilinear and the velocity of passing the coherence length by the particle is lower than the real particle velocity. For frequencies lower than optical frequencies, this leads to the suppression of the intensity of bremsstrahlung and this suppression is particularly strong for low frequencies for which the radiation formation length is larger. This effect was first considered by Landau and Pomeranchuk in 1953 in the framework of classical electrodynamics [4, 5], whereas quantum theory including the generalization to the pair production process was developed in 1954 by Migdal [6]. This effect was experimentally observed only in 1996 [7] (see also review [8]).

1.3 Radiation from a Heavy Charged Particle Colliding With an Atom

Considering the collision of a heavy charged particle with an atom, we take into account that radiation can appear due not only to the deviation of the fast particle, but also to change in the state of atomic electrons. The energy and momentum conservation laws forbid the emission of a transverse electromagnetic wave by the freely moving charge. For the appearance of radiation, momentum must be transferred to another particle or body. If the internal state of the atom remains unchanged in the emission process, momentum is transferred from the particle to the atom, the atom undergoes recoiling, and the particle is deviated and emits. In this case, the deviated particle is the main source of radiation. This radiation is called bremsstrahlung [9]. Its intensity is inversely proportional to the particle mass squared and is low for heavy particles. The effect of atomic electrons on bremsstrahlung was considered in [10, 11].

The second type of the process occurs when the deviation of the moving particle after collision is very small, whereas the atom changes its internal state. A part of the energy of the fast particle is transferred to atomic electrons, inducing polarization currents, which are sources of polarization radiation. In this case, the energy of electron oscillations is much lower than the energy of the fast particle; hence, the energy loss of the fast particle can be neglected and the velocity of this particle can be considered as constant. Cherenkov radiation appearing when the particle velocity is higher than the speed of light in the medium, transition radiation appearing when the particle intersect an interface between media, parametric x-ray radiation appearing when the particle moves through a crystal, and diffraction radiation appearing when the charged particle moves in vacuum near the surface of a medium and does not intersect it belong to radiation generated by polarization currents or polarization bremsstrahlung [12].

Transition radiation and diffraction radiation are most often considered in the uniform motion approximation. This approximation is applicable when the total emitted energy is much lower than the kinetic energy of the charged particle. Thus, we assume that the field of the heavy charged particle acting on an atomic electron coincides with the field of the charge Ze undergoing uniform rectilinear motion

$$\mathbf{E_0}(\mathbf{r},t) = \int d^3q \mathbf{E_0}(\mathbf{q}) \exp(i\mathbf{qr} - i\mathbf{qv}t); \qquad \mathbf{E_0}(q) = -i\frac{Ze}{2\pi^2}\frac{\mathbf{v}(\mathbf{qv}) - \mathbf{q}c^2}{q^2c^2 - (\mathbf{qv})^2}, \tag{1.42}$$

and the emission process is described by the transformation of the Fourier component of the self field of the charge, $\mathbf{E_0}(\mathbf{q})\exp(i\mathbf{qr} - i\mathbf{qv}t)$, to the emitted plane wave $\mathbf{E}\exp(i\mathbf{kr} - i\omega t)$. Since the problem is stationary, frequency cannot change in the framework of linear electrodynamics, so that $\omega = \mathbf{qv}$.

As an example, let us consider radiation appearing when charge e moves with velocity $\mathbf{v}$ near an electron bound in an oscillator. The field of the uniformly moving charge, $\mathbf{E_0}(\mathbf{r},t)$, excites the oscillations of the bound electron. Implying that the characteristic wavelength of the field is much larger than the amplitude of the oscillations of the bound electron, we use the dipole approximation. In this case, the field of the particle is almost the same for all the positions of the oscillating electron, and the field at the equilibrium position of the electron, $\mathbf{R}$, can be used as the field acting on the electron. Since the wavelength of the field is much larger than the atomic size, the atomic nucleus position can be taken as the equilibrium electron position.

In this approximation, the equation of motion of the electron in the field of the particle has the form

$$\frac{d^2\mathbf{r}}{dt^2} + \Gamma\frac{d\mathbf{r}}{dt} + \omega_0^2\mathbf{r} = \frac{e}{m}\mathbf{E_0}(\mathbf{R},t) = \frac{e}{m}\int d\omega\, \mathbf{E_0}(\mathbf{R},\omega)\exp(-i\omega t). \tag{1.43}$$

The dipole moment of this electron

$$\mathbf{d}(\mathbf{R}, t) = \int d\omega\, \mathbf{d}(\mathbf{R}, \omega) \exp(-i\omega t), \tag{1.44}$$

which is induced in the atom by the field of the particle, is related to the solution $\mathbf{r}(t)$ of Eq. (1.43) for forced oscillations as $\mathbf{d}(t) = e\mathbf{r}(t)$. Taking the time Fourier transforms of Eq. (1.44), one easily arrives at the following expression for the Fourier transform of the dipole moment:

$$\mathbf{d}(\mathbf{R}, \omega) = \frac{e^2}{m} \frac{\mathbf{E}_0(\mathbf{R}, \omega)}{\omega_0^2 - \omega^2 - i\Gamma\omega}. \tag{1.45}$$

In the dipole approximation, the Fourier transform of the atomic current density generated by the field of the particle has the form

$$\mathbf{j}(\mathbf{r}, \omega) = -i\omega \sum_s \mathbf{d}_s(\mathbf{R}, \omega)\, \delta(\mathbf{r}_s - \mathbf{R}), \tag{1.46}$$

where the summation is performed over all the atomic electrons. The Fourier transform of this current in space and time is written as

$$\mathbf{j}(\mathbf{q}, \omega) = \frac{-i\omega}{(2\pi)^3} \alpha(\omega)\, \mathbf{E}_0(\mathbf{R}, \omega) \exp(-i\mathbf{q}\mathbf{R}). \tag{1.47}$$

Here, the polarizability of the atom is introduced by the expression

$$\alpha(\omega) = \frac{e^2}{m} \sum_s \frac{f_s}{\omega_s^2 - \omega^2 - i\Gamma_s\omega}, \tag{1.48}$$

where f_s is the probability of excitation of an electron with the frequency ω_s in the atom.

The spectral–angular distribution of radiation energy is represented in the form

$$\frac{d^2 W}{d\omega d\Omega} = (2\pi)^6 \left(\frac{1}{c\varepsilon^{1/2}}\right) \left|[\mathbf{k}\mathbf{j}(\mathbf{k}, \omega)]\right|^2 = |\alpha(\omega)|^2 \frac{\omega^2}{c} |[\mathbf{k}\mathbf{E}_0(\mathbf{R}, \omega)]|^2. \tag{1.49}$$

Let us consider radiation appearing when the ultrarelativistic particle moves along the z axis at distance b from the equilibrium position of the bound electron that coincides with the origin of the coordinate system. The field component of the ultrarelativistic particle along the velocity is much smaller than the transverse field component. Hence, it is possible to take into account only the transverse component of the field

$$E_{0x}(\mathbf{R}, t) = \frac{Ze\gamma b}{\left(b^2 + \gamma^2 v^2 t^2\right)^{3/2}}. \tag{1.50}$$

The Fourier transform of the transverse component of the field is expressed as

$$E_{0x}(\mathbf{R}, \omega) = \frac{Ze}{\pi b v} \frac{\omega b}{\gamma v} K_1\left(\frac{\omega b}{\gamma v}\right), \tag{1.51}$$

where $K_1(u)$ is the modified Bessel function of the first order. In the limits of small and large values of the argument, this function has the form

$$K_1(u) \cong \begin{cases} 1/u & u \ll 1 \\ (\pi/2u)\exp(-u) & u \gg 1 \end{cases}. \tag{1.52}$$

The Fourier transform of the field given by Eq. (1.51) as a function of the variable $u = \omega b/\gamma v$ approaches a constant for small u values and decreases exponentially for large u values. In the case under consideration, the spectral–angular distribution of the emitted energy has the form

$$d^2W(\mathbf{n}, \omega) = \frac{Z^2\omega^4\left(1 - \sin^2\vartheta\cos^2\phi\right)}{\pi^2 c^3 v^2 b^2} |\alpha(\omega)|^2 \left\{\frac{\omega b}{\gamma v} K_1\left(\frac{\omega b}{\gamma v}\right)\right\}^2 d\omega d\Omega. \tag{1.53}$$

The expression in braces is on the order of one for $\omega b < \gamma v$ and decreases rapidly for $\omega b > \gamma v$. The emitted energy is inversely proportional to the square of the impact parameter b for $\omega b < \gamma v$ and decreases exponentially for $\omega b > \gamma v$. We emphasize that Eq. (1.53) does not contain the mass of the fast particle and includes only the mass of the atomic electron.

Thus, the conservation laws for bremsstrahlung are satisfied due to transferring the momentum from the fast particle to the medium (the fast particle is deviated). The conservation laws for transition or diffraction radiation are satisfied due to transferring the momentum from the field of the charged particle to the medium (the fast particle does not change its velocity).

In the general case, the particle—atom collision is certainly accompanied by the transfer of the moment to the medium from both the particle and the field. However, the momentum transfer from the field to the medium can be neglected in the electron—atom collision and the momentum transfer from the proton to the medium can be neglected in the fast proton—atom collision. Thus, ordinary bremsstrahlung and transition radiation are limiting cases of radiation appearing in collision with the atom. More complex cases, where the momentum transfers to the medium from both the field and the particle itself cannot be neglected, naturally exist. Other kinds of such processes are included in polarization bremsstrahlung that, in contrast to bremsstrahlung, involves both the deviation of the fast particle and change in the state of bound atomic electrons [11]. Such emission processes were considered in monograph [12], where references on polarization bremsstrahlung can be found.

1.4 Transition Radiation and Diffraction Radiation

The field of the fast charged particle moving near the surface of the medium excites oscillations of atomic electrons; i.e., polarization currents appear and produce a secondary electromagnetic field. For example, the appearance of such a field accompanies the motion of particles in the units of accelerators and the effect of this field on the motion of particles is of primary interest. In particular, the secondary field can give rise to instability in linear accelerators.

The secondary field produced by polarization currents is often called wakefield in works devoted to the theory of the processes in accelerators. However, the term wakefield is rarely used in the case of radiation generated by polarization currents; the notions of diffraction radiation, Smith–Purcell radiation, etc. are used more often. Both longitudinal and transverse components of the wave field at the points close to the trajectory of a particle play an important role in the action of the wakefield on the particle in an accelerator. Only the transverse component of the secondary field, i.e., the radiation field at long distances from the particle trajectory is of interest in the case of transition radiation and diffraction radiation.

As mentioned above, both transition radiation and diffraction radiation appear in the process of the emission of electromagnetic waves that accompanies oscillations of atomic electrons excited by the field of the moving charged particle.

The energy loss of the fast particle to radiation is much lower than the energy of the particle; hence, the particle velocity can be treated as constant with a good accuracy. The emission from the charge whose motion in the medium is specified by the law $\mathbf{r} = \mathbf{v}t$ can be described as the transformation of the Fourier component of the intrinsic field of the charge, $\mathbf{E_0}(\mathbf{q}) \exp\{i\mathbf{q}(\mathbf{r} - \mathbf{v}t)\}$, to the monochromatic plane wave $\mathbf{E} \exp(i\mathbf{kr} - i\omega t)$ in the process of interaction with the atoms of the medium. Quantity $\mathbf{q}$ does not generally coincide with $\mathbf{k}$ and quantity $\mathbf{qv}$ does not generally coincide with ω. Therefore, the momentum and energy of the atom change in the process of interaction by $\hbar(\mathbf{q} - \mathbf{k})$ and $\hbar(\omega - \mathbf{qv})$, respectively. An exception is the case, where the relative permittivity of the uniform stationary medium, $\varepsilon(\omega)$, and the particle velocity $\mathbf{v}$ satisfy the inequality $\varepsilon(\omega) \geq (c/v)^2$. Then, the equality $\mathbf{kv} = \omega(v/c)\sqrt{\varepsilon(\omega)}\cos\theta = \omega$ is valid at $\mathbf{q} = \mathbf{k}$ for a certain θ value. Such a radiation is called Cherenkov radiation.

Let us consider a case, where the conditions of existing Cherenkov radiation are not satisfied. In this case, radiation can appear if the momentum $\Delta p = \hbar(\mathbf{q} - \mathbf{k})$ is transferred from the field to the medium or the energy $\Delta E = \hbar(\omega - \mathbf{qv})$ is transferred from the medium to the field. The transfer of the momentum and energy is possible only in the inhomogeneous medium and nonstationary medium, respectively. The energy conservation law for the emission process in a nonstationary inhomogeneous medium has the form

$$\hbar\omega = \hbar\mathbf{kv} + \mathbf{v}\Delta\mathbf{p} + \Delta E. \tag{1.54}$$

Therefore, the transfer of only the longitudinal (i.e., directed along the particle velocity) component of the momentum p_v is important for the appearance of radiation. The minimum value of the quantity $\mathbf{v}\Delta\mathbf{p} + \Delta E$ below which radiation does not appear is given by the expression

$$(v\Delta p_v + \Delta E)^{\min} = \hbar\omega - \hbar\mathbf{k}\mathbf{v} = \hbar\omega\left\{1 - (v/c)\sqrt{\varepsilon(\omega)}\cos\theta\right\}. \tag{1.55}$$

For particles with the energy $E = \gamma mc^2 \gg mc^2$ for small angles θ and high frequencies, when the relative permittivity of the medium satisfies the relation $\varepsilon(\omega) = 1 - \left(\omega_p/\omega\right)^2$, it follows from Eq. (1.55) that

$$(v\Delta p_v + \Delta E)^{\min} = \frac{\hbar\omega}{2}\left\{\theta^2 + \left(\omega_p/\omega\right)^2 + \gamma^{-2}\right\}. \tag{1.56}$$

The transfer of momentum Δp_v from the field to the inhomogeneous medium is most efficient when the characteristic size of inhomogeneities along the velocity direction is on the order of $\hbar/\Delta p_v$. If this size is larger than $\hbar/\Delta p_v^{\min}$, the longitudinal momentum transfer is lower than the minimum longitudinal momentum necessary for the emission process and radiation is negligibly small. The threshold value of the inhomogeneity size for the nonrelativistic particle is on the order of $v/\omega \sim \lambda\,(v/c)$, i.e., is much smaller than the radiation wavelength. On the contrary, in the ultrarelativistic case for high frequencies, the threshold size of an inhomogeneity is on the order of $\lambda/\left[\theta^2 + \left(\omega_p/\omega\right)^2 + \gamma^{-2}\right]$, i.e., is much larger than the radiation wavelength.

Radiation is called transition radiation if the particle intersects the boundary of the medium and it is called diffraction radiation if the particle does not intersect this boundary. Note that diffraction radiation does no appear when the particle moves in parallel to the plane boundary of the homogeneous medium, because the transfer of the longitudinal momentum to the medium is impossible in such a geometry, but this transfer is a necessary condition for emission according to the conservation laws. However, diffraction radiation always appears in the presence of inhomogeneities on the surface of the medium or inside it (for instance, near an edge of surface).

Let us consider the problem of transition radiation in the framework of macroscopic electrodynamics. The particle generates its own macroscopic electromagnetic field in each medium. However, these fields do not satisfy the boundary conditions at the interface between the media. It is known that the general solution of the inhomogeneous wave equation for these fields is the sum of a partial solution of the inhomogeneous wave equation and the general solution of the homogeneous wave equation. The own field of the uniformly moving charge is described by the partial solution of the inhomogeneous equation. The radiation field is described by the general solution of the homogeneous equation. It contains two arbitrary constants, which are chosen so as to satisfy the boundary conditions.

Thus, the solution of Maxwell's equations for the semi-infinite medium differs from the solution of these equations for the infinite medium by the presence of the radiation field. Therefore, the uniformly moving charge intersecting the interface between two media emits electromagnetic waves. Such a radiation was theoretically predicted by Ginzburg and Frank in 1946 and was called transition radiation [13]. Transition radiation was investigated in many studies [2, 14–19].

We now consider the simplest example of transition radiation, radiation generated by a charged particle intersecting the vacuum—ideal conductor interface. Let the particle with charge Ze be emitted from the ideal conductor to vacuum with velocity $\mathbf{v}$ perpendicular to the surface of the conductor, $x = 0$.

According to the symmetry of the problem, the magnetic field has no component along the x axis; correspondingly, the boundary conditions for the normal components of the magnetic field H_x are satisfied automatically. According to the boundary conditions, the tangential components of the electric field must vanish on the conductor surface. This requirement can be satisfied by introducing "image charge" $-Ze$, moving along the x axis with velocity $-\mathbf{v}$ inside the conductor. In this case, the field generated by the real charge for $x > 0$ in vacuum coincides with the field that is produced in the half-space $x > 0$ by the real charge and image charge in infinite vacuum. Thus, the solution of the problem with one charge and boundary condition is reduced to the solution of the problem with two, real and image, charges.

Finally, in order to solve the problem of transition radiation, it is necessary to determine radiation generated by two charges $+Ze$ and $-Ze$ that move toward each other with velocities $+v$ and $-v$ and then stop abruptly at time $t = 0$. The current density corresponding to this motion of the charges has the form

$$\mathbf{j}(\mathbf{r}, t) = Ze\mathbf{v}\left\{\delta\left(\mathbf{r} - \mathbf{v}t\right) + \delta\left(\mathbf{r} + \mathbf{v}t\right)\right\}\ (t < 0)\,; \quad \mathbf{j}(\mathbf{r}, t) = 0\ (t > 0)\,. \tag{1.57}$$

The spectral–angular distribution of the emitted energy can be obtained with the use of Eq. (1.11) in the form similar to Eq. (1.13):

$$\frac{d^2W\left(\mathbf{n}, \omega\right)}{d\Omega d\omega} = \frac{Z^2e^2\omega^2}{4\pi^2c^3}\left[\mathbf{nv}\right]^2\left|\int_{-\infty}^{0} dt\left[\exp\left\{i\left(\omega - \mathbf{kv}\right)t\right\} + \exp\left\{i\left(\omega + \mathbf{kv}\right)t\right\}\right]\right|^2. \tag{1.58}$$

In vacuum $\mathbf{k} = (\omega/c)\,\mathbf{n}$, so that $(\omega - \mathbf{kv})$ and $(\omega + \mathbf{kv})$ are positive and integration yields

$$d^2W(\mathbf{n}, \omega) = \frac{Z^2e^2\omega^2}{4\pi^2c^3}\left[\mathbf{nv}\right]^2\left|\frac{1}{\omega - \mathbf{kv}} - \frac{1}{\omega + \mathbf{kv}}\right|^2 d\Omega\, d\omega \tag{1.59}$$

or, with the notation $\mathbf{nv} = v\cos\theta$ and $\beta = v/c$,

$$d^2W(\mathbf{n}, \omega) = \frac{Z^2e^2\beta^2}{\pi^2c}\frac{\sin^2\theta}{\left(1 - \beta^2\cos^2\theta\right)^2}d\Omega d\omega. \tag{1.60}$$

The integration of this expression with respect to the angles provides the radiation spectrum

$$dW(\omega) = \left(Z^2 e^2/\pi c\right) \left\{ \frac{1+\beta^2}{2\beta} \ln \frac{1+\beta}{1-\beta} - 1 \right\} d\omega. \tag{1.61}$$

In the ultrarelativistic case, where $\beta = \frac{v}{c} = 1 - (1/2)\left(mc^2/E\right)^2 = 1 - \gamma^{-2}/2$, we obtain

$$\frac{dW(\omega)}{d\omega} = \frac{Z^2 e^2}{\pi c} \left\{ \ln(2\gamma) - 1 \right\}. \tag{1.62}$$

In the nonrelativistic case, where $\beta \ll 1$, the radiation spectrum is given by the expression

$$\frac{dW(\omega)}{d\omega} = \frac{Z^2 e^2}{\pi c} \frac{4}{3} \beta^2. \tag{1.63}$$

The applicability of the above formulas is restricted by the condition of the large imaginary part of the relative permittivity at a given frequency.

It follows from Eq. (1.62) that energy loss spectra to transition radiation increases logarithmically with the particle energy.

Let us consider transition radiation generated by the charged particle intersecting the dielectric—vacuum interface for frequencies much higher than atomic frequencies. In this case, beyond narrow bands, where absorption is significant, the relative permittivity has the form

$$\varepsilon(\omega) = 1 - \frac{\omega_p^2}{\omega^2}, \quad \omega_p^2 = \frac{4\pi n e^2}{m} \ll \omega^2. \tag{1.64}$$

Since the difference $1 - \varepsilon(\omega)$ is small, the method of successive approximations can be used to solve the problem. Transition radiation can be considered as radiation appearing when the charge stops abruptly at the interface and suddenly begins to move at the same point and same time from the interface in the other medium. The spectral–angular distribution of radiation generated by the charge that moves according to law $\mathbf{r}_0(t)$ in the infinite medium and then stops abruptly is given by the expression ($\mathbf{v}_0(t) = d\mathbf{r}_0(t)/dt$):

$$\frac{d^2 W(\mathbf{n}, \omega)}{d\omega d\Omega} = \frac{Z^2 e^2 \omega^2}{4\pi^2 c^3} \varepsilon^{1/2}(\omega) \left| \int_{-\infty}^{0} dt \left[\mathbf{n}\mathbf{v}_0(t)\right] \exp\left\{ i\omega \left(t - \frac{\sqrt{\varepsilon}\mathbf{n}\mathbf{r}_0(t)}{c} \right) \right\} \right|^2. \tag{1.65}$$

For the case of the sudden stop of the charge at the interface between the semi-infinite medium and vacuum, this distribution changes due to refraction and reflection at the interface, so that corrections proportional to the small difference $1-\varepsilon(\omega)$ between the relative permittivities of vacuum and the medium appear for high frequencies. In the first approximation, quantity $1 - \varepsilon(\omega)$ can be neglected everywhere except the exponential in Eq. (1.65). In this case, the distributions of radiation appearing when the charge stops abruptly in the infinite and semi-infinite media are the same. The distribution of radiation appearing when the charge suddenly begins to move in infinite vacuum has the form

$$\frac{d^2W(\mathbf{n},\omega)}{d\omega d\Omega} = \frac{Z^2e^2\omega^2}{4\pi^2c^3}\left|\int_0^\infty dt\,[\mathbf{n}\mathbf{v}_0(t)]\exp\{i\omega(1-\mathbf{n}\mathbf{v}/c)\,t\}\right|^2. \tag{1.66}$$

Refraction and reflection at the medium—vacuum interface for high frequencies also change this distribution by a quantity proportional to the difference $1-\varepsilon(\omega)$ between the relative permittivities of the media. In the first nonvanishing approximation in $1-\varepsilon(\omega)$, the radiation distributions in the semi-infinite and infinite media are also the same.

When the energy distribution of transition radiation is calculated in the same approximation, the difference $1-\varepsilon(\omega)$ can be also neglected everywhere except the exponential. Then, the spectral–angular distribution of the radiation energy can be represented as

$$\begin{aligned}\frac{d^2W(\mathbf{n},\omega)}{d\omega d\Omega} = \frac{Z^2e^2\omega^2}{4\pi^2c^3}&\left|\int_0^\infty dt\,[\mathbf{n}\mathbf{v}_0(t)]\exp\{i\omega(t-\mathbf{n}\mathbf{r}_0(t))/c\}+\right.\\ &\left.+\int_{-\infty}^0 dt\,[\mathbf{n}\mathbf{v}_0(t)]\exp\left\{i\omega\left(t-\sqrt{\varepsilon}\mathbf{n}\mathbf{r}_0(t)/c\right)\right\}\right|^2.\end{aligned} \tag{1.67}$$

For the case of the motion of the charged particle with a constant velocity **v**, the spectral–angular distribution of relativistic electron in this approximation has the form

$$\begin{aligned}\frac{d^2W(\mathbf{n},\omega)}{d\omega d\Omega} = \frac{Z^2e^2\omega^2}{4\pi^2c^3}[\mathbf{n}\mathbf{v}]^2&\left|\int_{-\infty}^0 dt\exp\left\{i\omega\left(1-\sqrt{\varepsilon}\mathbf{n}\mathbf{v},c\right)t\right\}+\right.\\ &\left.+\int_0^\infty dt\exp\{i\omega(1-\mathbf{n}\mathbf{v}/c)\,t\}\right|^2.\end{aligned} \tag{1.68}$$

The integration with respect to time gives

$$\frac{d^2W(\mathbf{n},\omega)}{d\omega d\Omega} = \frac{Z^2e^2\omega^2}{4\pi^2c^3}[\mathbf{n}\mathbf{v}]^2\left|\frac{1}{1-\sqrt{\varepsilon}\mathbf{n}\mathbf{v}/c}-\frac{1}{1-\mathbf{n}\mathbf{v}c}\right|^2. \tag{1.69}$$

For $\omega \gg \omega_p$, one can take $\sqrt{\varepsilon(\omega)} = 1-\omega_p^2/2\omega^2$. In the ultrarelativistic case, the radiation emission angles are small, so that $\cos\theta = 1-\theta^2/2$. In this approximation, Eq. (1.69) is modified to the form

$$\frac{d^2W(\mathbf{n},\omega)}{d\omega d\Omega} = \frac{Z^2}{4}\frac{e^2}{c}\left(\frac{\omega_p}{\omega}\right)^4\frac{\theta^2}{\left(\theta^2+\gamma^{-2}+\omega_p^2/\omega^2\right)^2\left(\theta^2+\gamma^{-2}\right)^2}. \tag{1.70}$$

The integration of this expression with respect to the angles yields the radiation spectrum

$$\frac{dW}{d\omega} = \frac{Z^2 e^2}{\pi c} \left\{ \left[1 + 2\left(\frac{\omega}{\gamma\omega_p}\right)^2 \right] \ln\left[1 + \left(\frac{\omega}{\gamma\omega_p}\right)^2 \right] - 2 \right\}. \tag{1.71}$$

The integration of Eq. (1.71) with respect to the frequency provides the total energy of transition radiation:

$$\Delta W = \frac{Z^2}{3} \frac{e^2}{c} \omega_p \frac{E}{mc^2} = \frac{1}{3} Z^2 \alpha \gamma \hbar \omega_p. \tag{1.72}$$

It follows from Eq. (1.72) that energy loss to transition radiation increases linearly with the particle energy. This result was obtained theoretically in [16, 17] and experimentally confirmed in [18]. Owing to the linear dependence of radiation energy on the energy of the ultrarelativistic particle, transition radiation is a unique tool for detecting ultrarelativistic particles. The point is that the difference between the velocities of ultrarelativistic particles with noticeably different energies can be very small. In particular, the difference between the velocities of identical ultrarelativistic particles with energies E_1 and E_2 is expressed as

$$v_1 - v_2 = c\left(1 - \frac{m^2c^4}{2E_1^2}\right) - c\left(1 - \frac{m^2c^4}{2E_2^2}\right) = c\frac{m^2c^4}{E_1^2}\frac{E_1^2 - E_2^2}{E_2^2}. \tag{1.73}$$

Thus, if the velocities of ultrarelativistic particles differ by 1%, their energies differ by $\left(E/mc^2\right)^2$ %. This means that the error in the energy of the ultrarelativistic particle calculated in terms of its measured velocity is much larger than the velocity measurement error.

To detect fast particles on the basis of their electromagnetic interactions, the properties of the current density produced by a particle is used. However, since the current density is proportional to the particle velocity, the particle velocity is really measured. Thus, the error in the energy of the ultrarelativistic particle calculated in terms of its measured velocity is increased; thus, such a measurement method is complicated. The detection of the intensity of transition radiation allows direct measurement of the energy of the ultrarelativistic particle. This circumstance explains the wide application of transition radiation for detecting ultrarelativistic particles [19].

Diffraction radiation accompanies the uniform motion of the charged particle near the medium surface [20–28]. Microscopically, such a radiation is radiation generated by particle-excited electrons of atomic shells, which was discussed in the preceding section, where it was shown that only atoms whose distance b from the particle trajectory is smaller than $\gamma v/\omega$ are effectively involved in the emission process. If the distance to the medium surface from the trajectory of the particle moving in vacuum is larger than $\gamma v/\omega$, the intensity of diffraction radiation decreases exponentially. For this reason, to experimentally observe diffraction radiation, the particle must move at a sufficiently small distance from the surface. For nonrelativistic particles, this condition is satisfied only in the radio frequency range, but for ultrarelativistic particles, it can be satisfied for high frequencies. Such a restriction is absent for transition radiation, because many atoms are always located near the trajectory of a particle moving in condensed matter.

The small thickness of the surface layer of the molecules involved in the formation of diffraction radiation becomes an advantage in investigation of the surface properties of the medium. By measuring the radiation spectrum, one can study change in the properties of the medium with the thickness of the surface layer under investigation.

Another advantage of diffraction radiation over transition radiation is its weak effect both on the characteristics of the motion of the charged particle (interactions of the particle with the atoms of the medium, i.e., so-called "close collisions", are absent) and on the properties of the medium under investigation. In particular, almost nondestructive investigation of the parameters of charged particle beams becomes possible.

The process of the formation of diffraction radiation is characterized not only by the particle Lorentz factor γ and radiation wavelength λ, but also by several length scales. These are the distance from the particle to the medium surface h, the radius of the damping of the own field of the particle $\lambda\beta\gamma$, the typical size of surface inhomogeneities a (no smaller than the intermolecular distance $n_0^{-1/3}$), and the radiation formation length l_c. These parameters for various particular cases can be related by various inequalities, which allow one to choose a certain method to approximately solve a problem. Various relations between the characteristic parameters lead to various approaches to description of diffraction radiation with features

1.5 Wakefield in Linear Accelerators

An increase in the particle number density in an accelerated beam is an important problem of modern accelerator physics, because it improves the conditions of experiments on high-energy particle interactions. Beam instabilities due to the interaction of the accelerated particles between each other and with accelerating equipment hinder such an increase.

In the case of acceleration of bunches of ultrarelativistic particles, the direct interaction between particles in a bunch plays an insignificant role. It is known that the interaction force between two charges moving with the same velocities in the laboratory (fixed) reference frame decreases noticeably when the velocity approaches the speed of light. Indeed, the Lorentz force $\mathbf{F}$ with which the charge e_2 acts on the charge e_1 in the laboratory reference frame is given by the expression

$$\mathbf{F} = e_1\mathbf{E}_2 + \frac{e_1}{c}\,[\mathbf{v}\mathbf{H}_2]\,. \tag{1.74}$$

Let the beam particles move with the same velocity, **v**. In this case, the magnetic field is absent in the comoving reference frame; hence, in the laboratory reference frame,

$$\mathbf{H_2} = \frac{1}{c^2}\,[\mathbf{v}\mathbf{E_2}]\,. \tag{1.75}$$

We denote the radius vector from the charge e_2 to the charge e_1 in the comoving reference frame as **R** and the angle between **v** and **R** as ϑ. In terms of these variables, the field generated by the charge e_2 at the point, where the charge e_1 is located at a given time, in the laboratory reference frame is expressed as [1]

$$\mathbf{E_2} = \frac{e_2\mathbf{R}}{R^3}\,\frac{1-\frac{v^2}{c^2}}{\left(1-\frac{v^2}{c^2}\sin^2\vartheta\right)^{3/2}}. \tag{1.76}$$

Therefore, the force **F** has the form

$$\mathbf{F} = \frac{e_1e_2}{R^3}\,\frac{\left(1-\frac{v^2}{c^2}\right)}{\left(1-\frac{v^2}{c^2}\sin^2\vartheta\right)^{3/2}}\left\{\mathbf{R}+\frac{[\mathbf{v}\,[\mathbf{v}\mathbf{R}]]}{c^2}\right\}. \tag{1.77}$$

Let us decompose vector **R** into two components $\mathbf{R}_\parallel$ and $\mathbf{R}_\perp$ parallel and perpendicular to the velocity, respectively:

$$\mathbf{R} = \mathbf{R}_\parallel + \mathbf{R}_\perp \tag{1.78}$$

where

$$\mathbf{R}_\parallel = \frac{\mathbf{v}\,(\mathbf{v}\mathbf{R})}{v^2}, \qquad \mathbf{R}_\perp = \frac{[\mathbf{v}\,[\mathbf{R}\mathbf{v}]]}{v^2}. \tag{1.79}$$

Then, Eq. (1.77) is represented as

$$\mathbf{F} = \frac{e_1e_2}{R^3}\,\frac{\left(1-\frac{v^2}{c^2}\right)}{\left(1-\frac{v^2}{c^2}\sin^2\vartheta\right)^{3/2}}\left\{\mathbf{R}_\perp+\mathbf{R}_\parallel-\frac{v^2}{c^2}\mathbf{R}_\perp\right\}. \tag{1.80}$$

The last term in braces in Eq. (1.80) corresponds to the second term in Eq. (1.74), i.e., is a force generated by the magnetic field appearing in the laboratory reference frame (i.e., the magnetic field whose action on particles in the moving beam is observed by an observer at rest). According to Eq. (1.76), repulsive Coulomb forces increase with the velocity of the beam particles. Correspondingly, the first two terms in braces in Eq. (1.80) decrease with increasing the particle velocity for any ϑ values. However, the appearing magnetic forces not only compensate the increase in the Coulomb repulsion in the moving beam, but also lead to the decrease in the total repulsion between the beam electrons with increasing the velocity. Indeed, since

$$|\mathbf{R}_\perp| = R \sin \vartheta, \qquad \left|\mathbf{R}_\parallel\right| = R \cos \vartheta, \tag{1.81}$$

the force components parallel ($F_\parallel$) and perpendicular ($F_\perp$) to the velocity are expressed as

$$F_\parallel = \frac{e_1 e_2}{R^2} \frac{\left(1 - \frac{v^2}{c^2}\right) \cos \vartheta}{\left(1 - \frac{v^2}{c^2} \sin^2 \vartheta\right)^{3/2}}, \qquad F_\perp = \frac{e_1 e_2}{R^2} \frac{\left(1 - \frac{v^2}{c^2}\right)^2 \sin \vartheta}{\left(1 - \frac{v^2}{c^2} \sin^2 \vartheta\right)^{3/2}}. \tag{1.82}$$

Therefore, the electrons located in one cross section of the beam ($\vartheta = \pi/2$) repulse each other with the force

$$F_\perp = \frac{e^2}{R^2} \sqrt{1 - v^2/c^2} = \frac{e^2}{R^2} \gamma^{-1} \tag{1.83}$$

whereas the electrons following each other in one straight line parallel to the beam axis ($\vartheta = 0$) repulse each other with the force

$$F_\parallel = \frac{e^2}{R^2} \left(1 - v^2/c^2\right) = \frac{e^2}{R^2} \gamma^{-2}. \tag{1.84}$$

Expressions (1.83) and (1.84) give the maximum $F_\perp$ and $F_\parallel$ values, respectively. Thus, repulsion between the beam particles in a moving ultrarelativistic bunch in the laboratory reference frame decreases with increasing the velocity of the beam particles.

The repulsion between the charged beam particles is compensated by special methods of acceleration engineering, which make it possible to stabilize bunches in the process of acceleration. For this reason, the interaction between particles owing to secondary fields generated by polarization currents in the conducting units of accelerators plays the main role in the formation of beam instabilities in linear accelerators. A certain analogue is seen with the problem of diffraction radiation, although significant differences also exist. The stability of the beam is determined by the forces acting on a particle; thus, the total field, including longitudinal and transverse components, at short distances, rather than the transverse field of diffraction radiation at long distances, is of interest. The problem of the stability of particle beams in accelerators is usually characterized by axial symmetry, which is most often absent in the problem of diffraction radiation.

The appearance of the instabilities of particle beams in accelerators is primarily associated with the motion of particles through the accelerator units, where the shape or properties of the channel in which a particle beam moves change. This can be attributed both to the design of the accelerator and to the imperfection of the materials of the accelerator units. These causes are responsible for the appearance of polarization currents, which are the main sources of the secondary fields. These secondary fields lead to change in the momentum of the particles and this change usually gives rise to the instability of the beam. However, the asymmetry of the geometry of the accelerator units can promote the suppression of instabilities [29].

The importance of the problem of the instabilities of the particle beam in linear accelerators stimulated intensive investigation of this problem in numerous studies. The problems appearing in this case were specific; for this reason, new solution methods were required and new terminology appeared. Since the velocities of the particles in the beam are close to the speed of light, the field of the leading particle moving in the accelerator acts on the following particle, but the action of the field of the second particle on the first particle can be disregarded. Thus, the field acting only on the particles moving in the wake behind the leading particle plays the main role. This field was called wakefield [30–33].

Following classification presented in [34], the wakefields calculated disregarding the finite conductivity of the materials of the accelerator units are called geometric wakes, whereas the fields calculated with the use of the real characteristics of the materials are called resistive wakes.

The instabilities of the particle beam are associated with change in the momenta of these particles. Let two particles move in the accelerator. Neglecting the difference of the velocity of the ultrarelativistic particle from the speed of light, we assume that the first particle with charge q at time t is located at the point with the coordinates $x=0$, $y=0$, $z=ct$, whereas the second particle with unit charge is located at the point with the coordinates $x=\rho_x$, $y=\rho_y$, $z=s\equiv ct-l$. The field of the first particle changes the momentum of the second particle by the value

$$\Delta \mathbf{p} = \int_{-\infty}^{\infty} dt \left\{ \mathbf{E}\left(\rho_x, \rho_y, s, t\right) + \left[\mathbf{e}_z \mathbf{B}\left(\rho_x, \rho_y, s, t\right)\right]\right\}, \tag{1.85}$$

where $\mathbf{e}_z$ is the unit vector of the z axis and the integration is performed along the unperturbed orbit of the second particle.

In view of the cylindrical symmetry of the problem, it is convenient to consider separately change in the longitudinal momentum Δp_z directed along the particle path and change in the transverse momentum $\Delta \mathbf{p}_{\mathrm{tr}}$ perpendicular to the particle path. We represent these quantities in the form

$$\Delta p_z = -\left(q/c\right) w_l\left(\boldsymbol{\rho}, l\right); \qquad \Delta \mathbf{p}_{\mathrm{tr}} = \left(q/c\right) \mathbf{w}_{\mathrm{tr}}\left(\boldsymbol{\rho}, l\right), \tag{1.86}$$

$$w_l\left(\boldsymbol{\rho}, l\right) = -\frac{c}{q} \int_{-\infty}^{\infty} dt E_z\left(\rho_x, \rho_y, s, t\right), \tag{1.87}$$

$$\mathbf{w}_{\mathrm{tr}}\left(\boldsymbol{\rho}, l\right) = \frac{c}{q} \int_{-\infty}^{\infty} dt \left\{ \mathbf{E}_{\mathrm{tr}}\left(\rho_x, \rho_y, s, t\right) + \left[\mathbf{e}_z \mathbf{B}\left(\rho_x, \rho_y, s, t\right)\right]_{\mathrm{tr}} \right\}. \tag{1.88}$$

Quantities $w_l\left(\boldsymbol{\rho}, l\right)$ and $\mathbf{w}_{\mathrm{tr}}\left(\boldsymbol{\rho}, l\right)$ are usually called longitudinal and transverse wake functions, respectively. Since the action of the second particle on the first particle can be neglected for ultrarelativistic particles, $w_l(\boldsymbol{\rho}, l) = 0$ and $\mathbf{w}_{\text{tr}}(\boldsymbol{\rho}, l) = 0$ for $l < 0$. The longitudinal and transverse wake functions are related as (Panofsky—Wentzel theorem)

$$\frac{\partial \mathbf{w}_{\text{tr}}}{\partial l} = \frac{\partial w_l}{\partial \boldsymbol{\rho}}. \tag{1.89}$$

Therefore, the longitudinal and transverse wake functions can be expressed in terms of the derivatives of the same function $W(\boldsymbol{\rho}, l)$:

$$w_l(\boldsymbol{\rho}, l) = \frac{\partial W(\boldsymbol{\rho}, l)}{\partial l}, \qquad \mathbf{w}_{\text{tr}}(\boldsymbol{\rho}, l) = \frac{\partial W(\boldsymbol{\rho}, l)}{\partial \boldsymbol{\rho}}. \tag{1.90}$$

We emphasize that the wake functions $w_l(\boldsymbol{\rho}, l)$, $\mathbf{w}_{\text{tr}}(\boldsymbol{\rho}, l)$, and $W(\boldsymbol{\rho}, l)$ were defined above with respect to the orbit of the leading particle. These functions naturally depend also on the coordinates of the leading particle $\boldsymbol{\rho}'$; i.e., these functions should be written as $w_l(\boldsymbol{\rho}', \boldsymbol{\rho}, l)$, $\mathbf{w}_{\text{tr}}(\boldsymbol{\rho}', \boldsymbol{\rho}, l)$, and $W(\boldsymbol{\rho}', \boldsymbol{\rho}, l)$.

The longitudinal and transverse impedances are expressed in terms of the Fourier transforms of the wake functions:

$$\begin{aligned} Z_l(\boldsymbol{\rho}, l) &= \frac{1}{c}\int_0^\infty dl w_l(\boldsymbol{\rho}, l) \exp(i\omega l/c), \\ Z_{\text{tr}}(\boldsymbol{\rho}, l) &= -\frac{1}{c}\int_0^\infty dl \mathbf{w}_{\text{tr}}(\boldsymbol{\rho}, l) \exp(i\omega l/c). \end{aligned} \tag{1.91}$$

With the use of Eqs. (1.87) and (1.88), the impedances can be expressed in terms of the Fourier transforms of the field. For example, if the particle has only one velocity component v_z, the longitudinal impedance can be expressed as

$$Z_l(\boldsymbol{\rho}, l) = -\frac{1}{q}\int_0^\infty dz E_z(\boldsymbol{\rho}, z, \omega) \exp(-i\omega z/c). \tag{1.92}$$

Change in the longitudinal and transverse momenta of the particles is easily expressed in terms of the impedance. As an example, let us calculate the impedance for the case of ultrarelativistic particles moving in a semi-infinite, ideally conducting, cylindrical waveguide with radius a [33]. Let the waveguide axis coincide with the z axis, the charged particle enter into the open end of the waveguide in the $z = 0$ plane, and then move along the waveguide axis. Owing to the axial symmetry of the problem, the polarization current density has only one, z, component. In the limit $\gamma \to \infty$, the time Fourier transform of the electric field inside the waveguide can be represented in the form

$$\begin{aligned} &E_z(\rho, z, \omega) = -iQ \exp(ikz) K_0(\tau\rho) - \\ &-(q/(2cka)) \int_{-\infty}^{\infty} dp F(p) \chi_p^2 J_0(\chi_p \rho/a) H_0^{(1)}(\chi_p) \exp(ipz) \end{aligned} \tag{1.93}$$

where

$$k = \frac{\omega}{c},\ \tau = \frac{k}{\gamma},\ Q = \frac{qk}{\pi c \gamma^2},\ \chi_p = a\sqrt{k^2 - p^2 + 2ik\varepsilon} \tag{1.94}$$

$J_0(x)$ is the Bessel function of the first kind, $H_0(x)$ is the Hankel function of the first kind, and $K_0(x)$ is the modified Bessel function of the second kind. Function $F(p)$ is determined from the boundary conditions, which can be written in the form

$$E_z(a, z, \omega) = 0,\ \text{for} z > 0. \tag{1.95}$$

In terms of the notation

$$L(p) = \pi \chi_p^2 J_0(\chi_p) H_0^{(1)}(\chi_p), \quad \Lambda = -2iaK_0(\tau a)\left(\frac{k}{\gamma}\right)^2, \tag{1.96}$$

boundary condition Eq. (1.95) is represented in the form

$$\int_{-\infty}^{\infty} dpF(p) L(p) \exp(ipz) = \Lambda \exp(ikz),\ \text{for}\, z < 0. \tag{1.97}$$

Equations (1.93) and (1.97) form the system of linear integral equations for the function $F(p)$. We assume that the desired function $F(p)$ satisfies the following conditions:

1. the product of the functions, $F(p) L(p)$, in the upper half-plane of the complex variable p has only one pole at $p = k$, the residual of this product at this pole is equal to $\Lambda/(2i\pi)$, and product $F(p) L(p)$ is an analytic function at the other points of the upper half-plane;
2. $F(p)$ is an analytic function in the upper half-plane and vanishes in the limit $|p| \to \infty$.

As shown in [35], the solution of the system of Eq. (1.93), (1.94), (1.95), (1.96), (1.97) that satisfies conditions (1) and (2) can be obtained in the form

$$F(p) = \frac{i\Lambda}{2\pi \Gamma_+(ka)(ka)^{1/2}} \frac{\Gamma_-(pa)}{(ka - pa)^{3/2}}. \tag{1.98}$$

The functions $\Gamma_\pm$ are defined by the expression

$$\Gamma_\pm(x) = [2\eta I_0(\eta) K_0(\eta)]^{\pm 1/2} \times$$

$$\times \exp\left\{-\frac{x}{i\pi}\int_0^{ka} dt \frac{\ln[\pi\sigma I_0(\sigma) K_0(\sigma)]}{x^2 - t^2} + \frac{x}{i\pi} PV \int_{ka}^{\infty} dt \frac{\ln[2\zeta I_0(\zeta) K_0(\zeta)]}{t^2 - x^2}\right\} \tag{1.99}$$

where

$$\eta = \sqrt{x^2 - k^2a^2}, \quad \sigma = \sqrt{k^2a^2 - t^2}, \quad \zeta = \sqrt{t^2 - k^2a^2}. \tag{1.100}$$

With the use of Eq. (1.98), the longitudinal impedance for the ultrarelativistic case was obtained in [36] in the form

$$Z(k) = \frac{2kaK_0(ka/\gamma)}{c\gamma^2 I_0(ka/\gamma)} \left\{ \frac{\gamma I_1(ka/\gamma)}{I_0(ka/\gamma)} - \frac{\Gamma'_+(ka)}{\Gamma_-(ka)} - \frac{1}{4ka} \right\}. \tag{1.101}$$

In the limiting case $ka \gg 1$,

$$Z(k) \simeq \frac{2}{c} \ln\left(\frac{2\gamma}{ka}\right). \tag{1.102}$$

The above example is one of a few cases, where the impedance problem is solved analytically. In most cases, change in the shape or properties of the channel through which particles move is complicated so that simplifying assumptions must be introduced and numerical methods should be used to solve the problem. All these circumstances lead to difficulties in solving problems associated with the wakefield, but many particular problems in this field have been solved due to the urgent necessity of solving the problem of instabilities of particle beams in linear accelerators.

In particular, the effect of a finite conducting cylinder on the compression of a bunch of particles moving in a cylindrical channel of a linear accelerator was studied in [31]. Denoting the radius and thickness of the skin layer of the linear accelerator (a cylinder compressing the beam) as d and δ_d (b and δ_b), respectively, and assuming that cylinder thickness τ is much smaller than its radius b and that the inequality $\delta_b \ll d(\beta\gamma)^{-2}$ is satisfied, Gluckstern and Zotter [37] obtained the condition of the beam compression in the form

$$\tau \gg \frac{d\delta_b}{b\delta_d}. \tag{1.103}$$

The effectiveness of the beam compression was also considered in [38]. The wakefield for the case of the motion of particles in the cylindrical channel with dielectric accelerating units was studied in [39].

The methods developed in the theory of the wakefield for the linear accelerators were also applied for transition radiation.

Transition radiation generated by a particle beam passing through a thin metallic foil was analyzed in [40]. The cases, where the bunch length is comparable with its transverse sizes and where this length is much smaller than the transverse sizes of the bunch, were considered. The transverse-coordinate distribution of the particles in the bunch was assumed to be Gaussian.

The force acting on a charge moving in parallel to the surface of a semi-infinite dielectric was considered in [41] with allowance for the possibility of Cherenkov radiation in the dielectric.

It is useful to point to the similarity of the problems of the instability of particle beams in linear accelerators and the problem of diffraction radiation. The common feature of both kinds of the problems is the prime necessity of the calculation of the polarization current density in the medium. After the calculation of the polarization current density, the field at finite distances from the source must be found in accelerator physics, whereas the characteristics of diffraction radiation are determined by the field at infinity; i.e., different problems should be solved in these two cases. Hence, the application of the methods developed in accelerator physics for the calculation of diffraction radiation can be very useful, but some adaptation of these methods is necessary.

In the recent work [42] there was performed a comparative study of the backward diffraction radiation field and the wakefield for a point-like charge passing through a hole in a conducting disk. For the former case authors used the virtual diffraction model (see, Chaps. 2, 3) and for latter one — particle-in-cell code "MAGIC". As the result they showed that the geometric wakefield generated by diffraction radiation target and diffraction radiation field have the same spectrum and angular distribution. Authors concluded that the difference between two concepts only exists in the subjective terminology.

References

1. Landau, L.D., Lifshitz, E.M.: The Classical Theory of Fields. Pergamon Press, Oxford (1987)
2. Ter-Mikaelyan, M.L.: High-Energy Electromagnetic Processes in Condensed Media. Wiley, New York, NY (1972)
3. Feinberg, E.L.: Nonelastic diffractive processes at high energies. Phys. Uspekhi **58**, 193 (1956)
4. Landau, L.D., Pomeranchuk, I.Ya.: Electron-avalanche process at high energies. Dokl. Akad. Nauk. USSR **92**, 735 (1953) (See English translation in the book: The Collected Papers of L.D. Landau. Pergamon Press, New York, NY (1965))
5. Landau, L.D., Pomeranchuk, I.Ya.: The Collected Papers of L.D. Landau. Pergamon Press, New York, NY (1965)
6. Migdal, A.B.: Bremsstrahlung and pair production in condensed media at high energies. Phys. Rev. **103**, 1811 (1956)
7. Anthony, P.L., Becker-Srendy, R., Bosted, P.E. et al.: Measurement of dielectric suppression of bremsstrahlung. Phys. Rev. Lett. **76**, 3350 (1996)
8. Akhiezer, A.I., Shul'ga, N.F.: Influence of multiple scattering on the radiation of relativistic particles in amorphous and crystalline media. Sov. Phys. Usp. **30**, 197–219 (1987); Baier, V.M., Katkov, V.M.: Concept of formation length in radiation theory. Phys. Rep. **409**, 261 (2005)
9. Bethe, H., Heitler, W.: On the stopping of fast particles and on creation of positive electrons. Proc. Roy. Soc. London A. **146**, 83 (1934)
10. Wheeler, J., Lamb, W.: Influence of atomic electrons on radiation and pair production. Phys. Rev. **55**, 858 (1939)
11. Astapenko, V.A., Buimistrov, V.M., Krotov, Yu.A.: Bremsstrahlung accompanied by atom excitation and ionization JETP **66** (3), 464 (1987)

12. Amusia, M., Buimistrov, V., Zon, B. et al.: Polarization Bremsstrahlung of Particles and Atoms. Plenum, New York, NY (1992)
13. Ginzburg, V.L., Frank, I.M.: On the transition radiation theory. Sov. Phys. JETP **16**, 15 (1946)
14. Garibyan, G.M., Yang, C.: X-ray Transition Radiation. Yerevan, Armenia (1983)
15. Ginzburg, V.L., Tsytovich, V.N.: Transition Radiation and Transition Scattering. Higler, Bristol (1990)
16. Garibyan, G.M.: On the theory of transition radiation and ionization losses of particle energy. Sov. Phys. JETP **10**, 372 (1960)
17. Barsukov, K.A.: Transition radiation in a wave guide. Sov. Phys. JETP **37**, 787 (1960)
18. Yuan, L.C.L.: Energy dependence of X-ray transition radiation from ultrarelativistic charged particles. Phys. Rev. Lett. **31**, 603 (1970)
19. Dolgoshein, B.: Transition radiation detectors. Nucl. Instrum. Methods Phys. Res. A **326**, 434 (1993)
20. Smith, S.J., Purcell, E.M.: Visible light from localized surface charges moving across a grating. Phys. Rev. **92**, 1069 (1953)
21. Barnes, C.W., Dedrick, K.G.: Radiation by an electron beam interacting with a diffraction grating. J. Appl. Phys. **37**, 411 (1966)
22. Kazantsev, A.P., Surdutovich, G.I.: Radiation of a charged particle flying by metal screen. Sov. Phys. Dokl. **7**, 990 (1963)
23. Bolotovskiy, B.M., Voskresenskiy, G.V.: Radiation of charged string flying by metal screen. Sov. Phys. JETP **34**, 11 (1964)
24. Glass, S.J., Mendlowitz, H.: Quantum theory of smith-purcell experiment. Phys. Rev. **174**, 57 (1968)
25. Bolotovskiy, B.M., Voskresenskiy, G.V.: Diffraction radiation. Sov. Phys. Usp. **94**, 377 (1968)
26. Lalor, E.: Three-dimensional theory of smith-purcell effect. Phys. Rev. A. **8**, 435 (1973)
27. Potylitsyn, A.P.: Transition radiation and diffraction radiation. similarities and differences. Nucl. Instrum. Methods Phys. Res. B **145**, 169 (1998)
28. Ryazanov, M.I., Strikhanov, M.N., Tishchenko, A.A.: Diffraction radiation from an inhomogeneous dielectric film on the surface of a perfect conductor. JETP **126**, 349 (2004)
29. Burov, A., Danilov, V.: Suppression of transverse bunch instabilities by asymmetries in the chamber geometry. Phys. Rev. Lett. **82**, 2286 (1999)
30. Bane, K.L.F., Wilson, P.B., Weiland, T.: Wake fields and wake field acceleration in physics of high energy particle accelerators, AIP Conf. Proc., **127**, 876 (1985).
31. Palmer, R.B.: A qualitative study of wake fields for very short bunches. Part. Accelerators **25**, 97 (1990)
32. Chao, A.W.: Physics of Collective Beam Instabilities in High Energy Accelerators. Wiley, New York, NY (1993)
33. Heifets, S.A., Kheifets, S.A.: Coupling impedance in modern accelerators. Rev. Mod. Phys. **63**, 631 (1991)
34. Stupakov, G.V.: Geometrical Wake of a Smooth Taper. SLAC-PUB-95-7086 (1995)
35. Weinstein, L.A.: Theory of Diffraction and the Factorization Method. The Golden Press Boulder, Denver, CO (1969)
36. Kheifets, S.A., Palumbo, L.: Analytical calculation of the longitudinal impedance of a semi-infinite circular waveguide. European Organization for Nuclear Research, ReportCERN/LEP, note 580 (1987)
37. Gluckstern, R., Zotter, B.: Analysis of shielding charged particle beams by thin conductors. Phys. Rev. ST-AB **4**, 024402 (2001)
38. Al-khateeb, A.M., Boine-Frankenheim, O., Hasse, R.W., Hofmann, I.: Longitudinal impedance and shielding effectiveness of a resistive beam pipefor arbitrary energy and frequency. Phys. Rev. E **71**(2):026501 (2005)
39. Schachter, L., Byer, R.L. Siemann, R.H.: Wake field in dielectric acceleration structures. Phys. Rev. E **68**, 036502 (2003)

40. Bane, K.L.F., Stupakov, G.: Transition radiation wake-fields for a beam passing through a metallic foil. Phys. Rev. ST-AB **7**, 064401 (2004)
41. Schieber, D., Schachter, L.: Reaction forces on a relativistic point charge moving above a dielectric or a metallic half-space. Phys. Rev. E **57**, 6008 (1998)
42. Xiang, D., Huang, W-H., Lin,Y-Z., Park, S-J., Ko, I.S.: Wake of a beam passing through a diffraction radiation target. Phys. Rev. ST Accel. Beams **11**(2), 024001 (2008)

Chapter 2
General Properties of Diffraction Radiation

2.1 Diffraction Radiation as Radiation from Polarization Currents

As mentioned above, diffraction radiation can be considered as radiation generated by polarization currents induced in a medium by the field of a moving charge. The distance between the charge trajectory and medium surface is usually much larger than the mean intermolecular distance in the medium. At the same time, it is well known that the field of the charge moving in vacuum with velocity $\mathbf{v}$ and energy $E = \gamma mc^2$ decreases as $\exp(-h\omega/\gamma v)$ with distance h in the direction perpendicular to the velocity. Hence, polarization currents are located in a layer close to the surface and the properties of diffraction radiation depend strongly on the properties of this layer. In particular, radiation does not appear from a charged particle uniformly moving in parallel to the infinite plane surface of a homogeneous medium, because the conservation laws for radiation forbid the transfer of the longitudinal momentum to the medium in such a geometry. However, if the medium is inhomogeneous or its surface is not planar, the field can transfer the longitudinal momentum to the medium and radiation can appear.

As known, a medium with an electromagnetic field can be considered as homogeneous if not only the average density of the number of atoms is constant, but also the intermolecular distances are much smaller than the field wavelength. Therefore, the medium can be treated as homogeneous in the optical frequency range, but it should be considered as inhomogeneous in the high-frequency range. This means that diffraction radiation from the same surface must be considered in different ways in different frequency ranges.

Charge e whose motion in the homogeneous medium with relative permittivity $\varepsilon(\omega)$ is described by the law $x = 0$, $y = 0$, and $z = vt$ creates a field with the vector potential whose Fourier transform in space and time has the form

$$\mathbf{A}(\mathbf{q}, \omega) = \frac{e\mathbf{v}}{2\pi^2} \frac{\delta(\omega - \mathbf{qv})}{q^2 - (\omega/c)^2\varepsilon(\omega)} \tag{2.1}$$

A.P. Potylitsyn et al., *Diffraction Radiation from Relativistic Particles*,
STMP 239, 29–53, DOI 10.1007/978-3-642-12513-3_2,

In order to determine the distance dependence of the Fourier transform of the field of such a charge with frequency ω, we consider the time Fourier transform of the vector potential:

$$\mathbf{A}(\mathbf{r},\omega) = \frac{e\mathbf{v}}{2\pi^2}\int\limits_{-\infty}^{\infty}\int\limits_{-\infty}^{\infty} dq_x dq_y \frac{\exp\{iq_x x + iq_y y + i(\omega z/v)\}}{q_x^2 + q_y^2 + (\omega/v)^2[1-(v/c)^2(\varepsilon(\omega))]} \tag{2.2}$$

We consider only the case, where Cherenkov radiation is absent at the frequency under consideration and the following condition is satisfied:

$$(v/c)^2\varepsilon(\omega) < 1 \tag{2.3}$$

In this case, the denominator of the integrand in Eq. (2.2) is positive. The leading contribution to the integral comes from the d_x^2 and d_y^2 values smaller than or about $[1-(v/c)^2(\varepsilon(\omega))]$, when the denominator in Eq. (2.3) is close to its minimum value. However, if x or y values are sufficiently large so that $q_x x \gg 1$ or $q_y y \gg 1$, the exponent in the numerator of the integrand in Eq. (2.2) oscillates rapidly and significantly reduces the integral. This does not occur if the coordinates x and y are smaller than or about the limiting values x_0 and y_0 determined from the conditions

$$x_0^2(\omega/v)^2\left[1-(v/c)^2\varepsilon(\omega)\right] = 1, \qquad y_0^2(\omega/v)^2\left[1-(v/c)^2\varepsilon(\omega)\right] = 1. \tag{2.4}$$

Hence, the time Fourier component of the vector potential is large in the region, where the x and y coordinates are smaller than or about values x_0 and y_0, and is small beyond this region. The calculation of the integral shows that the x dependence has the form $\exp(-x/x_0)$. The dependence of the scalar potential, electric field, and magnetic field of the charge is the same. Thus, the ω-frequency Fourier component of the field of the charge uniformly moving in the medium decreases as $\exp\left\{-\frac{h\omega}{v}\sqrt{1-(v/c)^2\varepsilon(\omega)}\right\}$ in the direction transverse to the velocity. If relative permittivity $\varepsilon(\omega)$ is not close to one, the difference $1-(v/c)^2\varepsilon(\omega)$ is not small and is on the order of one. Therefore, the exponent is comparable to $-x\omega/v$; i.e., it depends only slightly on the energy of the fast particle. However, for frequencies much higher than atomic frequencies, the relative permittivity is close to one and has the form $\varepsilon(\omega) = 1-(\omega_p/\omega)^2$. For ultrarelativistic particles and high frequencies, $1-(v/c)^2\varepsilon(\omega) \approx (\omega_p/\omega)^2 + \gamma^{-2}$ and the ω-frequency Fourier component of the field of the charge moving in the medium decreases as $\exp\left\{-\frac{h\omega}{v}\sqrt{(\omega_p/\omega)^2+\gamma^{-2}}\right\}$ in the direction x transverse to the velocity.

Thus, the effective range of the field of the ultrarelativistic charged particle moving in vacuum increases linearly with the particle energy at frequency ω, whereas the energy dependence of the range of the field of the ultrarelativistic charged particle moving in the medium is more complex. Under condition (2.3), the range of the field increases only for frequencies much higher than atomic frequencies. This means that the polarization currents induced by the uniformly moving charged particle are primarily concentrated in a layer with thickness $\lambda\,(v/c)$ and they are concentrated in a wider layer with a thickness of about $\lambda\,(v/c)\left(\left(\omega_p/\omega\right)^2+\gamma^{-2}\right)^{-1/2}$ only for high frequencies.

The source of diffraction radiation is the polarization current that is generated by the field of the particle and whose time Fourier transform can be represented in the form

$$\mathbf{j}\,(\mathbf{r},\omega)=\frac{i\omega}{4\pi}\left\{1-\varepsilon\,(\mathbf{r},\omega)\right\}\mathbf{E}\,(\mathbf{r},\omega)\equiv\sigma\,(\mathbf{r},\omega)\,\mathbf{E}\,(\mathbf{r},\omega)\,, \tag{2.5}$$

where $\varepsilon\,(\mathbf{r},\omega)$ is the relative permittivity and $\sigma\,(\mathbf{r},\omega)$ is the conductivity of the medium.

If the layer in which polarization currents appear is sufficiently thin, the effect of the polarization currents can be considered as small perturbation and this circumstance allows one to solve the problem by the method of successive approximations. To this end, the fields are represented in the form of power series in the polarization current. In the zeroth approximation, the polarization currents can be disregarded in microscopic Maxwell's equations, so that the field in this approximation coincides with the field of the charge uniformly moving in vacuum. In the first approximation, the polarization currents in Maxwell's equations are considered to be generated by the zeroth approximation field, and exact field $\mathbf{E}\,(\mathbf{r},\omega)$ induced by the fast particle in the medium can be replaced in expression (2.5) for the polarization current by the field $\mathbf{E_0}\,(\mathbf{r},\omega)$ of this particle in vacuum.

Let us take into account that, if the homogeneous medium is bounded by the $x=0$ plane, the uniform motion of the charged particle in parallel to the z axis does not induce the radiation field. Diffraction radiation appears if the region occupied by the medium is specified by a more general condition of $x<\varsigma\,(y,z)$. Let us find the point with the minimum x coordinate on the medium surface $x=\varsigma\,(y,z)$ and chose the coordinate axes so that the $x=0$ plane pass through this point. In this case, all inhomogeneities are located inside the layer between the $x=\varsigma\,(y,z)$ and $x=0$ surfaces.

The diffraction radiation intensity depends strongly on the relation between the characteristic sizes of the problem: the range of the exponential decrease in the field of the fast particle $\lambda\,(v/c)\left[1-(v/c)^2\,\varepsilon\,(\omega)\right]^{-1/2}$, wavelength λ, and the thickness of the inhomogeneous layer $\varsigma\,(y,z)$. In the nonrelativistic case, the field of the particle decreases rapidly with the penetration depth to the medium. As a result, the thickness of the inhomogeneous layer can become much larger than the field penetration depth; i.e., $\varsigma\,(y,z)\gg\lambda\,(v/c)$. In this case, diffraction radiation is determined only by surface sections closest to the particle trajectory and information on the properties of the entire surface cannot be obtained from diffraction radiation. This information can be acquired only under the condition

$$\lambda\left(v/c\right) > \varsigma\left(y, z\right). \tag{2.6}$$

The investigation of diffraction radiation began with the nonrelativistic case [1]. The results of investigations for this case were summarized in [2].

For nonrelativistic particles, $\gamma \simeq 1$ and v/c can be so small that the penetration depth of the optical-frequency field is about or smaller than intermolecular distances. In this case, optical-frequency diffraction radiation cannot be described in the framework of macroscopic electrodynamics. However, macroscopic electrodynamics at the same particle velocity and medium surface can be applicable for diffraction radiation with lower frequencies, e.g., for cm wavelength range.

Diffraction radiation generated by relativistic particles began to be investigated slightly later, but in a wider range including optical frequencies and is actively studied by many authors [3–11].

Note that the intensity of diffraction radiation from surfaces of certain profiles was determined in many studies through approximate numerical calculations, because this problem is rather complicated. In particular, the calculations of the energy losses of the electron moving near an inhomogeneous dielectric were reported in [12]. The numerical calculations of the energy losses of the electron beam moving near a dielectric sphere and radiation appearing in this case were presented in [13]. The characteristics of radiation appearing when the electron moves near a dielectric surface on the shape of this surface were discussed in [14] with the use of the results of the numerical calculations.

2.2 Formation Length of Diffraction Radiation

The estimate presented in Sect. 1.2 for the formation length of radiation from the fast particle refers to the case, where the charged particle itself is a source of radiation. Strictly speaking, polarization currents generated in the medium by the field of the charge uniformly moving in vacuum are directly responsible for diffraction radiation. For this reason, it is useful to estimate the formation length with the inclusion of the features of diffraction radiation. Diffraction radiation is usually considered with the use of the equations of macroscopic electrodynamics with the boundary conditions at the interface between the media. If these inhomogeneities are small, phenomenological theory can be inapplicable for describing such radiation. In this case, microscopic theory should be used. Let us estimate the formation length of diffraction radiation with the use of this theory.

From the microscopic point of view, diffraction radiation appears due to the scattering of the field of the uniformly moving charge from the atoms of the medium. Such a scattering from one atom with the formation of the radiation field was considered in Sect. 1.3.

Let us consider the scattering of one Fourier component of the self field of the fast particle from two identical medium atoms at points $\mathbf{R}_1$ and $\mathbf{R}_2$ on the z axis. Let the fast charged particle uniformly move in vacuum in parallel to the z axis according to the law $\mathbf{r} = \mathbf{b} + \mathbf{v}t$ in the $x = b$ plane. Taking the x axis along $\mathbf{b}$, we can represent the field created in such a motion of the particle in the form

$$\mathbf{E}_0(\mathbf{r}, t) = \int d^3q \int d\omega \mathbf{E}_0(\mathbf{q}, \omega) \exp(i\mathbf{qr} - i\mathbf{qv}t),$$
$$\mathbf{E}_0(\mathbf{q}, \omega) = \mathbf{E}_0(\mathbf{q})\, \delta(\omega - q_z v), \tag{2.7}$$
$$\mathbf{E}_0(\mathbf{q}) = -\frac{ie}{2\pi^2} \frac{\mathbf{v}\omega c^2 - \mathbf{q}}{q^2 - \omega^2/c^2} \exp(-iq_x b).$$

Let us consider the Fourier component of the field of the particle, $\mathbf{E}_0(\mathbf{q}, \omega)\, e^{i\mathbf{qr} - i\mathbf{qv}t}$, as the incident wave. Repeating the consideration leading to Eq. (1.47), we can obtain the following expression for the Fourier transform of the current density generated in the atoms located at points $\mathbf{R}_1$ and $\mathbf{R}_2$ by the Fourier component of the field of the particle:

$$\mathbf{j}(\mathbf{k}, \omega) = \frac{-i\omega}{(2\pi)^3} \alpha(\omega)\, \mathbf{E}_0(\mathbf{q}, \omega) \left[\exp\{i(\mathbf{q} - \mathbf{k})\mathbf{R}_1\} + \exp\{i(\mathbf{q} - \mathbf{k})\mathbf{R}_2\}\right]. \tag{2.8}$$

Spectral–angular distribution of radiation created by the Fourier component of the field of the fast particle takes the form (where $\mathbf{k} = \frac{\omega}{c}\frac{\mathbf{r}}{r}$)

$$\frac{d^2W(\mathbf{n}, \omega)}{d\omega d\Omega} = \frac{\omega^2}{c} |\alpha(\omega)|^2 \left|\left[\mathbf{k}\mathbf{E}_0(\mathbf{q}, \omega)\right]\right|^2 2\left[1 + \cos\{(\mathbf{q} - \mathbf{k})(\mathbf{R}_1 - \mathbf{R}_2)\}\right]. \tag{2.9}$$

For the case under consideration, vector $\mathbf{R}_1 - \mathbf{R}_2$ is directed along the z axis; hence, $(\mathbf{q} - \mathbf{k})(\mathbf{R}_1 - \mathbf{R}_2) = (q_z - k_z)(Z_1 - Z_2) = L(\omega/v - k_z)$. The factor $2\left[1 + \cos\{L(\omega/v - k_z)\}\right]$ takes values from zero to four in dependence on the cosine argument $(\mathbf{q} - \mathbf{k})(\mathbf{R}_1 - \mathbf{R}_2) = L(\omega/v - k_z)$. For $L(\omega/v - k_z) \ll 1$, this factor is equal to four. In this case, the energy emitted by two atoms is four times higher than the energy emitted by one atom. This means that the waves emitted by both atoms are coherent, i.e., arrive at a detector with the same phases and their amplitudes are summed. As a result, the field near the detector is doubled and the energy reaching the detector is quadrupled.

If $L(\omega/v - k_z) \gg 1$, $\cos\{L(\omega/v - k_z)\}$ is a rapidly oscillating function. The detector is detected the radiation energy arriving in finite frequency and angular ranges. The integral of the expression with the rapidly oscillating function over these ranges is equal to zero. Therefore, if $L(\omega/v - k_z) \gg 1$, the radiation energy from two atoms is twice as high as the radiation energy from one atom. In this case, the waves arriving at the detector from different atoms are incoherent, i.e., have significantly different phases; for this reason, the interference term in the energy is negligibly small and intensities, rather than amplitudes, of the waves are summed.

Thus, the condition of the coherence of radiations from two atoms can be written in the form

$$L \ll l_c, \tag{2.10}$$

where the length

$$l_c \sim \frac{2\pi}{\omega/v - k_z} = \frac{\lambda}{\beta^{-1} - n_z} \tag{2.11}$$

is called the coherence length or the length of the formation of diffraction radiation. Note that length l_c varies from $\beta\lambda$ for nonrelativistic particles or emission in the direction perpendicular to the particle velocity to $\gamma^2\lambda$ for the case of emission along the velocity of the ultrarelativistic particle.

However, according to Eq. (2.9), radiations from two atoms can also be coherent under the condition

$$L\,(\omega/v - k_z) = 2\pi n, \quad n = 1, 2, 3 \ldots \tag{2.12}$$

In terms of the variables $\beta = v/c$ and $k_z = (\omega/c)\,n_z$, this condition can be represented in the form

$$\beta^{-1} - n_z = \frac{\lambda n}{L}, \tag{2.13}$$

or, in view of Eq. (2.11),

$$L = l_c n. \tag{2.14}$$

Thus, two atoms of the medium emit coherently if distance L between them is equal to an integer number of coherence lengths l_c.

We now consider the coherence conditions for the case of the ordered arrangement of atoms, e.g., in a single crystal. Let N atoms be located on a straight line at the same distance $|\mathbf{d}| = d$ from each other so that $\mathbf{R}_g = gd$, where $g = 0, 1, 2, \ldots, N-1$. The Fourier transform of the current density generated in these atoms by the Fourier component of the field of the particle is written similarly to Eq. (2.8) as

$$\mathbf{j}\,(\mathbf{k}, \omega) = \frac{-i\omega}{(2\pi)^3}\alpha\,(\omega)\,\mathbf{E}_0\,(\mathbf{q}, \omega) \sum_{g=0}^{N-1} \exp\left\{i\,(\mathbf{q} - \mathbf{k})\,g\mathbf{d}\right\}. \tag{2.15}$$

When vector $\mathbf{d}$ is directed along the z axis,

$$(\mathbf{q} - \mathbf{k})\,\mathbf{d} = d\left(\frac{\omega}{v} - \frac{\omega}{c}\cos\theta\right) = d\frac{\omega}{c}\left(\beta^{-1} - \cos\theta\right), \tag{2.16}$$

where θ is the angle between the photon emission direction and z axis. The ratio of the intensity of radiation generated by N atoms to the intensity of radiation generated by one atom is given by the expression

$$\frac{d^2 W_N(\mathbf{n},\omega)}{d^2 W_1(\mathbf{n},\omega)} = \left| \sum_{g=0}^{N-1} \exp\left\{ igd\frac{\omega}{c}\left(\beta^{-1} - \cos\theta\right)\right\} \right|^2 = \\ = \frac{\sin^2\left(\frac{dN}{2}\frac{\omega}{c}\left(\beta^{-1} - \cos\theta\right)\right)}{\sin^2\left(\frac{d}{2}\frac{\omega}{c}\left(\beta^{-1} - \cos\theta\right)\right)}. \tag{2.17}$$

The function on the right-hand side of Eq. (2.17) has a number of sharp peaks determined by the condition

$$d\frac{\omega}{c}\left(\beta^{-1} - \cos\theta\right) = 2\pi n, \quad n = 1, 2, 3, \ldots \tag{2.18}$$

The height and width of each peak corresponding to a given n value are proportional to N^2 and N^{-1}, respectively. Dispersion relation (2.18) describes Smith—Purcell radiation, which will be considered in detail in Chap. 6.

Note that, although we discuss radiation generated by individual atoms, the consideration is also applicable to radiation from individual inhomogeneities such as strips of a diffraction grating or target surface irregularities.

Formula (2.18) can be represented in the form of the requirement that the period of the structure is equal to an integer number of coherence lengths:

$$d = nl_c. \tag{2.19}$$

It is convenient to use radiation coherence condition (2.19) for periodic structures, whereas condition (2.10) is more convenient for qualitative analysis of phenomena caused by one irregularity or irregularities are located chaotically.

2.3 Radiation from Relativistic Particle Near a Screen

It is useful to consider the manifestations of the features of diffraction radiation generated by ultrarelativistic particles for a simple case of radiation generated by a particle moving near a flat screen. Let the screen of a homogeneous medium occupy a spatial region specified by the inequalities $-a < z < 0$, $x < 0$ and the particle with the charge e move in vacuum near the screen according to the law $z = vt$, $y = 0$, and $x = b > 0$. We consider diffraction radiation with optical and lower frequencies, i.e., when the wavelength is much larger than intermolecular distances and the medium of the screen can be treated as homogeneous. In this case, the Fourier transform of the polarization current given by Eq. (2.5) is represented as

$$\mathbf{j}(\mathbf{r},\omega) = \frac{i\omega}{4\pi}\left\{1 - \varepsilon(\mathbf{r},\omega)\right\}\mathbf{E}(\mathbf{r},\omega) = \sigma(\mathbf{r},\omega)\,\mathbf{E}(\mathbf{r},\omega), \\ \sigma(\mathbf{r},\omega) = \sigma(\omega)\,\theta(z+a)\,\theta(-z)\,\theta(-x); \quad \theta(x) = \frac{x+|x|}{2|x|}. \tag{2.20}$$

As pointed out above, diffraction radiation is formed not on the entire screen, but only in a small screen part that is the screen layer with a thickness of about of the $\lambda\,(v/c)\left[1-(v/c)^2\,\varepsilon\,(\omega)\right]^{-1/2}$ closest to the particle trajectory. The total field $\mathbf{E}\,(\mathbf{r},\omega)$ in Eq. (2.20) is the sum of the self field of the fast particle, $\mathbf{E}_0\,(\mathbf{r},\omega)$, and radiation field, $\mathbf{E}_1\,(\mathbf{r},\omega)$. Since the volume of the screen part in which the radiation field is formed is small, we can assume that $\mathbf{E}_1 \ll \mathbf{E}_0$ and use the method of successive approximations to solve the problem. In the first approximation, the polarization current can be treated as being generated only by the self field of the fast particle and the effect of the radiation field can be neglected. In this case, the polarization current density is known and the problem reduces to the calculation of radiation generated by the given current. In order to use expression (1.11) for the spectral–angular distribution of the emitted energy, it is necessary to determine the Fourier transform of the polarization current density given by Eq. (2.20) in space and time. We take into account that the field of the charged particle whose motion is specified by the law $z=vt$, $y=0$, $x=b$ has the form of Eq. (2.7). With the use of Eqs. (2.20) and (2.7), it is easy to obtain the Fourier transform of the polarization current in the coordinates and time in the form (where $Q\equiv\omega/v-q_z$)

$$\mathbf{j}\,(\mathbf{q},\omega)=\frac{\sigma\,(\omega)}{4\pi^2 v}\int_{-a}^{0}dz\int_{-\infty}^{0}dx\int_{-\infty}^{\infty}ds_x\exp\left(is_x x+iQz\right)\mathbf{E}\left(q_x+s_x,q_y,\omega/v\right). \tag{2.21}$$

First, we integrate with respect to s_x with the use of the relation [15]

$$\int_{-\infty}^{\infty}ds\frac{(1;\,s)}{s^2+G^2}\exp\left(isp\right)=(1;\;iG)\,\frac{\pi}{G}\exp\left(-pG\right). \tag{2.22}$$

Introducing the notation

$$G\left(q_y\right)=\left[q_y^2+\gamma^{-2}\,(\omega/c)^2\right]^{1/2} \tag{2.23}$$

and disregarding the γ^{-2} corrections, we can reduce Eq. (2.21) for the ultrarelativistic case to the form

$$\begin{aligned}\mathbf{j}\,(\mathbf{q},\omega)=&\frac{\sigma\,(\omega)}{4\pi^2 v}\left\{i\mathbf{e_x}+\frac{q_y\mathbf{e_y}}{G\left(q_y\right)}\right\}\times\\&\times\int_{-a}^{0}dz\int_{-\infty}^{0}dx\exp\left\{iQz-(b-x)\,G\left(q_y\right)-iq_x x\right\},\end{aligned} \tag{2.24}$$

where $\mathbf{e_x}$ and $\mathbf{e_y}$ are the orts of the x and y axes, respectively. Integration with respect to x and z yields

$$\mathbf{j}(\mathbf{q},\omega)=\frac{e\sigma(\omega)}{8\pi^3 v}\frac{\left[1-\exp(iQa)\right]\exp\left\{-bG\left(q_y\right)\right\}}{G\left(q_y\right)Q\left\{G\left(q_y\right)+iq_x\right\}}\left\{\mathbf{e_x}G\left(q_y\right)-i\mathbf{e_y}q_y\right\}. \quad (2.25)$$

The energy emitted by polarization current (2.25) in vacuum for the total observation time in frequency range $d\omega$ in solid angle element $d\Omega$ in the direction of vector $\mathbf{k}=\frac{\omega}{c}\frac{\mathbf{r}}{r}$ is given by the expression

$$d^2W(\mathbf{n},\omega)=\frac{1}{c}(2\pi)^6\left|\left[\mathbf{kj}(\mathbf{k},\omega)\right]\right|^2 d\omega d\Omega. \quad (2.26)$$

The substitution of Eq. (2.25) into Eq. (2.26) yields the spectral–angular distribution of diffraction radiation generated by the ultrarelativistic particle

$$\begin{aligned}\frac{d^2W(\mathbf{n},\omega)}{d\omega d\Omega}&=\frac{e^2}{v}|\sigma(\omega)|^2\frac{\left[\mathbf{ke_y}\right]^2k_y^2+\left[\mathbf{ke_x}\right]^2G^2\left(k_y\right)}{G^2\left(k_y\right)\left\{G^2\left(k_y\right)+k_x^2\right\}}\times\\&\times\frac{4\sin^2\left(\frac{a}{2}\left(\omega/v-k_z\right)\right)}{\left(\omega/v-k_z\right)^2}\exp\left\{-2bG\left(k_y\right)\right\}.\end{aligned} \quad (2.27)$$

Radiation in the ultrarelativistic case is concentrated in the region of small angles θ with respect to the particle velocity: $k_z\approx k\left(1-\theta^2/2\right)$, $k_y\approx k\theta\sin\varphi$, and $k_x\approx k\theta\cos\varphi$. Therefore, $(\omega/v)-k_z\approx(\omega/2)\left(\theta^2+\gamma^{-2}\right)$. The function $G(k)$ is on the order of k at $\theta\sim 1$ and $\sin^2\varphi\sim 1$, so that $\exp(-2bG)\sim\exp(-2bk)$, whereas $G(k)\sim k/y$ for small θ and $\sin^2\varphi$ values, so that $\exp(-2bG)\sim\exp(-2bk/y)$. Thus, the contribution to radiation from the angular range $\theta\sim 1$ and $\sin^2\varphi\sim 1$ is exponentially small and the main contribution to radiation comes from the region of small angles $\theta\le\gamma^{-1}$ and angles φ close to zero or π. This dependence on angle φ means that the radiation emission directions are primarily concentrated near the xz plane, i.e., near the symmetry plane of the problem.

The radiation intensity depends strongly on the ratio of screen thickness a to coherence length $\lambda\gamma^2$. If the screen thickness is much smaller than the coherence length, i.e., $a\left[(\omega/v)-k_z\right]/2\ll 1$, the radiation distribution given by Eq. (2.27) takes the form

$$\frac{d^2W(\mathbf{n},\omega)}{d\omega d\Omega}=\frac{e^2}{v}4a^2|\sigma(\omega)|^2\frac{\left[\mathbf{ke_y}\right]^2k_y^2+\left[\mathbf{ke_x}\right]^2G^2\left(k_y\right)}{G^2\left(k_y\right)\left\{G^2\left(k_y\right)+k_x^2\right\}}\exp\left\{-2b\,G\left(k_y\right)\right\}. \quad (2.28)$$

For optical frequencies and $\gamma\sim 10^3$, the coherence length can be equal to several centimeters, so that the condition $a\ll\lambda\gamma^2$ is easily satisfied. If $a\left[(\omega/v)-k_z\right]/2=\pi/2$, Eq. (2.27) is modified to the form

$$\frac{d^2W(\mathbf{n},\omega)}{d\omega d\Omega}=\frac{e^2}{v}4\left|\sigma(\omega)\right|^2\frac{\left[\mathbf{ke_y}\right]^2k_y^2+\left[\mathbf{ke_x}\right]^2G^2\left(k_y\right)}{G^2\left(k_y\right)\left\{G^2\left(k_y\right)+k_x^2\right\}\left(\omega/v-k_z\right)^2}\exp\left\{-2b\,G\left(k_y\right)\right\}. \tag{2.29}$$

However, if the frequency and direction of radiation satisfy the condition $a\,(\omega/v-k_z)=2K\pi$, where K is an integer number, the radiation intensity is equal to zero. This means that angles corresponding to the maxima and minima of the intensity exist at a given radiation frequency. The appearance of these maxima and minima is attributed to the coherence of radiation formed in various sections of the screen. We emphasize that the coherence length for diffraction radiation generated by the nonrelativistic particle is on the order of $\lambda\,(v/c)$ and the problem of creating the screen with the thickness comparable with the coherence length for optical and higher frequencies is impracticable.

Disregarding the γ^{-2} corrections, we can represent Eq. (2.28) for angles $\theta\le\gamma^{-1}$ in the form

$$\begin{aligned}\frac{d^2W(\omega,\theta,\varphi)}{\theta d\theta d\varphi d\omega}&=\frac{e^2}{v}\left|\sigma(\omega)\right|^2\frac{4a^2}{k^2}\frac{2\theta^2\sin^2\varphi+\gamma^{-2}}{\left(\theta^2\sin^2\varphi+\gamma^{-2}\right)\left(\theta^2+\gamma^{-2}\right)}\times\\&\times\exp\left\{-2bk\sqrt{\theta^2\sin^2\varphi+\gamma^{-2}}\right\}.\end{aligned} \tag{2.30}$$

According to this expression, the radiation intensity is low if the distance between the trajectory of the ultrarelativistic particle and the dielectric screen, b (impact parameter), is much larger than $\gamma\lambda$.

This result is obtained under the assumption that the radiation field is much lower than the self field of the particle. Although this assumption is always valid for the frequencies exceeding optical frequencies, it will be shown in Chap. 4 that a more accurate approach similar to that developed in Sect. 1.4 for calculating the characteristics of transition radiation at the frequencies exceeding optical frequencies is required in ultraviolet and soft X-ray frequency ranges.

The imaginary part of the relative permittivity should be taken into account for optical and lower frequencies. This is most substantial for conducting media when $\varepsilon(\omega)=\varepsilon'(\omega)+i\varepsilon''(\omega)$ and $\varepsilon'(\omega)\ll\varepsilon''(\omega)$. In this case, the field of the particle varies strongly in the medium at the skin-layer thickness $\delta\sim\dfrac{c}{\omega}\sqrt{\varepsilon''(\omega)}$; therefore, the method of successive approximations is inapplicable. Hence, Eqs. (2.27), (2.28), (2.29), (2.30) are applicable in a frequency region, where $\varepsilon''(\omega)\ll\varepsilon'(\omega)<1$.

2.4 Diffraction Radiation from Ultrarelativistic Particles

As known, radiation in the ultrarelativistic case is concentrated in the region of small θ angles near the particle velocity direction. In this case, the thickness of the surface layer with polarization currents for the ultrarelativistic particle can be on the order of or much smaller than the wavelength, whereas the formation length of diffraction radiation, $l_c \sim \lambda\left(\beta^{-1} - \cos\theta\right)^{-1}$, for the characteristic emission angles about $\theta \ll 1$ is much larger than the field wavelength λ. The diffraction radiation intensity is low when the linear sizes of inhomogeneities either much smaller or much larger than the radiation formation length. This means that diffraction radiation is most intense when the sizes of the surface inhomogeneities are on the order of the radiation formation length. Therefore, diffraction radiation generated by the ultrarelativistic particle moving near the medium whose surface irregularities have a linear size of b is most intense for the frequencies on the order of $4\pi\,(c/b)\left(\gamma^{-2} + \theta^2\right)^{-1}$. This is much different from diffraction radiation generated by the nonrelativistic particle, where the intensity maximum is in a frequency range of about $4\pi\,(v/b)$.

Let us consider diffraction radiation in the X-ray frequency range when the radiation frequency is much higher than atomic frequencies and the wavelength is smaller than or about the atomic size. Note that the consideration of X-ray diffraction radiation is meaningful only for the ultrarelativistic particles, because such a radiation for the nonrelativistic particles is negligibly small due to the fast decrease in the self field in the direction perpendicular to the velocity.

In the x-ray frequency range, macroscopic electrodynamics is inapplicable, averaging over the volume is not performed, but the electron number density is averaged over the quantum-mechanical states and thermal motion of the atoms. The polarization current density depending on the atomic coordinates is involved in microscopic Maxwell's equations in order to include the reverse effect of the polarization current on the field. This effect in the x-ray frequency range is small and can be considered as small perturbation. Since atomic frequencies are much lower than the field frequency, the coupling forces of the atomic electrons interacting with the field can be disregarded and the electrons can be considered as free particles when calculating the polarization current. In this approximation, microscopic Maxwell's equations can be represented in a form similar to the equations of the macroscopic electrodynamics of an inhomogeneous medium by introducing the following analogue of the relative permittivity:

$$\varepsilon\,(\mathbf{r}, \omega) = 1 - \chi\,(\mathbf{r}, \omega) \equiv 1 - \frac{4\pi e^2}{m\omega^2} \sum_a f\,(\mathbf{r} - \mathbf{R}_a)\,, \tag{2.31}$$

where the summation is performed over all the molecules of the medium, $\mathbf{R}_a$ is the radius vector of the center-of-mass of the molecule, and $f\,(\mathbf{r})$ is the electron number density in the molecule averaged over the quantum-mechanical electronic states and thermal motion. The microscopic electron number density in the medium can be represented in the form

$$n_e = \sum_a f\,(\mathbf{r} - \mathbf{R}_a)\,. \tag{2.32}$$

The electric displacement can be formally introduced by the expression

$$\mathbf{D}\,(\mathbf{r}, \omega) = \varepsilon\,(\mathbf{r}, \omega)\,\mathbf{E}\,(\mathbf{r}, \omega)\,. \tag{2.33}$$

However, this quantity is a microscopic quantity and depends on the coordinates of the atoms of the medium. The formal coincidence of Eq. (2.33) with the relation between the displacement and field of the inhomogeneous medium in macroscopic electrodynamics is natural, because the medium cannot be treated as homogeneous for a field with the wavelength of about atomic sizes. When the charged particle moves in parallel to the plane surface of the inhomogeneous medium, the field can transfer momentum to the irregularities of the medium and thereby generate diffraction radiation. This phenomenon is the simplest example of X-ray diffraction radiation.

In this region, the relative permittivity is close to one and, hence, perturbation theory in the small quantity $\chi(\mathbf{r},\omega) = 1 - \varepsilon(\mathbf{r},\omega)$ can be used. In the first approximation, only the expansion terms linear in $\chi(r,\omega)$ can be retained. In the zeroth approximation, the quantity $\chi(\mathbf{r},\omega)$, i.e., polarization currents in the medium, can be neglected. For the problem concerning diffraction radiation, this means that the zeroth approximation field is the field of the charged particle uniformly moving in infinite vacuum. The field of the charged particle whose motion is described by the law $x = b,\ y = 0,\ z = vt$ has the form of Eq. (2.7).

A source of diffraction radiation is the polarization current that is induced by the self field $\mathbf{E}$ in the medium and whose time Fourier transform can be represented in the form

$$\mathbf{j}(\mathbf{r},\omega) = \frac{i\omega}{4\pi}\chi(\mathbf{r},\omega)\,\mathbf{E}_0(\mathbf{r},\omega) = \frac{ie^2}{m\omega}\mathbf{E}_0(\mathbf{r},\omega)\sum_a f(\mathbf{r}-\mathbf{R}_a). \tag{2.34}$$

Passing to the space Fourier transforms in Eq. (2.34) with the use of Eq. (2.7) and the relation

$$f(\mathbf{r}) = \int d^3q f(\mathbf{q})\exp(i\mathbf{qr}), \tag{2.35}$$

we easily obtain

$$\mathbf{j}(\mathbf{k},\omega) = \frac{ie^2}{m\omega}\int d^3p\,\mathbf{E}_0(\mathbf{k}-\mathbf{p})\,\delta(\omega - k_z v + p_z v)\,f(\mathbf{p})\sum_a \exp\left(-i\mathbf{pR}_a\right). \tag{2.36}$$

For a crystal, $\mathbf{R}_a = \mathbf{e_x}a_x l + \mathbf{e_y}a_y m + \mathbf{e_z}a_z s$, where $l = 1,2,\dots L$, $m = 1,2,\dots M$, $s = 1,2,\dots S$, so that

$$\begin{aligned}\sum_a \exp\left(-i\mathbf{pR}_a\right) = {} & \frac{\exp(-ip_x a_x L) - 1}{\exp(-ip_x a_x) - 1} \times \frac{\exp\left(-ip_y a_y M\right) - 1}{\exp\left(-ip_y a_y\right) - 1} \\ & \times \frac{\exp(-ip_z a_z S) - 1}{\exp(-ip_z a_z) - 1}.\end{aligned} \tag{2.37}$$

Each factor in Eq. (2.37) is a rapidly oscillating function of vector **p** and almost does not contribute to the integral except for the **p** magnitudes for which the arguments of the exponentials are small, $p_x a_x L \ll 1$, $p_y a_y M \ll 1$, and $p_z a_z S \ll 1$. Thus, the right-hand side of Eq. (2.37) is equal to the product $L \cdot M \cdot S$, i.e., to the number of molecules, N, in the entire crystal volume. If $L \gg 1$, $M \gg 1$, and $S \gg 1$, the **p** vector magnitudes for which sum (2.37) is not small coincide with the reciprocal lattice vectors

$$\mathbf{g}\left(\frac{2\pi}{a_x}n_1, \frac{2\pi}{a_y}n_2, \frac{2\pi}{a_z}n_3\right), \tag{2.38}$$

where n_1, n_2, n_3 are arbitrary integers. Thus,

$$\sum_a \exp\left(-i\mathbf{p}\mathbf{R}_a\right) = N\delta\left(\mathbf{p}-\mathbf{g}\right). \tag{2.39}$$

The substitution of Eq. (2.39) into Eq. (2.36) yields

$$\mathbf{j}\left(\mathbf{k},\omega\right) = \frac{ie^2}{m\omega}N\sum_{\mathbf{g}}\mathbf{E}_0\left(\mathbf{k}-\mathbf{g}\right)\delta\left(\omega - k_z v + g_z v\right) f\left(\mathbf{g}\right). \tag{2.40}$$

Here, the Dirac delta function means that the Fourier transform of the polarization current is nonzero only at certain values of angle θ between vector **k** and particle velocity **v**:

$$\cos\theta = \frac{c}{v}\left(1 + \frac{vg_z}{\omega}\right). \tag{2.41}$$

The energy emitted by a certain current in vacuum for the total observation time in the frequency range $d\omega$ to the solid angle element $d\Omega$ in the direction of vector **k** is given by the expression

$$d^2W\left(\mathbf{n},\omega\right) = \frac{(2\pi)^6}{c}\left|\left[\mathbf{k}\mathbf{j}\left(\mathbf{k},\omega\right)\right]\right|^2 d\omega d\Omega. \tag{2.42}$$

The substitution of Eq. (2.40) into Eq. (2.42) gives rise to the appearance of the Dirac delta function squared

$$\delta^2\left(\omega - k_z v + g_z v\right) = \frac{T}{2\pi}\delta\left(\omega - k_z v + g_z v\right), \tag{2.43}$$

where T is the total observation time. This means that the angular distribution of diffraction radiation from the crystal consists of a set of narrow peaks near the angles satisfying inequality (2.41) for various g_z values. For this reason, each peak can be analyzed independently. X-ray diffraction radiation generated by the ultrarelativistic particle moving near a single crystal is discussed in detail in Sect. 4.5.

The properties of the electromagnetic waves with the frequencies between optical frequencies and x-ray frequencies differ from the properties of the optical and X-ray waves. This frequency range is the range of ultraviolet and soft x-ray radiations. In this range, the field wavelength is larger than atomic sizes:

$$\lambda \gg \frac{\hbar^2}{me^2}, \quad \text{i.e.} \quad \hbar\omega \ll \alpha mc^2 \tag{2.44}$$

(where $\alpha = e^2/\hbar c \simeq 1/137$ is the fine structure constant) and the frequencies are higher than atomic frequencies:

$$\hbar\omega \gg \alpha^2 mc^2. \tag{2.45}$$

In the frequency range, where both conditions (2.44) and (2.45) are satisfied,

$$\alpha mc^2 \gg \hbar\omega \gg \alpha^2 mc^2, \tag{2.46}$$

macroscopic electrodynamics is applicable; i.e., the properties of the medium are described by the usual relative permittivity. Since the wavelength is much larger than atomic sizes, the homogeneous medium approximation is applicable. Owing to inequality (2.45), the binding forces of the electrons in the atom can be disregarded. Hence, the relative permittivity of the medium in the frequency range specified by inequalities (2.46) can be represented as

$$\varepsilon(\omega) = 1 - \omega_p^2/\omega^2, \quad \omega_p = \sqrt{4\pi NZe^2/m}, \tag{2.47}$$

where N is the number of atoms per unit medium value and Z is the nuclear charge number. The thickness of the medium layer, where the polarization currents induced by an ultrarelativistic particle with the energy $E = \gamma Mc^2$ in the frequency range specified by inequalities (2.46) are located, is given by the expression (where $\beta = v/c$)

$$a \sim \frac{\lambda\beta}{\sqrt{(\omega_p/\omega)^2 + \gamma^{-2}}}. \tag{2.48}$$

The thickness of this layer is much larger than the radiation wavelength.

Let an ultrarelativistic charged particle move near the nonplanar surface of a homogeneous medium occupying the spatial region specified by the inequality $\varsigma(y, z) > x$ and changes in the surface profile occur in the region $x > 0$. In this case, the coordinate dependence of the relative permittivity is determined only by the medium surface profile and, hence,

$$1 - \varepsilon(\mathbf{r}, \omega) = (\omega_p/\omega)^2 \theta[\varsigma(y, z) - x], \qquad \theta(u) = \frac{u + |u|}{2|u|}. \tag{2.49}$$

As shown in Sect. 2.1, diffraction radiation in this case is generated by polarization currents in the layer between the $x = 0$ plane and $x = \varsigma(y, z)$ surface. The Fourier transform of the polarization current density can be represented as

$$\mathbf{j}(\mathbf{r}, \omega) = \frac{i\omega_p^2}{4\pi\omega}\mathbf{E}(\mathbf{r}, \omega)\,\theta(x)\,\theta\left[\varsigma(y, z) - x\right], \tag{2.50}$$

where $\mathbf{E}(\mathbf{r}, \omega)$ is the Fourier transform of the field generated by the moving particle in the medium. In order to determine this field by usual methods of macroscopic electrodynamics, it is necessary to find the general solution of Maxwell's equations in the medium and in vacuum and to match these solutions on the surface of the homogeneous medium, $x = \varsigma(y, z)$. The exact solution can be obtained for the surfaces of the simplest profile, whereas approximate methods should be used for other cases. When $\omega_p^2/\omega^2 \ll 1$, this ratio can be used as a small parameter in the method of successive approximations. This problem is analyzed in detail in Chap. 4.

2.5 Effect of the Excitation of the Medium on Diffraction Radiation

Diffraction radiation appearing when the charged particle moves near the surface of the stationary medium is considered above. The properties of such a medium are time independent and its energy is conserved. If the relative permittivity of the medium is $\varepsilon(\omega) < (c/v)^2$, Cherenkov radiation is impossible, the energy and momentum conservation laws in the emission process are satisfied due to the transfer of momentum to the medium, but the transfer of momentum is possible only in the inhomogeneous medium. If the properties of the medium vary in time, the energy of the medium is not conserved. The properties of the nonstationary medium in macroscopic electrodynamics were considered in [16, 17]. In such a medium, the energy exchange between the medium and field is possible; in particular, the energy and momentum conservation laws for radiation can be satisfied without the transfer of momentum to the medium due only to the energy transfer from the medium to the field. This means that diffraction radiation appears when the charged particle uniformly moves in parallel to the planar surface of the homogeneous nonstationary medium [18]. If the medium is nonstationary and inhomogeneous, momentum transfer $\Delta\mathbf{p}$ to the medium and energy transfer ΔE from the medium can exist in the process of emission. In this case, the energy conservation law for the emission of the transverse wave with frequency ω and wave vector $\mathbf{k}$ has the form of Eq. (1.54):

$$\hbar\omega = \hbar\mathbf{k}\mathbf{v} + \mathbf{v}\Delta\mathbf{p} + \Delta E. \tag{2.51}$$

If $\mathbf{v}\Delta\mathbf{p} \ll \Delta E$, the transfer of the energy from the medium to the field obviously plays the leading role in the emission process and the momentum transfer can be disregarded in the first approximation. In this approximation, the problem reduces to the calculation of diffraction radiation from the homogeneous nonstationary medium.

As an example, we consider the medium excited by a certain interaction. Our consideration is limited to the case, where the ionization of the medium in the process of excitation can be disregarded; i.e., excitation energy per atom is lower than the ionization potential. This means that excitation is weak. After the termination of such an interaction, the electronic excitation energy migrates in the medium in the form of long-lived elementary excitations of the medium, i.e., electromagnetic longitudinal plane waves.

As known, the relation between frequency ω and wave vector $\mathbf{q}$ of the longitudinal plane electromagnetic wave is determined by the condition of vanishing the relative permittivity as a function of the frequency and wave vector: $\varepsilon(\mathbf{q}, \omega) = 0$. For small $\mathbf{q}$ magnitudes, the solution of this equation has the form

$$\omega(\mathbf{q}) = \omega_p + (\alpha/2)\, q^2, \tag{2.52}$$

where α is a constant and ω_p is the frequency at which the relative permittivity is equal to zero at $q = 0$ [19] : $\varepsilon\left(\omega_p, q = 0\right) = 0$. Hence, the electron field of the longitudinal plane wave for small q values has the form

$$\mathbf{E}(\mathbf{r}, t) = \mathbf{E} \exp\left\{i\mathbf{qr} - i\omega_p t - i(\alpha/2)\, q^2 t\right\}. \tag{2.53}$$

The propagation velocity of such a wave is $\mathbf{u} = \alpha\mathbf{q}$.

Let us consider diffraction radiation appearing when the ultrarelativistic charged particle moves in parallel to the planar surface of the homogeneous medium whose excitation is described by longitudinal wave (2.53). In such a geometry of the problem, diffraction radiation is impossible in the absence of the longitudinal wave. In the presence of the longitudinal wave, the medium becomes nonstationary and inhomogeneous. The energy conservation law in such a medium has the form of Eq. (2.51). The typical momentum transferred to the medium is on the order of $\hbar\mathbf{q}$ and the typical energy acquired from the medium is on the order of $\hbar\omega(q)$. The inequality $\mathbf{v}\Delta\mathbf{p} \ll \Delta E$ takes the form $qv \ll \omega_p + (\alpha/2)\, q^2$ and, under this inequality, radiation in the first approximation can be considered as diffraction radiation from the homogeneous nonstationary medium by changing the field of the longitudinal waves to $\mathbf{E}_p \exp\left(-i\omega_p t\right)$. However, since the momentum transfer from the field to the medium is disregarded, $\mathbf{q} = \mathbf{k}$ and the equality $\mathbf{kv} \ll \omega_p + \alpha k^2/2$ should be satisfied. This equality is always satisfied for the nonrelativistic particles and is satisfied for the ultrarelativistic particles only at the frequencies $\omega \ll \omega_p$. This condition determines the region of applicability of the approximate solution method.

In order to avoid insignificant complications, we assume that the distance from the particle trajectory to the medium surface is not small and the self field in the medium is low, so that the layer in which the polarization current exists is sufficiently thin. In this case, the effect of the polarization current can be treated by the method of successive approximations in the powers of the polarization current. In the zeroth approximation, the polarization currents can be disregarded in microscopic Maxwell's equations, so that the field in this approximation coincides with the field of the charged particle uniformly moving in vacuum. In the first approximation, the terms linear in the polarization current are taken into account. In this approximation, it can be assumed that the polarization current in the equations for the first approximation field is induced by the zeroth approximation field, i.e., by the field $\mathbf{E}_0(\mathbf{r}, t)$ of the charged particle uniformly moving in vacuum. Thus, the calculation of the first approximation field reduces to the problem of the field generated by a given current in vacuum.

Let us find the polarization current density in the medium for the case of the sufficiently thin layer with the polarization current. Assuming that the fields acting on a bound atomic electron are much lower than atomic fields, the electron can be treated as quasielastically bound. The equations of motion of the electron quasielastically bound in an atom under the action of the field of the fast particle,

$$\mathbf{E}_0(\mathbf{r}, t) = \int d^3q \int d\omega \mathbf{E}_0(\mathbf{q}, \omega) \exp(i\mathbf{qr} - i\omega t), \tag{2.54}$$

and field $\mathbf{E}_p \exp\left(-i\omega_p t\right)$ have the form

$$\frac{d^2\mathbf{r}(t)}{dt^2} + \omega_\mu^2 \mathbf{r}(t) = \frac{e}{m}\left\{\mathbf{E}_p \exp\left(-i\omega_p t\right) + \int d^3q \int d\omega \mathbf{E}_0(\mathbf{q}, \omega)\, e^{i\mathbf{qr}(t)}\right\}, \tag{2.55}$$

where e is the elementary charge and m, and ω_μ are the mass, and oscillation frequency of the bound atomic electron, respectively. Taking into account that the field wavelength is much larger than the oscillation amplitude, we can solve this equation by the method of successive approximations in the field powers:

$$\mathbf{r}(t) = \mathbf{r}_0(t) + \mathbf{r}_1(t) + \mathbf{r}_2(t) + \ldots, \tag{2.56}$$

where $\mathbf{r}_0(t)$ is field independent, $\mathbf{r}_1(t)$ is linear in the field, $\mathbf{r}_2(t)$ is square in the field, etc. In the zeroth approximation in the field, this equation is the equation of the free oscillations of the electron. In the first approximation, only terms linear in field hold in Eq. (2.55). In this case, since the field wavelength is much larger than atomic sizes, $\mathbf{r}_0(t)$ can be replaced by the radius vector of the atomic nucleus, $\mathbf{R}_a$:

$$\frac{d^2\mathbf{r}_1(t)}{dt^2} + \omega_\mu^2 \mathbf{r}_1(t) = \frac{e}{m}\left\{\mathbf{E}_p \exp\left(-i\omega_p t\right) + \int d^3q \int d\omega \mathbf{E}_0(\mathbf{q}, \omega)\, e^{i\mathbf{q}\mathbf{R}_a - i\omega t}\right\}. \tag{2.57}$$

The solution of this equation makes it possible to determine the polarization current density $\mathbf{J}(\mathbf{r}, t)$ appearing in the case of the homogeneous stationary medium. However, as shown above, diffraction radiation does not appear in the problem geometry under consideration for the case of the homogeneous stationary medium.

In this approximation, when the solution linear in the polarization current is sought, current $\mathbf{J}(\mathbf{r}, t)$ can be disregarded and the additional polarization current $\mathbf{j}(\mathbf{r}, t)$ appearing due to the excitation of the medium can be calculated. To this end, it is necessary to solve the second-approximation equation for motion of the electron quasielastically bound in the atom in the presence of the field of the fast particle, $\mathbf{E}_0(\mathbf{r}, t)$, in the excited medium. This equation contains the field-squared terms:

$$\frac{d^2\mathbf{r}_2(t)}{dt^2} + \omega_\mu^2 \mathbf{r}_2(t) = \frac{ie}{m} \int d^3p \left(\mathbf{p}\mathbf{r}_1(t)\right) \int d\omega \, \mathbf{E}_0(\mathbf{p}, \omega) \exp\left(i\mathbf{p}\mathbf{R}_a - i\omega t\right). \tag{2.58}$$

The transition to the time Fourier transforms of the coordinates in Eq. (2.57) provides

$$\left(\omega_\mu^2 - \omega\right) \mathbf{r}_1(\omega) = \frac{e}{m} \left\{ \mathbf{E}_p \delta\left(\omega - \omega_p\right) + \int d^3p \, \mathbf{E}_0(\mathbf{p}, \omega) \exp\left(i\mathbf{p}\mathbf{R}_a\right) \right\}, \tag{2.59}$$

so that

$$\mathbf{r}_1(t) = \mathbf{r}_{1p}(t) + \mathbf{r}_{10}(t); \qquad \mathbf{r}_{1p}(t) = \frac{e}{m} \mathbf{E}_p \frac{\exp\left(-i\omega_p t\right)}{\omega_\mu^2 - \omega_p^2}, \tag{2.60}$$

$$\mathbf{r}_{10}(t) = \frac{e}{m} \int d^3p \int d\omega \frac{\mathbf{E}_0(\mathbf{p}, \omega)}{\omega_\mu^2 - \omega^2} \exp\left(i\mathbf{p}\mathbf{R}_a - i\omega t\right). \tag{2.61}$$

The substitution of the expression for $\mathbf{r}_1(t)$ into Eq. (2.58) results in the appearance of two terms on the right-hand side of Eq. (2.58). The first term is squared in $\mathbf{E}_0$ and the second term is proportional to both self field $\mathbf{E}_0$ and longitudinal wave field $\mathbf{E}_p$. Correspondingly, the solution of Eq. (2.58) consists of two terms, $\mathbf{r}_2(t) = \mathbf{r}_{2p}(t) + \mathbf{r}_{20}(t)$. Radiation of interest is associated with the solution $\mathbf{r}_{2p}(t)$ of the equation following from Eq. (2.58):

$$\frac{d^2\mathbf{r}_{2p}(t)}{dt^2} + \omega_\mu^2 \mathbf{r}_{2p}(t) = \frac{ie}{m} \int d^3p \left(\mathbf{p}\mathbf{r}_{1p}(t)\right) \int d\omega \mathbf{E}_0(\mathbf{p}, \omega) e^{i\mathbf{p}\mathbf{R}_a - i\omega t}. \tag{2.62}$$

The substitution of Eq. (2.60) into Eq. (2.62) provides the expression

$$\mathbf{r}_{2p}(t) = i \left(\frac{e}{m}\right)^2 \left(\omega_\mu^2 - \omega_p^2\right) \int d^3q \left(\mathbf{p}\mathbf{E}_p\right) \int d\omega \frac{\mathbf{E}_0(\mathbf{p}, \omega - \omega_p)}{\omega_\mu^2 - \omega^2} e^{i\mathbf{p}\mathbf{R}_a - i\omega t}. \tag{2.63}$$

From this expression, the current density in the atom for such a motion of the electrons is easily obtained in the dipole approximation (summation is performed over all the atomic electrons):

$$\mathbf{j}_a(\mathbf{r}, t) = e \sum_{\mu} \frac{d\mathbf{r}_{2p}(t)}{dt} \delta(\mathbf{r} - \mathbf{R}_a). \tag{2.64}$$

Summing this expression over all the atoms of the medium, we can arrive at the following expression for the polarization current density, which is responsible for diffraction radiation:

$$\mathbf{j}(\mathbf{r}, t) = \sum_a \delta(\mathbf{r} - \mathbf{R}_a) \int d^3p \mathbf{E}_0(\mathbf{p}) (\mathbf{p}\mathbf{E}_p) \int d\omega U(\omega) \delta(\omega - \omega_p - p_z v) e^{i\mathbf{p}\mathbf{R}_a - i\omega t}, \tag{2.65}$$

where

$$U(\omega) = \sum_{\mu} \frac{e^3 \omega}{m^2 \left(\omega_{\mu}^2 - \omega_p^2\right)\left(\omega_{\mu}^2 - \omega^2\right)} \tag{2.66}$$

and it is taken into account that the field of the charged particle whose motion is described by the law $y = 0$, $x = b$, and $z = vt$ is given by Eq. (2.7). Assuming that the field wavelength is much larger than interatomic distances, we can average the current density over the coordinates of the medium atoms and arrive at the expression

$$\mathbf{j}(\mathbf{r}, t) = n_0 \int d^3R\, \theta(-X) \int d^3p\, (\mathbf{p}\mathbf{E}_p) \int d\omega\, U(\omega)\, \delta(\omega - \omega_p - p_z v)\, \mathbf{E}_0(\mathbf{p})\, e^{i\mathbf{p}\mathbf{R} - i\omega t}, \tag{2.67}$$

where n_0 is the average number of atoms per unit volume and $\theta(x)$ is the Heaviside step function (see, Eq. (2.49)). The Fourier transform of polarization current (2.67) in space and time is easily obtained in the form

$$\begin{aligned} \mathbf{j}(\mathbf{k}, \omega) = U(\omega)\, \delta(\omega - \omega_p - k_z v) \int_0^{\infty} dX \int_{-\infty}^{\infty} dp_x \left\{\mathbf{k}\mathbf{E}_p + (p_x - k_x)\, \mathbf{e_x}\mathbf{E}_p\right\} \times \\ \times \mathbf{E}_0(p_x, k_y, k_z) \exp\{i(k_x - p_x) X\}, \end{aligned} \tag{2.68}$$

where $\mathbf{e_x}$ is the ort of the x axis. The substitution of Eq. (2.7) into Eq. (2.68) yields (where $\gamma = 1/\sqrt{1 - v^2/c^2}$)

$$\begin{aligned} \mathbf{j}(\mathbf{k}, \omega) = \beta(\omega)\, \delta(\omega - \omega_p - k_z v) \int_0^{\infty} dX \exp(ik_x X) \int_{-\infty}^{\infty} dp_x \exp\{-ip_x(X + b)\} \times \\ \times \left\{\mathbf{k}\mathbf{E}_p + (p_x - k_x)\, \mathbf{e_x}\mathbf{E}_p\right\} \frac{ie}{2\pi^2 c^2} \frac{\mathbf{e_x}(p_x - k_x) + \mathbf{e_y} k_y + \mathbf{e_z}(\omega - \omega_p)/v\gamma^2}{p_x^2 + k_y^2 + \left[(\omega - \omega_p)/v\gamma\right]^2}. \end{aligned} \tag{2.69}$$

Introducing the notation

$$G\left(k_y\right)=\sqrt{k_y^2+\left(\frac{\omega-\omega_p}{v\gamma}\right)^2} \tag{2.70}$$

and integrating with respect to p_x with the use of the known relation [15]

$$\int_{-\infty}^{\infty} du\,\left(1;u;u^2\right)\frac{\exp(iua)}{u^2+G^2}=\left(1;\ iG\mathrm{sign}(a);\ -G^2\right)\frac{\pi}{G}\exp\left(-\left|a\right|G\right), \tag{2.71}$$

we reduce Eq. (2.69) to the form

$$\begin{aligned}\mathbf{j}\left(\mathbf{k},\omega\right)=U\left(\omega\right)\delta\left(\omega-\omega_p-k_z v\right)\int_0^{\infty} dX\exp\left\{ik_xX-(X+b)G\right\}\times\\ \times\left\{\mathbf{k}\mathbf{E}_p+\left(iG-k_x\right)\mathbf{e}\mathbf{E}_p\right\}\frac{ie}{2\pi G}\left\{\mathbf{e_x}\left(iG-k_x\right)+\mathbf{e_y}k_y+\mathbf{e_z}\frac{\omega-\omega_p}{v\gamma^2}\right\}.\end{aligned} \tag{2.72}$$

The integration with respect to X yields

$$\begin{aligned}\mathbf{j}\left(\mathbf{k},\omega\right)=U\left(\omega\right)\delta\left(\omega-\omega_p-k_z v\right)\frac{\exp\left(-bG\right)}{ik_x-G}\times\\ \times\left\{\mathbf{k}\mathbf{E}_p+\left(iG-k_x\right)\mathbf{e}\mathbf{E}_p\right\}\frac{ie}{2\pi G}\left\{\mathbf{e_x}\left(iG-k_x\right)+\mathbf{e_y}k_y+\mathbf{e_z}\frac{\omega-\omega_p}{v\gamma^2}\right\}.\end{aligned} \tag{2.73}$$

The spectral–angular distribution of the energy emitted by this current at long distances has the form

$$\frac{d^2W}{d\omega\,d\Omega}=(2\pi)^6\,\frac{1}{c}\left|\left[\mathbf{kj}\left(\mathbf{k},\omega\right)\right]\right|^2. \tag{2.74}$$

The substitution of Eq. (2.73) into Eq. (2.74) provides the distribution of the diffraction radiation energy from the homogeneous medium excited by one longitudinal wave (where T is the total observation time):

$$\begin{aligned}\frac{d^2W}{d\omega\,d\Omega}=(2\pi)^3\,T\,e^2\left|U\left(\omega\right)\right|^2\exp\left(-2bG\right)\delta\left(\omega-\omega_p-k_zv\right)\times\\ \times\frac{\left\{\left(\left(\mathbf{k}-k_x\mathbf{e_x}\right)\mathbf{E}_p\right)^2+\left(G\mathbf{e}\mathbf{E}_p\right)^2\right\}\left\{\left[\mathbf{k}\mathbf{e_x}\right]^2G^2+\left[\mathbf{k}\left(\mathbf{e_y}k_y-\mathbf{e_x}k_x+\mathbf{e_z}(\omega-\omega_p)/v\gamma^2\right)\right]^2\right\}}{G^2\left(k_y^2+G^2\right)}.\end{aligned} \tag{2.75}$$

The excitation of the medium is often distributed homogeneously and isotropically. This corresponds to the isotropic distribution of the longitudinal waves in the medium. In this case, the distribution of the energy of diffraction radiation can be obtained from Eq. (2.75) by integrating over the directions of the field of the longitudinal wave $\mathbf{E}_p$ and the result has the form

$$\frac{d^2W}{d\omega\, d\Omega} = (2\pi)^3\, Te^2\, |U(\omega)|^2 \exp(-2bG)\, \delta\left(\omega - \omega_p - k_z v\right) \times$$
$$\times \frac{1}{2}\mathbf{E}_p^2 \left\{G^2 + k_y^2 + k_z^2\right\} \frac{\left\{[\mathbf{ke}]^2\, G^2 + \left[\mathbf{k}\left(\mathbf{e_y} k_y - \mathbf{e_x} k_x + \mathbf{e_z}(\omega - \omega_p)/v\gamma^2\right)\right]^2\right\}}{G^2\left(k_x^2 + G^2\right)}. \tag{2.76}$$

The medium can be excited by an acoustic wave. In this case, the homogeneous stationary initial medium becomes both inhomogeneous and nonstationary. If a surface acoustic wave is generated in the medium, the surface profile of the medium also changes. Radiation generated by the charged particle moving near the medium surface along which the surface acoustic wave propagates was investigated in [21, 22]. Considering the effect of the acoustic wave as small perturbation, the authors of those works obtained the spectral–angular distribution of the diffraction-radiation energy and showed that such a radiation can be observed in the range of millimeter and submillimeter waves.

2.6 Diffraction Radiation from a Charged Particle Reflected from a Single Crystal

The fast charged particle incident on the surface of a single crystal at small grazing angle ς (angle between the particle velocity and surface) undergoes mirror reflection from the surface if $\varsigma < \theta_L = (U/E)^{1/2}$ [23] (here, θ_L is the Lindhard angle, U is the potential barrier of the surface, and E is the particle energy).

Change in the velocity in the process of reflection leads to bremsstrahlung. At the same time, the polarization of the surface by the charged particle results in the appearance of polarization currents also leading to radiation. Therefore, such a radiation appears due to the join action of the mechanisms of bremsstrahlung and diffraction radiation [24].

Estimation of the intensity of such a radiation is of interest, because the effective surface of the single crystal from which electrons are reflected can differ from the effective surface from which the electromagnetic field is reflected. Indeed, the electrons are reflected from the surface atomic layer, whereas the electromagnetic field is reflected from the electrons of the medium. Meanwhile, the conduction electron density near the surface in metals undergoes small Fridel oscillations and vanishes outside the surface ion layer at distance $z \sim \hbar/p_F$, where $\hbar$ is Planck's constant, p_F is the Fermi momentum, and the $z = 0$ plane coincides with the surface ion layer. The effective crystal surfaces determined from the reflection of particles and light are generally spaced by a certain distance b from each other. The intensity of the radiation under consideration is a function of this distance and can provide information on it.

It should be taken into account that radiation is formed in a finite time of about $\tau = 1/(\omega - \mathbf{kv})$ (where ω and $\mathbf{k}$ are the frequency and wave vector of radiation, respectively, and v is the particle velocity). The distance from the particle to the surface changes by $\tau\varsigma v$ in time τ.

When $b \ll \tau\varsigma v$, the distance between the effective reflection surface for particles and light does not affect the radiation intensity. We emphasize that the case, where $\varsigma > \theta_L$ and the charged particle penetrates into the crystal rather than is reflected from it, is not considered below.

Let us consider radiation appearing when the particle with charge e is reflected from the planar surface of the semi-infinite ($z < 0$) cubic crystal with relative permittivity ε_1. The particle moves in the homogeneous isotropic medium with relative permittivity ε_2. The Fourier transform of the current density of the charge undergoing mirror reflection from the $z = 0$ surface can be represented in the form

$$\begin{aligned} \mathbf{j}\left(q_x, q_y, z, \omega\right) &= (2\pi)^{-3} \iiint dxdydt\mathbf{j}\left(\mathbf{r}, t\right) \exp\left(-iq_y y + i\omega t\right) = \\ &= \frac{e}{4\pi^3 u} \left\{\mathbf{v} \cos\left[(\omega - \mathbf{qv})\, zu\right] + i\mathbf{u} \sin\left[(\omega - \mathbf{qv})\, z/u\right]\right\}. \end{aligned} \tag{2.77}$$

Here, $\mathbf{v}$ and $\mathbf{u}$ are the velocity components tangential and normal to the surface, respectively.

The partial solution of Maxwell's equations for the self field of the particle has the form

$$E_{0z}\left(\mathbf{q}, z, \omega\right) = \frac{e}{\pi^2 \varepsilon_2 c^2} \frac{\omega \varepsilon_2 u^2 - c^2\left(\omega - \mathbf{qv}\right)}{u^2 k_2^2 - (\omega - \mathbf{qv})^2} \sin\left[(\omega - \mathbf{qv})\, z/u\right], \tag{2.78}$$

$$H_{0z}\left(\mathbf{q}, z, \omega\right) = \frac{ieu}{\pi^2 c} \times \frac{\left[\mathbf{qv}\right]}{u^2 k_2^2 - (\omega - \mathbf{qv})^2} \cos\left[(\omega - \mathbf{qv})\, z/u\right], \tag{2.79}$$

$$k_{1(2)} = \sqrt{(\omega/c)^2\, \varepsilon_{1(2)} - q^2}. \tag{2.80}$$

The total field outside the crystal consists of the self field of the particle and the field $\mathbf{E}_2$ of the transverse waves leaving the surface. The field inside the crystal consists only of the field $\mathbf{E}_1$ of the transverse waves leaving the surface:

$$\mathbf{E}_{1(2)}\left(\mathbf{q}, z, \omega\right) = E_{1(2)}\left(q, \omega\right) \exp\left\{-(+)\, ik_{1(2)} z\right\}. \tag{2.81}$$

From the condition that fields $\mathbf{E}_1$ and $\mathbf{E}_2$ are transverse and the boundary conditions on the crystal surface $z = 0$, it is easy to derive the relations determining the normal components of the radiation fields:

$$(\varepsilon_1 k_2 + \varepsilon_2 k_1) E_{1(2)z}(\mathbf{q}, \omega) = +(-)\varepsilon_{2(1)} k_{2(1)} E_{0z}(\mathbf{q}, 0, \omega) + \varepsilon_{2(1)} (\mathbf{q}\mathbf{E}_0(\mathbf{q}, 0, \omega)), \tag{2.82}$$

$$(k_2 + k_1) H_{1(2)z}(\mathbf{q}, \omega) = +(-) k_{2(1)} H_{0z}(\mathbf{q}, 0, \omega) + \left(\mathbf{q}\mathbf{H}_0(\mathbf{q}, 0, \omega)\right). \tag{2.83}$$

The other components are expressed in terms of the normal components with the use of Maxwell's equations. The spectral–angular distribution in the medium, where the particle moves, can be obtained in the form

$$\frac{d^2 W(\mathbf{n}, \omega)}{d\Omega\, d\omega} = \frac{4\pi^2 \omega^4 \varepsilon_2^{3/2}}{q^2 c^3} \left(\varepsilon_2 |E_{2z}|^2 + |H_{2z}|^2\right) \cos^2\theta. \tag{2.84}$$

The substitution of the explicit expressions of the field provides the distribution of the emitted energy in the form

$$\frac{d^2 W}{d\omega\, d\Omega} = \frac{4e^2 \varepsilon_2^{3/2} u^2 \cos^2\vartheta}{\pi^2 c^3 \left| \left(1 - (v/c)\,\varepsilon_2^{1/2} \sin\vartheta \cos\varphi\right)^2 - (u/c)^2 \varepsilon_2 \cos^2\vartheta \right|^2} \times$$
$$\times \left\{ \frac{\left|\varepsilon_1 - \varepsilon_2 \sin^2\vartheta\right| (v/c)^2 \sin^2\varphi}{\left|\varepsilon_2^{1/2} \cos\vartheta + \left[\varepsilon_1 - \varepsilon_2 \sin^2\vartheta\right]^{1/2}\right|^2} + \frac{|\varepsilon_1|^2 \left|\sin\vartheta - (v/c)\,\varepsilon_2 \cos^2\varphi\right|^2}{\left|\varepsilon_1 \varepsilon_2^{1/2} \cos\vartheta + \varepsilon_2 \left[\varepsilon_1 - \varepsilon_2 \sin^2\vartheta\right]^{1/2}\right|^2} \right\}. \tag{2.85}$$

The radiation distribution in the crystal from which the particle is reflected differs noticeably from that given by Eq. (2.85). The calculations are similar and at $\mathrm{Im}\varepsilon_1 = 0$ yield the expression

$$\frac{d^2 W}{d\omega\, d\Omega} = \frac{4e^2 \varepsilon_2^{3/2} u^2 \cos^2\vartheta}{\pi^2 c^3 \left| \left(1 - (v/c)\,\varepsilon_2^{1/2} \sin\vartheta \cos\varphi\right)^2 - (u/c)^2 \left(\varepsilon_2 - \varepsilon_1 \sin^2\vartheta\right) \right|^2} \times$$
$$\times \left\{ \frac{\left|\varepsilon_2 - \varepsilon_1 \sin^2\vartheta\right| (v/c)^2 \sin^2\varphi}{\left|\varepsilon_1^{1/2} \cos\vartheta + \left[\varepsilon_2 - \varepsilon_1 \sin^2\vartheta\right]^{1/2}\right|^2} + \frac{\left|\varepsilon_1 \sin\vartheta - (v/c)\,\varepsilon_1^{1/2} \varepsilon_2 \cos^2\varphi\right|^2}{\left|\varepsilon_2 \varepsilon_1^{1/2} \cos\vartheta + \varepsilon_1 \left[\varepsilon_2 - \varepsilon_1 \sin^2\vartheta\right]^{1/2}\right|^2} \right\}. \tag{2.86}$$

The problem of the generation of the surface waves in the process of the mirror reflection of the fast charged particle from the surface of the single crystal is considered similarly [25].

The dependence of the radiation intensity on the difference between the surfaces of the effective reflection of particles and field can be illustrated by an example of reflection from a metal when it is convenient to use the image method. Let the electromagnetic field be reflected from the $z = 0$ plane. If the reflection planes of particles and field coincide, the spectral—angular distribution of the emitted energy has the form

$$\frac{d^2W}{d\omega\, d\Omega} = \frac{4}{\pi^2}\frac{e^2}{c}\frac{u^2}{c^2} \times \frac{(v/c)^2\cos^2\vartheta\sin^2\varphi + \left[\sin\vartheta - (v/c)\cos\varphi\right]^2}{\left(\left[1-(v/c)^2\sin\vartheta\cos\varphi\right]^2 - \left[(u/c)\cos\vartheta\right]^2\right)^2}. \tag{2.87}$$

If the particle is reflected not reaching the effective field reflection planes at the distance b from it, the laws of motion of the actual charge and charge image have the form $\left(\theta(t) = \dfrac{t+|t|}{2|t|}\right)$

$$\begin{aligned} \mathbf{r}_+(t) &= \mathbf{b} + (\mathbf{v}+\mathbf{u})\, t\theta(t) + (\mathbf{v}-\mathbf{u})\ t\theta(-t)\,; \\ \mathbf{r}_-(t) &= -\mathbf{b} + (\mathbf{v}-\mathbf{u})\, t\theta(t) + (\mathbf{v}+\mathbf{u})\ t\theta(-t)\,. \end{aligned} \tag{2.88}$$

The ratio of the spectral–angular distributions of the emitted energy at a finite positive b value and $b = 0$ can be obtained in the form

$$\frac{d^2W(b)}{d^2W(b=0)} = \cos^2\left(b\frac{\omega}{c}\cos\vartheta\right). \tag{2.89}$$

When the particle is reflected from "inner" layers of the medium, i.e., at negative b values, the laws of the motion of the charge and charge image have the form

$$\begin{aligned} \mathbf{r}_+(t) &= (\mathbf{v}+\mathbf{u})\, t\theta(t - |b|/u) + (\mathbf{v}-\mathbf{u})\ t\theta(-t - |b|/u)\,; \\ \mathbf{r}_-(t) &= (\mathbf{v}-\mathbf{u})\, t\theta(t - |b|/u) + (\mathbf{v}+\mathbf{u})\ t\theta(-t - |b|/u)\,. \end{aligned} \tag{2.90}$$

The ratio of the distributions of the emitted energy is written as

$$\frac{d^2W(-|b|)}{d^2W(b=0)} = \cos^2\{(\omega - \mathbf{kv})\,|b|/u\}\,. \tag{2.91}$$

Thus, the measurement of the angular distribution of the radiation under consideration provides the possibility of measuring the magnitude and sign of the displacement of the effective reflection planes of light and particles from the surface.

References

1. Smith, S.J., Purcell, E.M.: Visible light from localized surface charges moving across a grating. Phys. Rev. **92**, 1069 (1953)
2. Shestopalov, V.P.: Diffraction electronics. Kharkov, Ukraine (1976)
3. Barnes, C.W., Dedrick, K.G.: Radiation by an electron beam interacting with a diffraction grating. J. Appl. Phys. **37**, 411 (1966)
4. Kazantsev, A.P., Surdutovich, G.I.: Radiation of a charged particle flying by metal screen. Sov. Phys. Dokl. 7, 990 (1963)
5. Bolotovskiy, B.M., Voskresenskiy, G.V.: Radiation of charged string flying by metal screen. Sov. Phys. JETP **34**, 11 (1964)
6. Glass, S.J., Mendlowitz, H.: Quantum theory of Smith-Purcell experiment. Phys. Rev. **174**, 57 (1968)
7. Bolotovskiy, B.M., Voskresenskiy, G.V.: Diffraction radiation. Sov. Phys. Usp. **94**, 377 (1968)
8. Lalor, E.: Three-dimensional theory of Smith-Purcell effect. Phys. Rev. A. **8**, 435 (1973)
9. Potylitsyn, A.P.: Transition radiation and diffraction radiation. Similarities and differences. Nucl. Instrum. Methods Phys. Res. B **145**, 169 (1998)
10. Potylitsyn, A.P.: Resonant diffraction radiation and Smith-Purcell effect. Phys. Lett. A **238**, 112 (1998)
11. Ryazanov, M.I., Strikhanov, M.N., Tishchenko, A.A.: Diffraction radiation from an inhomogeneous dielectric film on the surface of a perfect conductor. JETP **126**, 349 (2004)
12. Garsia de Abajo, F.J., Aizpurua, J.: Numerical simulation of electron energy loss near inhomogeneous dielectrics. Phys. Rev. B. **56**, 15873 (1997)
13. Garsia de Abajo, F.J.: Relativistic energy loss and induced photon emission in the interaction of a dielectric sphere with an external electron beam. Phys. Rev. B **59**, 3095 (1999)
14. Garsia de Abajo, F.J., Howie, A.: Retarded field calculation of electron energy loss in inhomogeneous dielectrics. Phys. Rev. B **65**, 115418 (2002)
15. Gradshtein, I.S., Ryzhik, I.M.: Tables of Integrals, Series. Products Academic Press, New York, NY (1980)
16. Pitaevskiy, L.P.: Electric forces in a transparent media with dispersion. Sov. Phys. JETP **39**, 1450 (1960)
17. Bollshov, L.A., Reshetin, V.P.: Dispersion phenomena at radiation passing through media with refraction index changing slowly in time. Sov. Zh. Eksp. Teor. Fiz. **77**, 1911 (1979)
18. Osipov, V.A., Ryazanov, M.I.: Emission of a charge moving uniformly parallel to the surface of a nonstationary medium. Laser Phys. **8**, 1007 (1998)
19. Landau, L.D., Lifshitz, E.M.: Electrodynamics of Continuous Media. Addison-Wesley, Reading, MA (1984)
20. Weinstein, L.A.: Theory of Diffraction and the Factorization Method. The GoldenPress Boulder, Denver, CO (1969)
21. Mkrtchyan, A.R., Grigoryan, L.Sh., Didenko, A.N., Saharyan, A.A.: Radiation of a charged particle fluing over a surface acoustic wave. Acustica **75**, 184 (1991)
22. Mkrtchyan, A.R., Grigoryan, L.Sh., Didenko, A.N., Saharyan, A.A.: Radiation of a charged particle moving over raleigh acoustic wave. Sov. Zhurnal Tekhnicheskoi Fiziki **61**, 21 (1991)
23. Lindhard, J.: Influence of crystal lattice on motion of energetic charged particles. Mat.- Fys. Medd. Dan. Vid. Selsk. **34**, 1 (1965)
24. Ryazanov, M.I., Safronov, A.N.: On quasi-transition radiation at reflecting of a charged particle from a crystal surface. Sov. Zh. Eksp. Teor. Fiz. **103**, 3114 (1993)
25. Ryazanov, M.I., Safronov, A.N.: On quasi-transition radiation of surface waves. Sov. Zh. Eksp. Teor. Fiz. **104**, 3512 (1993)

Chapter 3
Diffraction Radiation at Optical and Lower Frequencies

3.1 Diffraction Radiation from a Circular Hole in an Opaque Screen

Calculation of the characteristics of radiation generated when a charged particle moves through a circular hole in an infinitely thin perfectly conducting screen is one of the most investigated problems in the theory of diffraction radiation. Such a screen can be a thin metal plate; in this case, radiation should be considered at frequencies below plasmon frequencies, i.e., at optical, infrared, millimeter, etc. frequencies.

This problem was considered for the first time likely by Bobrinev and Braginskii [1] in 1958. However, their results are applicable only for nonrelativistic particles. In 1959, Dnestrovskii and Kostomarov obtained analytical expressions for the nonrelativistic case [2] and, for the first time, for the ultrarelativistic case [3].

We consider a circular hole in an infinitely thin perfectly conducting screen. Let the screen be located on the xy plane and a charged particle moves through the center of the circle along the z axis. The problem of calculation of the characteristics of generated radiation can be formulated in the framework of classical electrodynamics: we divide space into three regions (two half-spaces and the screen between them) and specify the matching conditions for the fields at the boundaries. A field in each half-space satisfies the system of Maxwell's equations (where $k = \omega/c$)

$$\begin{aligned} \mathrm{rot}\mathbf{H}(\mathbf{r},\omega) &= -ik\mathbf{E}(\mathbf{r},\omega)\,, \quad \mathrm{div}\mathbf{E}(\mathbf{r},\omega) = 0 \\ \mathrm{rot}\mathbf{E}(\mathbf{r},\omega) &= ik\mathbf{H}(\mathbf{r},\omega)\,, \quad \mathrm{div}\mathbf{H}(\mathbf{r},\omega) = 0. \end{aligned} \tag{3.1}$$

Here, $\mathbf{E}$ and $\mathbf{H}$ are the differences between the total fields and self fields of the charged particle.

This system of equations can be solved with the use of the vector Green formula: two arbitrary functions $\mathbf{A}$ and $\mathbf{B}$ continuous together with their first and second derivatives in volume V and on the surface S bounding this volume satisfy the relation

$$\int_V (\mathbf{A}\mathrm{rotrot}\mathbf{B} - \mathbf{B}\mathrm{rotrot}\mathbf{A})\,dV = \int_S ([\mathbf{B}\mathrm{rot}\mathbf{A}] - [\mathbf{A}\mathrm{rot}\mathbf{B}])\,d\mathbf{s}. \tag{3.2}$$

A.P. Potylitsyn et al., *Diffraction Radiation from Relativistic Particles*,
STMP 239, 55–103, DOI 10.1007/978-3-642-12513-3_3,

Here, $d\mathbf{s} = \mathbf{n}ds$, where $\mathbf{n}$ is the unit vector normal to surface S.

For investigating radiation, we are interested in the Fourier transforms of electric, $\mathbf{E}(\mathbf{r},\omega)$, and magnetic, $\mathbf{H}(\mathbf{r},\omega)$, fields. Correspondingly, vector $\mathbf{E}(\mathbf{r},\omega)$ or $\mathbf{H}(\mathbf{r},\omega)$ is taken as vector $\mathbf{A}$. The following product of arbitrary unit vector $\mathbf{e}$ by the Green's function of the corresponding wave equation for the field is taken as vector $\mathbf{B}$:

$$\mathbf{B} = G(R)\,\mathbf{e} \equiv \frac{e^{ik|\mathbf{r}-\mathbf{r}'|}}{|\mathbf{r}-\mathbf{r}'|}\mathbf{e}, \quad R = \left|\mathbf{r}-\mathbf{r}'\right|. \tag{3.3}$$

Here, $\mathbf{r}$ is the radius vector of the observation point and $\mathbf{r}'$ is the radius vector of the points over which integration is performed. The half-space in which radiation is measured is taken as V and the $z = 0$ plane with the infinite hemisphere based on this plane is taken as S. Since the radiation field decreases with increasing the distance to the radiation source, surface S reduces to only the $z = 0$ plane. Using Eq. (3.2), we can arrive at the following expressions for the radiation fields [4]:

$$\begin{aligned}\mathbf{E}(\mathbf{r},\omega) = & -\frac{1}{4\pi}\int_S \left(ik\left[\mathbf{nH}\left(\mathbf{r}',\omega\right)\right]G(R) + \left[\left[\mathbf{nE}\left(\mathbf{r}',\omega\right)\right]\mathrm{grad}G(R)\right]\right)dS' - \\ & -\frac{1}{4\pi}\int_S \left(\mathbf{nE}\left(\mathbf{r}',\omega\right)\right)\mathrm{grad}G(R)\,dS' + \frac{i}{4\pi k}\oint_C \mathrm{grad}\,G(R)\left(\mathbf{H}_1 - \mathbf{H}_2, dl\right) \\ \mathbf{H}(\mathbf{r},\omega) = & \frac{1}{4\pi}\int_S \left(ik\left[\mathbf{nE}\left(\mathbf{r}',\omega\right)\right]G(R) - \left[\left[\mathbf{nH}\left(\mathbf{r}',\omega\right)\right]\mathrm{grad}G(R)\right]\right)dS' - \\ & -\frac{1}{4\pi}\int_S \left(nH\left(r',\omega\right)\right)\mathrm{grad}G(R)\,dS\prime - \frac{i}{4\pi k}\oint_C \mathrm{grad}G(R)\left(\mathbf{E}_1 - \mathbf{E}_2, dl\right).\end{aligned} \tag{3.4}$$

In the terms with contour integrals, integration is performed over contour C separating screen S_1 and hole S_2. Fields $\mathbf{E}_i$ and $\mathbf{H}_i$ for $i = 1, 2$ refer to S_i.

The surface integrals in Eq. (3.4) are obtained immediately from the Green formula (detailed derivation was given in [5]). As shown by Khachatryan [4], expressions (3.4) satisfy the system of Eq. (3.1) only in the presence of the contour integrals. The physical meaning of the contour integrals is that they reflect stepwise change in the surface current density on contour C, which leads to the presence of a certain linear current density. Note that a similar situation arises in diffraction theory [6, 7]. For the case of the normal passage through the center of the hole, these contour integrals are equal to zero [4]. Indeed, in the presence of axial symmetry in the problem, the linear current density along the edge of the hole (contour C) is not induced. However, according to the same argumentation, in the presence of strong symmetry breaking, e.g., for the case of eccentric passage or asymmetric hole, the contribution from the contour integrals in Eq. (3.4) can be significant (this circumstance was not taken into account in, e.g., [8, 9]).

In the ultrarelativistic case $\gamma \gg 1$, radiation propagates at very small angles with respect to the trajectory and the self field of the particle given by the expression

$$\mathbf{E}^0(\mathbf{r},\omega) = \frac{e\omega}{\pi v^2\gamma}\left(\frac{\boldsymbol{\rho}}{\rho}K_1\left(\frac{\omega\rho}{v\gamma}\right) - \frac{i}{\gamma}\frac{\mathbf{v}}{v}K_0\left(\frac{\omega\rho}{v\gamma}\right)\right)\exp\left(i\frac{\omega}{v}z\right) \simeq$$
$$\simeq \frac{e\omega}{\pi v^2\gamma}\frac{\boldsymbol{\rho}}{\rho}K_1\left(\frac{\omega\rho}{v\gamma}\right)\exp\left(i\frac{\omega}{v}z\right). \tag{3.5}$$

is almost transverse (where $\boldsymbol{\rho} = (x, y)$). Owing to this fact, the terms proportional to $(\mathbf{nE})$ and $(\mathbf{nH})$ can be neglected on the right-hand side of Eq. (3.4).

We are interested in the radiation field at the distances much larger than the wavelength, i.e., $r \gg \lambda$. For these distances, it is easy to obtain

$$\mathrm{grad}G(R) \simeq -i\mathbf{k}G(R) \tag{3.6}$$

and, therefore,

$$\begin{aligned}\big[[\mathbf{nH}]\,\mathrm{grad}G\big] &\simeq -ik\,G\,[\mathbf{nE}]\\ \big[[\mathbf{nE}]\,\mathrm{grad}G\big] &\simeq ik\,G\,[\mathbf{nH}].\end{aligned} \tag{3.7}$$

Thus, formulas (3.4) for the case under consideration reduce to the comparatively simple expressions

$$\begin{aligned}\mathbf{E}(\mathbf{r},\omega) &= -\frac{ik}{2\pi}\int_S \left[\mathbf{nH}(\mathbf{r}',\omega)\right]G(R)\,dS'\\ \mathbf{H}(\mathbf{r},\omega) &= \frac{ik}{2\pi}\int_S \left[\mathbf{nE}(\mathbf{r}',\omega)\right]G(R)\,dS'.\end{aligned} \tag{3.8}$$

In order to finally formulate the problem, boundary conditions should be specified. The most obvious boundary condition is the condition of zero tangential component of the total electric field on the screen surface. In our notation, this condition can be expressed as

$$E_\rho|_{S_1} = -\left.E^0_\rho\right|_{S_1}. \tag{3.9}$$

As the boundary condition on the hole surface S_2, we can take the Kirchhoff conditions

$$\mathbf{E}|_{S_2} = 0, \quad \mathbf{H}|_{S_2} = 0; \tag{3.10}$$

i.e., the total electromagnetic field in the hole should be equal to the field of the moving particle, i.e., should be the same as that in the absence of the screen. In this case, in view of Eqs. (3.9) and (3.10), it follows from Eq. (3.8) that

$$\begin{aligned} \mathbf{E}(\mathbf{r},\omega) &= \frac{ik}{2\pi}\int_{S_1}\left[\mathbf{n}\mathbf{H}^0(\mathbf{r},\omega)\right]G(R)\,dS_1 \\ \mathbf{H}(\mathbf{r},\omega) &= -\frac{ik}{2\pi}\int_{S_1}\left[\mathbf{n}\mathbf{E}^0(\mathbf{r},\omega)\right]G(R)\,dS_1 \end{aligned} \tag{3.11}$$

These expressions coincide with the formulation of Huygens's principle in the Kirchhoff method. Calculation of the characteristics in this method is relatively simple and was performed in [8–11]. For the sake of completeness, we consider diffraction radiation generated when the ultrarelativistic charged particle moves through a ring with inner radius a and outer radius b. The exact theory of radiation from the ring that is a plane screen between two radii was developed in [12].

According to Eq. (3.11), the electric component of the radiation field is determined as

$$E_{x,y}(\mathbf{r},\omega) = -\frac{ik}{2\pi}\int_a^b \rho d\rho \int_0^{2\pi} d\varphi' E^0_{x,y}(\mathbf{r},\omega)\, G(R), \tag{3.12}$$

where $E^0_{x,y}$ is found from Eq. (3.5) at $z = 0$. The function $G(R)$ in the Fresnel zone has the form

$$G(R) = \frac{e^{ik|\mathbf{r}-\mathbf{r}'|}}{|\mathbf{r}-\mathbf{r}'|} \xrightarrow[kr\gg 1]{r\gg r'} \frac{e^{ikr}}{r} e^{-i\mathbf{k}\mathbf{r}'}, \tag{3.13}$$

where $\mathbf{k} = k\mathbf{r}/r$ is the wave vector of a plane wave arriving at the observation point. In the cylindrical coordinate system,

$$\mathbf{k}\mathbf{r}' = k\rho\sin\theta\cos\left(\varphi - \varphi'\right), \tag{3.14}$$

where θ is the polar angle of radiation measured from the z axis and φ is the angle between the x axis and the projection of the wave vector of radiation on the xy plane. Thus, field Eq. (3.12) is given by the expression

$$E_{x,y}(\mathbf{r},\omega) = -\frac{ik}{\pi}\frac{e\omega}{v^2\gamma}\frac{e^{ikr}}{r}\int_a^b \rho d\rho K_1\left(\frac{\omega\rho}{v\gamma}\right)\int_0^{2\pi} d\varphi' \begin{Bmatrix}\cos\varphi' \\ \sin\varphi'\end{Bmatrix} e^{-ik\rho\sin\theta\cos(\varphi-\varphi')}. \tag{3.15}$$

The inner integral is calculated by the formula

$$\int_0^{2\pi} d\varphi' \begin{Bmatrix}\cos\varphi' \\ \sin\varphi'\end{Bmatrix} e^{-ik\rho\sin\theta\cos(\varphi-\varphi')} = \begin{Bmatrix}\cos\varphi \\ \sin\varphi\end{Bmatrix} J_1(k\rho\sin\theta). \tag{3.16}$$

Then, the outer integral with respect to ρ is calculated analytically as

$$\text{int}_a^b \rho d\rho K_1\left(\frac{\rho\omega}{v\gamma}\right) J_1\left(\rho k \sin\theta\right) = \left(\frac{v}{\omega}\right)^2 \frac{F\left(a\right) - F\left(b\right)}{1-\beta^2\cos^2\theta}, \tag{3.17}$$

where

$$F\left(a\right) = \frac{a\omega}{v\gamma} J_1\left(ak\sin\theta\right) K_2\left(\frac{a\omega}{v\gamma}\right) - ak\sin\theta J_2\left(ak\sin\theta\right) K_1\left(\frac{a\omega}{v\gamma}\right). \tag{3.18}$$

Using the following formulas reducing the order of cylindrical functions:

$$J_2\left(x\right) = \frac{2J_1\left(x\right)}{x} - J_0\left(x\right), \qquad K_2\left(x\right) = K_0\left(x\right) + \frac{2K_1\left(x\right)}{x}, \tag{3.19}$$

we obtain the following formula for $F\left(a\right)$:

$$F\left(a\right) = ak\sin\theta J_0\left(ak\sin\theta\right) K_1\left(\frac{a\omega}{v\gamma}\right) + \frac{a\omega}{v\gamma} J_1\left(ak\sin\theta\right) K_0\left(\frac{a\omega}{v\gamma}\right). \tag{3.20}$$

As a result,

$$E_{x,y}\left(\mathbf{r},\omega\right) = \frac{e^{ikr}}{r}\frac{ie}{\pi c\gamma}\left\{\begin{matrix}\cos\varphi\\ \sin\varphi\end{matrix}\right\}\frac{F\left(b\right) - F\left(a\right)}{1-\beta^2\cos^2\theta}. \tag{3.21}$$

Note that

$$F\left(0\right) = \gamma\sin\theta, \quad F\left(\infty\right) = 0. \tag{3.22}$$

Formula (3.21) with $a = 0$, $b = \infty$ provides the following standard formula for transition radiation from the infinite screen:

$$E_{x,y}^{\text{TR}}\left(\mathbf{r},\omega\right) = -\frac{e^{ikr}}{r}\frac{ie}{\pi c}\left\{\begin{matrix}\cos\phi\\ \sin\phi\end{matrix}\right\}\frac{\sin\theta}{1-\beta^2\cos^2\theta}. \tag{3.23}$$

For diffraction radiation from the hole in the infinite screen ($b \to \infty$), we obtain

$$E_{x,y}\left(\mathbf{r},\omega\right) = -\frac{e^{ikr}}{r}\frac{ie}{\pi c}\left\{\begin{matrix}\cos\varphi\\ \sin\varphi\end{matrix}\right\}\frac{\sin\theta}{1-\beta^2\cos^2\theta}\left[\frac{ak}{\gamma}K_1\left(\frac{a\omega}{v\gamma}\right)\right] J_0\left(ak\sin\theta\right), \tag{3.24}$$

where it is taken into account that the second term in Eq. (3.20) for $\theta \sim \gamma^{-1}$ is much smaller than the first term. In the ultrarelativistic case at $a \ll \gamma\lambda$, the expression in the square brackets is on the order of one. In this case, for the characteristic radiation angles $\theta \sim \gamma^{-1}$, $ka\theta \ll 1$ and $J_0\left(ka\theta\right) \simeq 1$, so that Eq. (3.24) reduces to Eq. (3.23).

Note that Eq. (3.24) coincides with the results of a more accurate theory developed in [2, 3] (see Eq. (3.35) below). Moreover, Eq. (3.24) coincides with good accuracy with the formula presented by Ter-Mikaelyan and Khachatryan [8, 9] (however, the expressions for the diffraction radiation fields from the circular hole that were presented in [8] and [9] contain a misprint, because the dimension of the radiation fields is incorrect in them).

Let us estimate the total losses on diffraction radiation by the formula

$$W = \int \frac{d^2W}{d\omega d\Omega} d\omega d\Omega = cr^2 \int |\mathbf{E}(\mathbf{r}, \omega)|^2 d\omega d\Omega, \tag{3.25}$$

where field $\mathbf{E}(\mathbf{r}, \omega)$ is taken from Eq. (3.24). As mentioned above, the expression in the square brackets is on the order of $\gamma \sin\theta$ for frequencies $\omega < \omega_c$, where the characteristic spectral frequency ω_c is given by the expression

$$\omega_c = \frac{c\gamma}{a}. \tag{3.26}$$

For this reason, for qualitative estimate of the total losses, it is sufficient to set

$$\left[\frac{ak}{\gamma} K_1 \left(\frac{ak}{\gamma}\right)\right] J_0(ak\theta) \simeq 1 \tag{3.27}$$

in the integrand; after that, the remaining integral with respect to $d\omega$ gives ω_c, whereas the integral with respect to the polar angle is on the order of one. Integration with respect to the azimuth angle yields 2π; as a result,

$$W \approx cr^2 \left(\frac{1}{r}\frac{e}{\pi c\gamma}\gamma\right)^2 2\pi\omega_c = \frac{2}{\pi}\frac{e^2}{c}\omega_c \sim \frac{e^2\gamma}{a} = \alpha\hbar c\frac{\gamma}{a}. \tag{3.28}$$

This rather rough estimate coincides up to a constant with more accurate estimates [3, 13].

Calculations are performed with the use of not only the vector theory of the Kirchhoff method [4, 8, 9], but also scalar theory [14]. The scalar theory is a good approximation for relativistic particles and small angles of radiation:

$$\theta \sim \gamma^{-1} \ll 1. \tag{3.29}$$

This range is most interesting for investigating diffraction radiation. The vector theory provides more accurate results and, in addition, makes it possible to determine the polarization of the radiation field. However, both approaches lead to approximately the same result for the spectral–angular radiation density under condition (3.29) [14].

The applicability of the Huygens—Kirchhoff method is limited by the condition

$$\lambda \ll a \leq \gamma\lambda. \tag{3.30}$$

The left-hand side of this inequality is attributed to the standard diffraction theory [15], whereas the right-hand side is characteristic of diffraction radiation and is associated with the fact that the charged particle field bounded in space by the quantity $\gamma\lambda$ is scattered at the edges of the hole. Note that Huygens' principle is applicable only for describing the forward scattering of the incident field (see in [15, section 98]) and can be used only for forward diffraction radiation. This conclusion is also supported by the fact that the transformation of Eq. (3.4) to Eq. (3.11) is performed under the assumption that radiation propagates at very small angle to the trajectory. Hence, it should be expected that formulas such as Eq. (3.11) can be applicable for calculation of Smith—Purcell radiation (propagating at large angles to the trajectory) only in a rather rough approximation.

The above discussion explains why more rigorous methods of solution are of interest. First, it is necessary to refuse condition (3.10) and to replace it by the condition

$$H_\varphi\big|_{S_2} = 0. \tag{3.31}$$

Condition (3.31) is a consequence of the accepted model: the screen is infinitely thin, so that the screen-transverse component of the density of the polarization currents induced on the screen by the field of the moving charged particle vanishes. However, the real metal screen has a finite thickness. The density of the polarization currents is qualitatively proportional to the conductivity of the target material, σ, and to the field of the moving charged particle, $\mathbf{E}^0$:

$$j_i \propto \sigma E_i^0, \tag{3.32}$$

where subscript i takes values $\{x, y, z\}$. Quantity j_z can be disregarded, because the longitudinal component of the self field of the ultrarelativistic charged particle is much lower than the transverse component: $E_z^0 \ll E_x^0, E_y^0$. It is seen that accepted condition (3.31) is approximate. Error associated with the use of this condition can be particularly noticeable, e.g., in the nonrelativistic case, for oblique motion, and grazing motion of the ultrarelativistic particle through the hole (the smallest angle between the particle trajectory and screen plane is comparable with γ^{-1}), because component H_φ in the hole plane is significantly nonzero for these cases.

Thus, under condition (3.31), it follows from Eq. (3.8) with condition (3.9) that

$$\mathbf{H}(\mathbf{r}, \omega) = \frac{ik}{2\pi} \int_{S_2} [\mathbf{nE}(\mathbf{r}, \omega)]\, G(R)\, dS_2 - \frac{ik}{2\pi} \int_{S_1} \left[\mathbf{nE}^0(\mathbf{r}, \omega)\right] G(R)\, dS_1. \tag{3.33}$$

The substitution of this expression into Eq. (3.31) for the points of surface S_2 yields

$$\int_{S_2} \cos(\varphi - \varphi_2)\, E_\rho(\mathbf{r}_2, \omega)\, G(R)\, dS_2 = \int_{S_1} \cos(\varphi - \varphi_1)\, E_\rho^0(\mathbf{r}_1, \omega)\, G(R)\, dS_1. \tag{3.34}$$

Here, φ is the angle in the cylindrical coordinate system and cosines appear due to scalar products $\mathbf{e}_\varphi \mathbf{e}_{\varphi_1}$ and $\mathbf{e}_\varphi \mathbf{e}_{\varphi_2}$ when taking the projection on $\mathbf{e}_\varphi$. Equation (3.34) is a first-kind Fredholm integral equation. Its solution makes it possible to determine field $E_\rho(\mathbf{r}_2, \omega)$; then, the radiation field is obtained by Eq. (3.33). Equation (3.34) was numerically solved by Dnestrovskii and Kostomarov [2]. Analytical expressions were obtained for the nonrelativistic and ultrarelativistic limiting cases. For the nonrelativistic case (see [2]), the results coincide with the result obtained in [1]. For the ultrarelativistic case, the following expression was obtained in [2, 3] at $\theta \ll 1$:

$$H_\varphi(\theta, \omega) = -\frac{e^{ikr}}{r}\frac{2e}{c}\left[\frac{ka}{\gamma}K_1\left(\frac{ka}{\gamma}\right)\right] J_0(ka\theta)\frac{1}{\gamma^{-2}+\theta^2}. \tag{3.35}$$

This expression provides the same result as Eq. (3.24) up to a factor of $(2\pi)^{-1}$, which is associated with another form of Fourier transforms and does not change the final result.

Another method for solution was proposed by Gianfelice et al. [13]. They also used Eq. (3.34), but solved it analytically rather than numerically. The essence of their method is the transition to the spheroidal coordinate system with the subsequent expansion of all desired functions in the system of special functions. After that, the coefficients of this expansion can be determined from Eq. (3.34). The resulting expansions for the fields are rather lengthy and are not presented here. We only note that the results of a more accurate theory developed in [13] lead to the expression for the total losses on radiation differs from that obtained by Dnestrovskii and Kostomarov by a factor of $3\pi/8 \approx 1,18$:

$$W \simeq \frac{3\pi}{8}\frac{e^2\gamma}{a}. \tag{3.36}$$

We also point to a number of works [16–18], where transition radiation generated by the charged particle moving through a continuous thin perfectly conducting disc was considered. When disc radius b is smaller than the characteristic size of decreasing the self field of the relativistic charge, $\gamma\lambda$, the transition radiation spectrum changes. This change can be interpreted as a result of the diffraction of a part of the self field of the charge at the edges of the disc. Note that Eq. (3.21) at $a = 0$ and finite b values is in good qualitative agreement with the results of the exact theory developed in [16–18].

For the infinitely thin perfectly conducting screen, the so-called Babinet theorem is valid [19, 5] (in [15], it is also referred to as the duality principle). It relates the radiation fields diffracted at the edges of two screens, which, being imposed on each other, form a continuous plane. Such screens are called supplemented. The scattered field can also be the field of the moving charge, so that the Babine theorem is also valid for transition radiation and diffraction radiation from supplemented, infinitely thin, perfectly conducting screens [14].

Let S_1 and S_2 be supplemented screens, so that $S_1 + S_2$ is a plane.

In the scalar theory, where the field is described by a certain scalar function ψ, the Babinet principle has the very simple form

$$\psi_1 + \psi_2 = \psi_0, \tag{3.37}$$

where ψ_1 is the field scattered from screen S_1 and ψ_2 is the field scattered from supplemented screen S_2.

n the vector theory, the formulation of the Babinet theorem is somewhat more complicated. Let the field with electric $\mathbf{E}_1^0$ and magnetic $\mathbf{H}_1^0$ components be incident on screen S_1. The supplemented diffraction system consists of fields $\mathbf{E}_2^0 = -\mathbf{H}_1^0$ and $\mathbf{H}_2^0 = \mathbf{E}_1^0$ incident on screen S_2. The theorem can be formulated as follows: if E_1 and H_1 are the fields scattered by screen S_1, fields $\mathbf{E}_2$ and $\mathbf{H}_2$ scattered by supplemented screen S_2 can be represented as

$$\begin{aligned} \mathbf{E}_2 &= \mathbf{H}_1 + \mathbf{H}_1^0, \\ \mathbf{H}_2 &= -\left(\mathbf{E}_1 + \mathbf{E}_1^0\right). \end{aligned} \tag{3.38}$$

The proof of this theorem can be found in [5] and in [15]. In principle, the use of this theorem makes it possible to determine the diffraction radiation field from the circular hole in the infinitely thin, perfectly conducting screen in terms of the transition radiation field from the thin, perfectly conducting disc that was found in [16, 17].

In conclusion, we point to a number of problems that arise in investigating radiation generated by the charge moving through the hole and seem to be insufficiently studied.

First, the strict solution of the problem has not yet been obtained in a compact analytical form for arbitrary particle energies. Works [2, 3] were based on the numerical solution of the system of integral equations. The solution in [13] was obtained in the form of an expansion in the system of special functions and is rather complicated. Meanwhile, the problem should be likely solved in a relatively compact form with the use of the Wiener—Hopf method developed for solving pair systems of integral equations by Fock and Weinstein [20, 21].

Second, the eccentric passage of the charged particle beam through the hole is of interest. To solve this problem, e.g., with the use of the vector Green formula, contour integrals should be accurately taken into account. The inclusion of these integrals corresponds to the inclusion of polarization currents induced at the edge of the hole in the process of eccentric passage.

Third, it is desirable to go beyond the scope of the perfectly conducting target model, because the conductivity of real targets is large but finite. The Leontovich boundary condition (see in [15]) instead of condition (3.9) of zero tangential component on the conductor surface seems to be a consistent generalization in the way of the inclusion of the real properties of the metal screen. The Leontovich boundary condition on the surface of the real conductor with large but finite conductivity σ has the form

$$[\mathbf{nE}] = -\zeta \; [\mathbf{n}\,[\mathbf{nH}]] \,. \tag{3.39}$$

Here, **n** is the unit vector normal to the surface and $\zeta = \sqrt{\mu/\varepsilon}$ is the wave impedance, where μ and ε are the relative permeability and relative permittivity of the metal, respectively. For good conductors that are not ferromagnets, $\mu = 1,\ \varepsilon = i4\pi\sigma/\omega$ and the wave impedance has the form

$$\zeta = \frac{1-i}{2}\sqrt{\frac{\omega}{2\pi\sigma}}. \tag{3.40}$$

For good conductors, $\zeta \ll 1$. The perfect conductivity approximation corresponds to the condition $\zeta = 0$, which corresponds to zero tangential component of the electric field, $[\mathbf{nE}] = 0$.

3.2 Diffraction Radiation from an Inclined, Perfectly Conducting Half-Plane

One of the exactly solvable problems is investigation of the process of generating diffraction radiation that accompanies the motion of the charged particle near inclined, perfectly conducting half-plane [22]. Note that such a description can be used with a good accuracy for wavelengths longer than 0.5 μm if a polished metal surface is used as a target near which the charge moves.

Let us consider the case where the charged particle moves near the semi-infinite, perfectly conducting screen so that the projection of the particle trajectory on the screen (target) is perpendicular to the edge of the target (see Fig. 3.1). We will refer to this case as the perpendicular geometry.

For this problem, we choose the following coordinate system (see Fig. 3.1): the z axis coincides with the screen edge and the x axis is located in the screen plane. In this coordinate system, particle velocity **v** is written as

$$\mathbf{v} = \boldsymbol{\beta} c, \qquad \boldsymbol{\beta} = \left\{\beta_x, \beta_y, \beta_z\right\} = \beta\ (\sin\psi_0 \cos\varphi_0, \sin\psi_0 \sin\varphi_0, \cos\psi_0)\,. \tag{3.41}$$

For the perpendicular geometry under consideration ($\psi_0 = \pi/2$), vector $\boldsymbol{\beta}$ has only two components:

$$\boldsymbol{\beta} = \beta\ (\cos\varphi_0, \sin\varphi_0, 0) = \left\{\beta_x, \beta_y, 0\right\}. \tag{3.42}$$

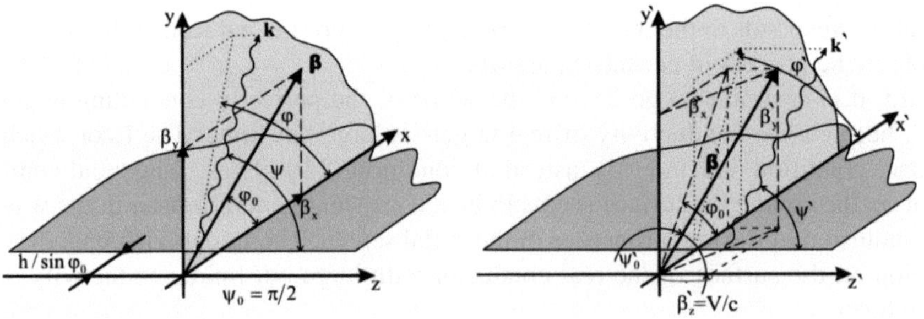

Fig. 3.1 Generation of diffraction radiation in the process of the oblique motion of the charged particle over the conducting semi-plane (perpendicular geometry)

For the geometry considered, the following notation is introduced:

- h is the minimal distance between the particle trajectory and screen edge (impact parameter);
- ψ_0, φ_0 are the polar and azimuth angles of the particle velocity, respectively;
- ψ, φ are the polar and azimuth angles of the wave vector, respectively;
- $\mathbf{k}$ is the wave vector,

$$\mathbf{k} = \left\{k_x, k_y, k_z\right\} = \omega\mathbf{n} = \omega\left\{\sin\psi\cos\varphi,\ \sin\psi\sin\varphi,\ \cos\psi\right\}, \tag{3.43}$$

- ω is the radiation frequency.

For this particular case, the solution of Maxwell's equations was found in [22] by the Wiener—Hopf method and potentials $\mathbf{A}$ and φ describing radiation in vacuum above and below the screen were expressed in terms of the charge, ρ, and current, $\mathbf{j}$, densities induced by the moving charge on the screen. For the case under consideration, $A_y = 0$ for the infinitely thin perfect screen. In this section, the system of units where $\hbar = m = c = 1$ is used until the resulting formulas.

Following [22], we write expressions for the strength of the diffraction radiation field. In that work, formulas for the Fourier components of ρ and $\mathbf{j}$ were derived. For the sake of convenience, these formulas are expressed in terms of the velocity of the particle and its Lorentz factor γ:

$$\begin{aligned}\rho = &-\frac{B_{q\omega}}{\sqrt{\omega\sin\psi\,(1+\cos\varphi)}}\times\\ &\times\left\{\frac{1}{\sin\psi} - \frac{\alpha_0\cos\varphi_0 + i\gamma^{-2}\sin\varphi_0}{\alpha_0\beta}\,\frac{\sin\psi\,(1+\cos\varphi)}{-\sin\psi\cos\varphi + \frac{\cos\varphi_0}{\beta} + i\frac{\alpha_0}{\beta}\sin\varphi_0}\right\},\end{aligned} \tag{3.44}$$

$$\begin{aligned}j_x &= \frac{B_{q\omega}}{\sqrt{\omega\sin\psi\,(1+\cos\varphi)}}\times\frac{\sin\psi + \frac{\cos\varphi_0}{\beta} + i\frac{\alpha_0}{\beta}\sin\varphi_0}{-\sin\psi\cos\varphi + \frac{\cos\varphi_0}{\beta} + i\frac{\alpha_0}{\beta}\sin\varphi_0},\\ j_y &= 0,\end{aligned} \tag{3.45}$$

where

$$\begin{aligned}B_{q\omega} &= \frac{e}{4\pi^2}\,\frac{\beta\sqrt{\omega\left[\sin\psi - \frac{\cos\varphi_0}{\beta} - i\frac{\alpha_0}{\beta}\sin\varphi_0\right]}}{\omega\left(\alpha_0\cos\varphi_0 + i\sin\varphi_0\right)}\exp\left(-\kappa\right),\\ \alpha_0 &= \sqrt{\gamma^{-2} + \beta^2\cos^2\psi} = \sqrt{1-\beta^2\sin^2\psi},\\ \kappa &= \frac{2\pi h}{\beta\lambda}\sqrt{1-\beta^2\sin^2\psi}.\end{aligned} \tag{3.46}$$

Here, e is the particle charge. Induced-current component j_z at $\mathbf{k} = \omega\mathbf{n}$ is found from continuity equation $\mathbf{nj} = \rho$:

$$j_z = \frac{\rho - n_x j_x}{n_z}. \tag{3.47}$$

The Fourier components of the electric field of diffraction radiation in the wave zone (the Fresnel zone) are easily found from Maxwell's equations:

$$E_i = \omega\left(j_i - \rho\, n_i\right), \quad i = x,\ y,\ z \tag{3.48}$$

or, in the explicit form, by expressing them in terms of ρ and j_x:

$$\begin{aligned} E_x &= \omega\left(j_x - \rho \sin\psi \cos\phi\right), \\ E_y &= -\omega\rho \sin\psi \sin\phi, \\ E_z &= \omega\left(\frac{\rho - j_x \sin\psi\cos\phi}{\cos\psi} - \rho\cos\psi\right). \end{aligned} \tag{3.49}$$

For solving the problems associated with the radiation polarization, it is necessary to pass to the coordinate system including the wave vector. For this reason, we choose the system of unit vectors as follows:

$$\begin{aligned} \mathbf{e}_2 &= \frac{[\mathbf{z}_0\mathbf{n}]}{\sin\psi} = \{-\sin\varphi,\ \cos\varphi,\ 0\}, \\ \mathbf{e}_1 &= -[\mathbf{n}\mathbf{e}_2] = \{\cos\varphi\cos\psi,\ \sin\varphi\cos\psi,\ -\sin\psi\}, \\ \mathbf{n} &= \frac{\mathbf{k}}{\omega}, \end{aligned} \tag{3.50}$$

Unit vector $\mathbf{z}_0$ is directed along the positive direction of the z axis. The components of the field strength in this coordinate system are written in the form

$$\begin{aligned} E_1 &= \mathbf{E}\mathbf{e}_1 = \omega\frac{j_x\cos\varphi - \rho\sin\psi}{\cos\psi}, \\ E_2 &= \mathbf{E}\mathbf{e}_2 = -\omega j_x \sin\varphi, \\ E_n &= \mathbf{E}\mathbf{n} = 0. \end{aligned} \tag{3.51}$$

It is obvious that, in order to obtain the spectral–angular distribution of diffraction radiation, the expression for the field strength in any of the above coordinate systems can be used:

$$\frac{d^2W}{d\omega d\Omega} = 4\pi^2 |\mathbf{E}|^2 = 4\pi^2 \left\{ |E_x|^2 + \left|E_y\right|^2 + |E_z|^2 \right\} = 4\pi^2 \left\{ |E_1|^2 + |E_2|^2 \right\} =$$

$$= 4\pi^2\omega^2 \left\{ |j_x \sin\varphi|^2 + \left| \frac{j_x \cos\varphi - \rho \sin\varphi}{\cos\psi} \right|^2 \right\} =$$

$$= 4\pi^2\omega^2 \left\{ |j_x|^2 - |\rho|^2 + \left| \frac{\rho - \sin\psi \cos\varphi \cdot j_x}{\cos\psi} \right|^2 \right\} =$$

$$= 4\pi^2\omega^2 \left\{ |j_x|^2 + |j_z|^2 - |\rho|^2 \right\} = 4\pi^2\omega^2 \left\{ |\mathbf{j}|^2 - |\rho|^2 \right\}. \tag{3.52}$$

Up to the notation, the resulting expression coincides with Eq. (1.16) for radiation in vacuum (where $\varepsilon(\omega) = 1$, $|\mathbf{k}| = \omega$)

$$\frac{d^2W}{d\omega d\Omega} = const \cdot \omega^2 \left\{ |\mathbf{j}(\mathbf{n},\ \omega)|^2 - |\rho(\mathbf{n},\ \omega)|^2 \right\}. \tag{3.53}$$

Indeed, it follows from Eq. (3.48) that

$$|E_x|^2 + \left|E_y\right|^2 + |E_z|^2 = \omega^2 \left\{ |\mathbf{j}|^2 + |\rho|^2 - 2\rho\mathbf{n}\mathbf{j} \right\}. \tag{3.54}$$

Taking into account continuity equation (1.16) $\mathbf{nj} = \rho$, the last expression reduces to the form

$$|\mathbf{E}|^2 = \omega^2 \left\{ |\mathbf{j}|^2 - |\rho|^2 \right\}. \tag{3.55}$$

The substitution of the expressions for ρ, j_x, and j_z into Eq. (3.52) yields the following expression for the spectral–angular distribution of radiation in the perpendicular geometry:

$$\frac{d^2W}{d\omega d\Omega} = \frac{e^2}{4\pi^2} \frac{\beta}{\sin\psi} \exp\left(-\frac{4\pi h}{\lambda\beta} \sqrt{1 - \beta^2 \sin^2\psi} \right) \times$$

$$\frac{\left(1 - \beta^2 \sin^2\psi\right) (1 + \beta_x \sin\psi) (1 - \cos\phi) + \cos^2\psi\ (1 - \beta_x \sin\psi) (1 + \cos\phi)}{\left(1 - \beta^2 \sin^2\psi\right) \left[(1 - \beta_x \sin\psi \cos\phi)^2 - \beta_y^2 \sin^2\psi \sin^2\phi \right]}. \tag{3.56}$$

Here, $\lambda = 2\pi c/\omega$ is the radiation wavelength. According to this formula, the diffraction radiation density decreases exponentially with the wavelength rise.

The dependence on angle φ_0 (between the velocity of the initial particle, $\boldsymbol{\beta}$, and target) is expressed in terms of the components $\beta_x = \beta \cos\varphi_0$ and $\beta_y = \beta \sin\varphi_0$. As shown in [23], the diffraction radiation intensity is maximal along the directions for which the denominator in Eq. (3.56) is minimal. These directions are specified for the ultrarelativistic case ($\gamma \gg 1$) by the conditions

$$\begin{aligned} \psi &= \frac{\pi}{2}, \\ \varphi &= \pm\varphi_0. \end{aligned} \tag{3.57}$$

Sign "+" corresponds to the radiation direction coinciding with the motion direction of the initial particle, i.e., to "forward" diffraction radiation and sign "−" corresponds to the direction of the mirror reflection of the particle momentum with respect to the target plane, i.e., to "backward" diffraction radiation. Let us use new angular variables (for the case of backward diffraction radiation) defined in the coordinate system associated with unit vectors $\mathbf{e}_{1,2}$:

$$\begin{aligned} \theta_x &= \left(\frac{\pi}{2} - \psi\right) \sim \gamma^{-1}, \\ \theta_y &= (\varphi + \varphi_0) \sim \gamma^{-1}. \end{aligned} \tag{3.58}$$

In terms of these variables, lengthy formula (3.56) reduces to the following simple formula for the diffraction radiation intensity generated by ultrarelativistic particles, where the $\sim \gamma^{-2}$ terms are disregarded [23]:

$$\begin{aligned} \frac{d^2W}{\hbar d\omega d\Omega} &= \frac{\alpha}{4\pi^2} \exp\left(-\frac{4\pi h}{\gamma\lambda}\sqrt{1+\gamma^2\theta_x^2}\right) \frac{\gamma^{-2}+2\theta_x^2}{\left(\gamma^{-2}+\theta_x^2\right)\left(\gamma^{-2}+\theta_x^2+\theta_y^2\right)} = \\ &= \frac{\alpha}{4\pi^2} \exp\left(-\frac{\omega}{\omega_c}\sqrt{1+\gamma^2\theta_x^2}\right) \frac{\gamma^{-2}+2\theta_x^2}{\left(\gamma^{-2}+\theta_x^2\right)\left(\gamma^{-2}+\theta_x^2+\theta_y^2\right)}. \end{aligned} \tag{3.59}$$

Expression (3.59) is written in a more convenient (dimensionless) form after the passage to the radiation photon energy, $\hbar\omega$, and with the use of the fine structure constant, $\alpha = e^2/(\hbar c) = 1/137$. In addition, Eq. (3.59) includes the characteristic frequency of the diffraction radiation spectrum,

$$\omega_c = \frac{\gamma c}{2h}, \tag{3.60}$$

which, for the perfectly conducting target, is expressed in terms of the Lorentz factor of the particle, γ, and impact parameter h. It can be assumed that the characteristic frequency, ω_c, for the case of the motion of the particle near the screen with a finite relative permittivity, $\varepsilon(\omega)$, also depends generally on the relative permittivity.

Note that angles θ_x and θ_y are measured in the transverse directions either from the mirror reflection direction (for backward diffraction radiation) or from the electron momentum (for forward diffraction radiation) and are on the order of the characteristic angles of radiation generated by the relativistic particle, γ^{-1}. Note also that $d\Omega = \sin\psi\, d\psi\, d\varphi = d\theta_x d\theta_y$.

From Eq. (3.59), it is easy to calculate radiation losses for each cone as

$$W = \int_{-\infty}^{+\infty} d\theta_x \int_{-\infty}^{+\infty} d\theta_y \int_0^{+\infty} d\omega \frac{d^2 W}{d\omega d\Omega} = \frac{3}{8}\alpha\, \hbar\omega_c. \qquad (3.61)$$

The total energy of diffraction radiation (for both radiation cones) is obtained by doubling this expression:

$$W_\Sigma = \frac{3}{4}\alpha\hbar\omega_c. \qquad (3.62)$$

The basic characteristics of diffraction radiation in the relativistic case for the geometry under consideration follow from Eq. (3.62) and are:

1. the radiation losses are independent of the target inclination φ_0;
2. when the spectrum cutoff frequency, ω_c, is lower than the maximum frequency for which the perfect conductor model (Lengmuir frequency of a metal) is still valid, the radiation losses are proportional to the Lorentz factor γ (similarly to the total losses on transition radiation);
3. the radiation intensity is proportional to the charge squared ($\alpha = \frac{e^2}{\hbar c}$).

In a more general case, the charged particle with charge Z and arbitrary energy (including nonrelativistic) moves over the perfectly conducting half-plane at arbitrary angles (i.e., with all three nonzero momentum components shown in Fig. 3.2),

$$\boldsymbol{\beta} = \{\beta_x,\ \beta_y,\ \beta_z\} = \beta\,\{\sin\psi_0\cos\varphi_0,\ \sin\psi_0\sin\varphi_0,\ \cos\psi_0\}\,. \qquad (3.63)$$

An approach that makes it possible to determine the characteristics of diffraction radiation in the general case was developed in [24] and is based on the application of the Lorentz transformation to the kinematics corresponding to the perpendicular geometry. The general case of the generation of diffraction radiation, where all three components of the particle velocity are nonzero (see Fig. 3.2) in the laboratory reference frame, corresponds to the perpendicular geometry of diffraction radiation in the reference frame moving with velocity $\beta_z c$ along the z axis (see Fig. 3.3).

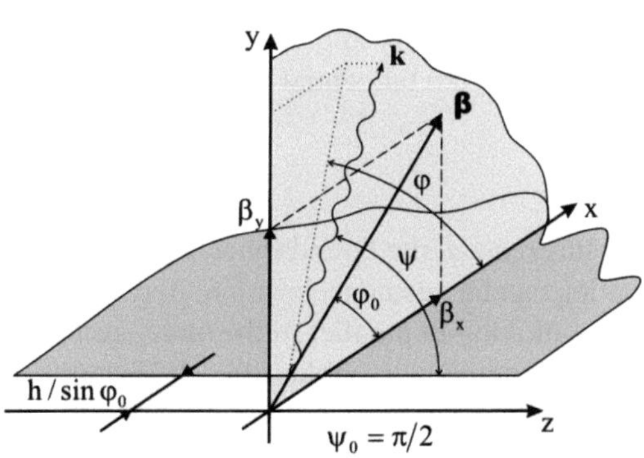

Fig. 3.2 Generation of diffraction radiation for the general case

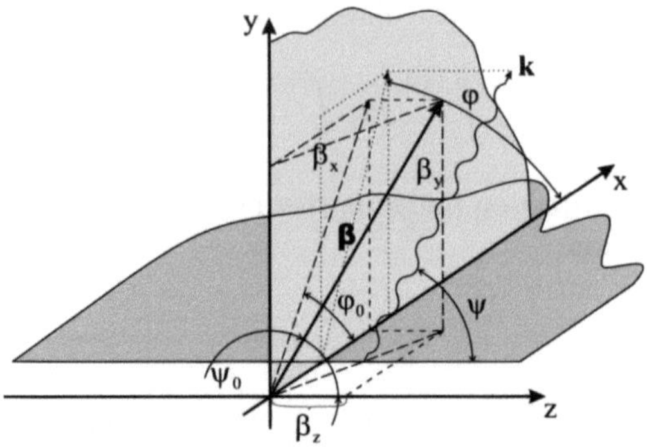

Fig. 3.3 Lorentz transformation from the inertial frame corresponding to the perpendicular geometry to the frame moving with velocity v along the z axis

In the general case, the application of the Lorentz transformations implies change in the boundary conditions on the target surface in the moving reference frame. As known (see, e.g., [20]), the boundary conditions on the surface of the actual conductor with large, but finite conductivity σ have the form

$$[\mathbf{nE}] = -\zeta\,[\mathbf{n}\,[\mathbf{nH}]]\,. \tag{3.64}$$

Here, $\mathbf{n}$ is the unit vector normal to the surface and $\zeta = \sqrt{\mu/\varepsilon}$ is the wave impedance, where μ and ε are the relative permeability and relative permittivity of the metal, respectively. This condition is called the Leontovich boundary condition. For good conductors that are not ferromagnets, $\mu = 1$, $\varepsilon = i4\pi\sigma/\omega$, and the wave impedance has the form

$$\zeta = \frac{1-i}{2}\sqrt{\frac{\omega}{2\pi\sigma}} = \frac{\omega}{c\delta}\frac{1-i}{\sqrt{2}}, \tag{3.65}$$

where δ is the thickness of the skin layer. It is seen that the boundary conditions on the surface of the conducting screen are not Lorentz invariant and this fact should generally be taken into account. However, for real good conductors, $\zeta \ll 1$. The perfect conductivity approximation corresponds to the condition $\zeta = 0$, which corresponds to zero tangential component of the electric field:

$$[\mathbf{nE}] = 0. \tag{3.66}$$

For this reason, the boundary conditions in the perfectly conducting target approximation can be treated as invariant under the Lorentz transformations in the reference frame moving in parallel to the target surface.

The kinematic variables after the Lorentz transformation (after the passage from the moving reference frame to laboratory frame) will be marked by primes. Thus, the particle velocity in the primed reference frame is written as follows:

$$\boldsymbol{\beta}' = \left\{\beta_x' = \beta_x\sqrt{1-\beta_z'^2}, \beta_y' = \beta_y\sqrt{1-\beta_z'^2}, \beta_z' = \frac{v}{c}\right\} = \left\{\gamma^{-1}\beta_x, \gamma^{-1}\beta_y, \beta\right\}. \tag{3.67}$$

Maxwell's equations are invariant under the Lorentz transformations. Hence, the solution of these equations has the form of Eq. (3.48) and, in the primed reference frame,

$$\mathbf{E}' = \omega'\left(\omega'\mathbf{j}' - \mathbf{k}'\rho'\right). \tag{3.68}$$

In this case, owing to the Lorentz invariance of Maxwell's equations and the invariance of the boundary conditions on the surface of the perfectly conducting screen, the relation of ρ' and $\mathbf{j}'$ to the characteristics of the particle motion remains the same as in the laboratory frame and is given by formulas similar to Eqs. (3.44) and (3.45). The relation of primed components in the new system to the unprimed components is given by known formulas (see, e.g., [25])

$$\begin{aligned}
\sin\psi' &= \frac{\sqrt{1-\beta_z'^2}\,\sin\psi}{1+\beta_z'\cos\psi},\\
\cos\psi' &= \frac{\cos\psi+\beta_z'}{1+\beta_z'\cos\psi},\\
\varphi' &= \varphi,\\
\omega' &= \omega\frac{\sqrt{1-\beta_z'^2}}{1-\beta_z'\cos\psi},
\end{aligned} \tag{3.69}$$

and 4-vector $j_\mu = (\rho, \mathbf{j})$ is transformed as follows:

$$\begin{aligned}
\rho' &= \frac{\rho+\beta_{z j_z}'}{\sqrt{1-\beta_z'^2}},\\
j_x' &= j_x,\\
j_y' &= j_y = 0,\\
j_z' &= \frac{j_z+\beta_z'\rho}{\sqrt{1-\beta_z'^2}}.
\end{aligned} \tag{3.70}$$

Since the components of 4-vector j_μ depend on angle ψ (see Fig. 3.1) and with the use of Eq. (3.69), ρ' and j_x' are expressed in the following explicit form in terms of the emission angles of a diffraction radiation photon in the laboratory reference frame:

$$\rho' = -\frac{B'_{q\omega'}}{\sqrt{\omega'}\sin\psi'(1+\cos\varphi)}\times$$
$$\times\left\{\frac{1-\beta'_x}{\sqrt{1-\beta'^2_z}\,\sin\psi'} - \frac{\alpha^*\beta'_x + i\gamma^{-2}\beta'_y\left(1-\beta'_z\cos\psi'\right)}{\alpha^*}\times\right.$$
$$\left.\times\frac{\sin\psi'\left(1+\cos\varphi\right)\sqrt{1-\beta'^2_z}}{-\left(\beta'^2_x+\beta'^2_y\right)\sin\psi'\cos\varphi + \beta'_x\left(1-\beta'_z\cos\psi'\right) + i\alpha^*\beta'_y}\right\}. \tag{3.71}$$

$$j'_x = -\frac{B'_{q\omega'}}{\sqrt{\omega'}\sin\psi'\left(1+\cos\varphi\right)}\times$$
$$\times\frac{\left(\beta'^2_x+\beta'^2_y\right)\sin\psi' + \beta'_x\left(1-\beta'_z\cos\psi'\right) + i\alpha^*\beta'_y}{-\left(\beta'^2_x+\beta'^2_y\right)\sin\psi'\cos\varphi + \beta'_x\left(1-\beta'_z\cos\psi'\right) + i\alpha^*\beta'_y}. \tag{3.72}$$

Here,

$$B'_{q\omega'} = \frac{e}{4\pi^2}\,\frac{\beta'_\perp\sqrt{\omega\left[\beta'^2_\perp\sin\psi' - \beta'_x\left(1-\beta'_z\cos\psi'\right) - i\alpha^*\beta'_y\right]}}{\omega'\left(\alpha^*\beta'_x + i\beta'_y\left(1-\beta'_z\cos\psi'\right)\right)}\exp\left(-\kappa'\right)$$

$$\alpha'_0 = \frac{\sqrt{\left(1-\beta'_z\cos\psi'\right)^2 - \beta'^2_\perp\sin^2\psi'}}{1-\beta'_z\cos\psi'} = \frac{\alpha^*}{1-\beta'_z\cos\psi'},$$

$$\kappa' = \frac{2\pi h}{\lambda'\beta'_\perp}\sqrt{\left(1-\beta'_z\cos\psi'\right)^2 - \beta'^2_\perp\sin^2\psi'}. \tag{3.73}$$

In Eq. (3.72) and below, $\beta'_\perp = \sqrt{\beta'^2_x + \beta'^2_y}$. Expressions for fields $E'_{1,2}$ have the same form as Eq. (3.51)

$$E'_1 = \omega'\frac{j'_x\cos\varphi - \rho'\sin\psi'}{\cos\psi'},$$
$$E'_2 = -\omega' j'_x\sin\varphi, \tag{3.74}$$
$$E'_n = 0.$$

The spectral–angular distribution of diffraction radiation in the general case can be obtained by summing the squares of field components (3.74), but calculations in this case are very cumbersome. In order to obtain an expression for the spectral–angular density of diffraction radiation in the general case, let use take into account that the spectral–angular density of radiation is proportional to the product of ω^2 by the magnitude squared of the current density 4-vector (see Eq. 1.16). The last quantity is Lorentz invariant and, therefore, the quantity

$$\frac{1}{\omega^2}\frac{d^2W}{d\omega d\Omega}=\frac{1}{\omega'^2}\frac{d^2W'}{d\omega' d\Omega'}=inv \tag{3.75}$$

is also Lorentz invariant. Then, transforming the angles, frequency, and velocity components when passing to the primed frame, we obtain

$$\frac{d^2W}{\hbar d\omega d\Omega}=Z^2\frac{\alpha}{4\pi^2}\frac{\beta_\perp(1-\beta_z\cos\psi)}{\sin\psi}\exp\left(-\frac{4\pi h}{\lambda\beta_\perp}\sqrt{(1-\beta_z\cos\psi)^2-\beta_\perp^2\sin^2\psi}\right)\times$$
$$\times\frac{1}{\left[(1-\beta_z\cos\psi)^2-\beta_\perp^2\sin^2\psi\right]\left[(1-\beta_z\cos\psi-\beta_x\sin\psi\cos\varphi)^2-\beta_y^2\sin^2\psi\sin^2\varphi\right]}\times$$
$$\times\left[\left((1-\beta_z\cos\psi)^2-\beta_\perp^2\sin^2\psi\right)\left(1+\frac{\beta_x\sin\psi}{1-\beta_z\cos\psi}\right)(1-\cos\varphi)+\right.$$
$$\left.+(\cos\psi-\beta_z)^2\left(1-\frac{\beta_x\sin\psi}{1-\beta_z\cos\psi}\right)(1+\cos\varphi)\right] \tag{3.76}$$

Since Eq. (3.76) is derived for the main (laboratory) frame, primes in this formula and below are omitted in order to simplify the expressions.

Formula (3.76) explicitly involves the dependence on the charge of the moving particle, Z.

Note that, in the particular case where $\varphi_0=\frac{\pi}{2}$ and $\beta_x=0$ $(\beta_\perp=\beta_y)$, i.e., when the particle velocity projection on the target plane is parallel to its edge (the so-called parallel geometry), Eq. (3.76) coincides completely with the results obtained in [26, 27], where the parallel geometry is considered.

Radiation losses for the general case are obtained by integrating Eq. (3.76) with respect to the azimuth and polar angles and frequency. It is interesting that the integral can be calculated analytically although the expression is complicated.

For the integration with respect to the azimuth angle, we represent Eq. (3.76) in the form

$$\frac{d^2W}{\hbar d\omega d\Omega}=\frac{Z^2\alpha}{4\pi^2}\frac{\beta_\perp(1-\beta_z\cos\psi)}{\sin\psi}\exp\left(-\frac{4\pi h}{\lambda\beta_\perp}\sqrt{(1-\beta_z\cos\psi)^2-\beta_\perp^2\sin^2\psi}\right)\times$$
$$\times\frac{1}{\left[(1-\beta_z\cos\psi)^2-\beta_\perp^2\sin^2\psi\right]}\times\frac{A+B\cos\varphi}{\left[(a\cos\varphi-\mu)^2+\nu^2\right]}, \tag{3.77}$$

where

$$A = \left[(1-\beta_z\cos\psi)^2 - \beta_\perp^2\sin^2\psi\right]\left(1+\frac{\beta_x\sin\psi}{1-\beta_z\cos\psi}\right)+ \\ +(\cos\psi-\beta_z)^2\left(1-\frac{\beta_x\sin\psi}{1-\beta_z\cos\psi}\right)= \\ =\left[(1-\beta_z\cos\psi)^2-\beta_\perp^2\sin^2\psi\right]\left(1+\frac{\beta_x\sin\psi}{1-\beta_z\cos\psi}\right)+ \\ \left[(1-\beta_z\cos\psi)^2-\beta_\perp^2\sin^2\psi-\gamma^{-2}\sin^2\psi\right]\cdot\left(1-\frac{\beta_x\sin\psi}{1-\beta_z\cos\psi}\right); \tag{3.78}$$

$$B = -\left[(1-\beta_z\cos\psi)^2 - \beta_\perp^2\sin^2\psi\right]\left(1+\frac{\beta_x\sin\psi}{1-\beta_z\cos\psi}\right)+ \\ +\left[(1-\beta_z\cos\psi)^2-\beta_\perp^2\sin^2\psi-\gamma^{-2}\sin^2\psi\right]\left(1-\frac{\beta_x\sin\psi}{1-\beta_z\cos\psi}\right) \tag{3.79}$$

$$a=\beta_\perp\sin\psi,\quad \mu=\frac{\beta_x}{\beta_\perp}(1-\beta_z\cos\psi) \\ \nu=\frac{\beta_y}{\beta_x}\sqrt{\mu^2-\frac{\beta_x^2}{\beta_\perp^2}a^2}=\frac{\beta_y}{\beta_\perp}\sqrt{(1-\beta_z\cos\psi)^2-\beta_\perp^2\sin^2\psi} \tag{3.80}$$

First, we reduce the degree in the integrand:

$$\frac{A+B\cos\varphi}{\left[(a\cos\varphi-\mu)^2+\nu^2\right]}=\frac{\lambda_1}{a\cos\varphi-\mu+i\nu}+\frac{\lambda_2}{a\cos\varphi-\mu-i\nu}= \\ =\frac{\lambda_1}{a\cos\varphi-\delta}+\frac{\lambda_1^*}{a\cos\varphi-\delta^*}, \\ \lambda_1=\frac{i}{2\nu}\left(A+\frac{\delta}{a}B\right),\quad \delta=\mu-i\nu. \tag{3.81}$$

Using the theory of residues , we obtain

$$\int_0^{2\pi}d\varphi\frac{A+B\cos\varphi}{\left[(a\cos\varphi-\mu)^2+\nu^2\right]}=-4\pi\,\mathrm{Re}\left(\frac{\lambda_1}{\sqrt{\delta^2-a^2}}\right). \tag{3.82}$$

Note that the square root is calculated exactly:

$$\sqrt{\delta^2 - a^2} = \sqrt{\left[\frac{\beta_x}{\beta_\perp}\sqrt{(1-\beta_z\cos\psi)^2 - \beta_\perp^2\sin^2\psi} - \frac{i\beta_y}{\beta_\perp}(1-\beta_z\cos\psi)\right]^2} =$$
$$= \frac{\beta_x}{\beta_y}\nu - i\frac{\beta_y}{\beta_x}\mu. \tag{3.83}$$

Therefore, taking the real part, we obtain

$$Re\left(\frac{\lambda_1}{\sqrt{\delta^2-\alpha^2}}\right) = -\frac{1}{2\nu}\frac{\left(A+\mu\frac{B}{a}\right)\mu\frac{\beta_y}{\beta_x} - \frac{\beta_x}{\beta_y}\nu^2\frac{B}{a}}{\left(\frac{\beta_y}{\beta_x}\mu\right)^2 + \left(\frac{\beta_x}{\beta_y}\nu\right)^2}. \tag{3.84}$$

The substitution of Eqs. (3.78), (3.79), (3.80) into the last formula provides the following expressions for the numerator,

$$\frac{\beta_y}{\beta_\perp}\frac{(1-\beta_z\cos\psi)^2 - \beta_x^2\sin^2\psi}{(1-\beta_z\cos\psi)}\left\{2(1-\beta_z\cos\psi)^2 - \left(1+\beta_\perp^2-\beta_z^2\right)\sin^2\psi\right\}, \tag{3.85}$$

and for the denominator,

$$-2\frac{\beta_y}{\beta_\perp}\left[(1-\beta_z\cos\psi)^2 - \beta_x^2\sin^2\psi\right]\sqrt{(1-\beta_z\cos\psi)^2 - \beta_\perp^2\sin^2\psi}. \tag{3.86}$$

Thus, the result of the azimuthal integration of Eq. (3.77) is given by the sufficiently simple formula

$$\frac{d^2W}{\hbar d\omega\sin\psi d\psi} = Z^2\frac{\alpha}{2\pi}\frac{\beta_\perp}{\sin\psi}\frac{2(1-\beta_z\cos\psi)^2 - (1+\beta_\perp^2-\beta_z^2)\sin^2\psi}{\left[(1-\beta_z\cos\psi)^2 - \beta_\perp^2\sin^2\psi\right]^{3/2}}\times$$
$$\times\exp\left(-\frac{4\pi h}{\lambda\beta_\perp}\sqrt{(1-\beta_z\cos\psi)^2 - \beta_\perp^2\sin^2\psi}\right). \tag{3.87}$$

Note that radiation vanishes when a particle moves parallel to the target edge (when $\beta_\perp = 0, \beta_z = \beta$). The integration with respect to the energy yields the expression

$$\frac{dW}{d\psi} = Z^2\frac{\alpha\hbar c}{2\pi}\frac{\beta_\perp^2}{2h}\frac{2(1-\beta_z\cos\psi)^2 - (1+\beta_\perp^2-\beta_z^2)\sin^2\psi}{\left[(1-\beta_z\cos\psi)^2 - \beta_\perp^2\sin^2\psi\right]^2}. \tag{3.88}$$

In order to calculate the total energy losses, we consider the integral

$$\int_0^{\pi} d\psi \frac{2(1-a\cos\psi)^2 - \left(1+b^2-a^2\right)\sin^2\psi}{\left[(1-a\cos\psi)^2 - b^2\sin^2\psi\right]^2} = \left(1 - \frac{1-a^2-b^2}{4b}\frac{\partial}{\partial b}\right) I\,(a,b), \tag{3.89}$$

where

$$I\,(a,b) = \int_0^{2\pi} \frac{d\psi}{(1-a\cos\psi)^2 - b^2\sin^2\psi}. \tag{3.90}$$

When a and b are real and $a^2 + b^2 < 1$, integral $I\,(a,b)$ is easily calculated with the use of residues

$$I\,(a,b) = \frac{2\pi}{1-a^2}\frac{1}{\sqrt{1-a^2-b^2}}. \tag{3.91}$$

Substituting the expression for $I\,(a,b)$ into Eq. (3.89), differentiating with respect to b, and setting $a = \beta_z$ and $b = \beta_\perp$, we arrive at the following expression for the total energy losses:

$$W = \frac{3}{4} Z^2 \alpha \hbar c \beta^2 \frac{\gamma}{2h} \frac{\beta_\perp^2}{\gamma^{-2}\beta_z^2 + \beta_\perp^2}. \tag{3.92}$$

For the perpendicular geometry ($\beta_z = 0$), Eq. (3.92) provides the expression

$$W_\perp = \frac{3}{4} Z^2 \alpha \beta^2 \hbar \omega_c, \tag{3.93}$$

which is independent of inclination φ_0 and coincides with previously obtained formula (3.62) with an accuracy of γ^{-2}. For the other particular case, the parallel geometry ($\beta_x = 0$, $\beta_\perp = \beta_y$, $\beta_z/\beta_y = \cot\psi_0$, Eq. (3.92) reduces to the expression

$$W_\parallel = \frac{3}{4} Z^2 \alpha \hbar c \beta^2 \frac{\gamma}{2h} \frac{\beta_y^2}{\beta_y^2 + \gamma^{-2}\beta_z^2} = W_\perp \frac{1}{1+\gamma^{-2}\cot^2\psi_0}. \tag{3.94}$$

Therefore, in the general case, $W_\perp > W_\parallel$. At $\psi_0 \ll 1$,

$$W_\parallel = W_\perp \frac{\psi_0^2}{\psi_0^2 + \gamma^{-2}}. \tag{3.95}$$

Thus, $W_\parallel$ varies from $W_\perp$ at $\psi_0 \sim 1$ to $W_\perp/2$ at $\psi_0 \sim \gamma^{-1}$. In the ultrarelativistic case for not too small angles of the particle motion near the target ($\psi_0 \gg \gamma^{-1}$),

$$W_\parallel = W_\perp \tag{3.96}$$

with an accuracy of γ^{-2}; i.e., the radiation losses are independent of angle ψ_0. In the ultrarelativistic case, the radiation losses are determined by the charge of the particle Z, its Lorentz factor γ, and impact parameter h. In other words, the radiation losses in this case coincide with the losses of the same particle moving with the same impact parameter near the semi-infinite perfect half-plane perpendicular to the particle trajectory (i.e., coinciding with the projection of the initial target on the plane perpendicular to the particle velocity).

Figure 3.4 shows the two-dimensional distribution of the diffraction radiation intensity for the general case as calculated by Eq. (3.76). It is seen that two radiation maxima coincide with the directions of the electron momentum and its mirror reflection from the target (forward diffraction radiation and backward diffraction radiation, respectively). For the moderate relativistic case with $\gamma = 50$, the radiation maximum coincides with the electron momentum direction with an accuracy better than γ^{-1}.

Figure 3.4 shows the distribution for an impact parameter of $h =$ (a) 50 and (b) 10 μm and a wavelength of (a) $\lambda = 1500\,\mu$m and $\lambda = 0.5\,\mu$m. Since the impact parameter and wavelength appear in the argument of the exponential in the form of a dimensionless variable of $z = \frac{4\pi h}{\gamma\beta\lambda}$, it is more convenient to use below this parameter for calculating the characteristics of diffraction radiation. This parameter is $z = 0.008$ and $z = 5$ for cases (a) and (b), respectively.

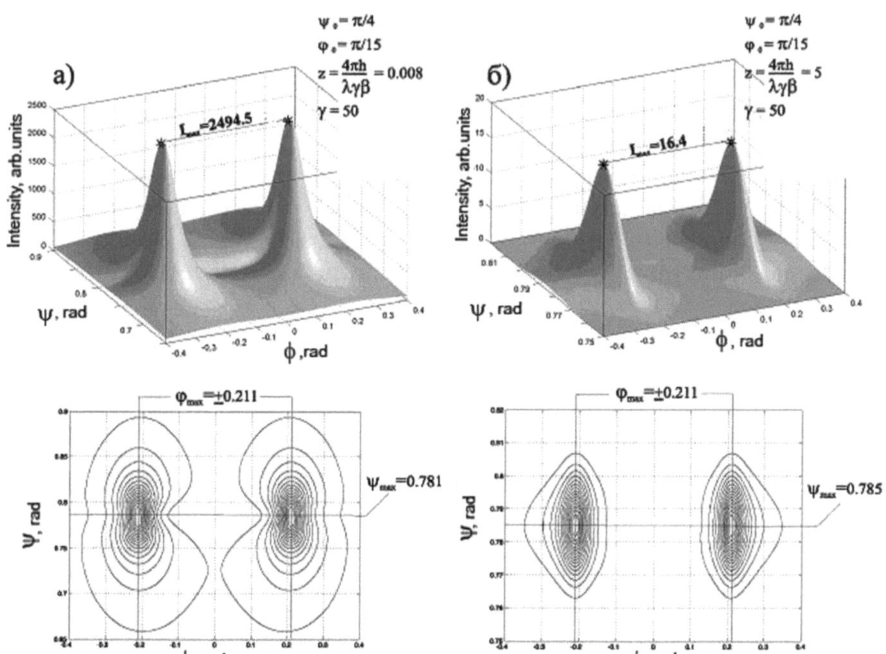

Fig. 3.4 (*Upper panels*) Angular distributions and (*lower levels*) contours of the diffraction radiation intensity for various values of dimensionless variable $z = 4\pi h/\gamma\beta\lambda$

According to the figure, the maxima in the distribution are insignificantly shifted from the directions corresponding to the electron momentum (ψ_0, φ_0) and mirror reflection $(\psi_0, -\varphi_0)$:

$$\frac{\psi_0 - \psi_{\max}}{\psi_0} = 0.006, \quad \frac{\Delta\phi}{\phi_0} = 0.008 \quad \text{for } z = 0.008, \text{ and}$$
$$\frac{\psi_0 - \psi_{\max}}{\psi_0} = 0.001, \quad \frac{\Delta\phi}{\phi_0} = 0.0075 \quad \text{for } z = 5. \tag{3.97}$$

It is obvious that the accuracy with which the positions of the maxima coincide with the directions (ψ_0, φ_0) (forward diffraction radiation) and $(\psi_0, -\varphi_0)$ (backward diffraction radiation) increases with γ. Thus, similar to the particular cases considered previously [23], the maxima in the general case coincide with the directions $\psi_0, \pm\varphi_0$ with an accuracy better than γ^{-1}.

In complete analogy with the case of the perpendicular geometry, we introduce the new angular variables

$$\begin{aligned} \theta_x &= \psi - \psi_0, \\ \theta_y &= (\varphi \mp \varphi_0) \sin \psi_0. \end{aligned} \tag{3.98}$$

In the latter formula, the upper and lower signs correspond to the forward and backward diffraction radiations, respectively. In most papers, this notation of the variables is used; i.e., variable θ_x characterizes the distribution along the edge of the semi-plane (see Fig. 3.5). For this reason, this direction will be used as the x axis. The cunbersome expression (3.76) is considerably simplified in terms of the variables given by Eq. (3.98).

Let us transform the argument of the exponential in Eq. (3.76) by expressing the velocity components in terms of ψ_0: $\beta_z = \beta \cos \psi_0$, $\beta_\perp = \beta \sin \psi_0$. Then, the radicand in Eq. (3.76) can be written in the form (where terms above γ^{-2} are neglected):

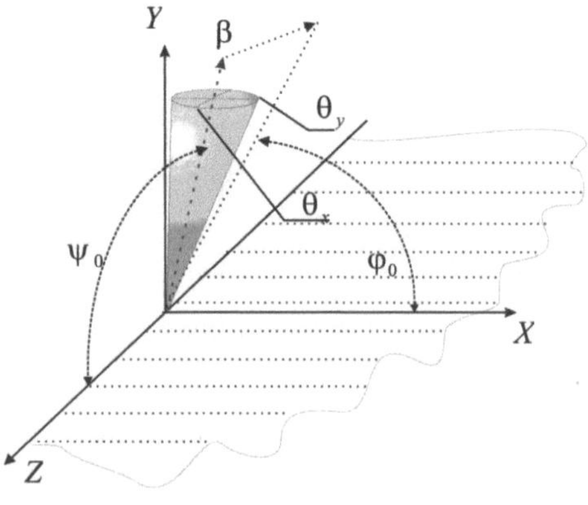

Fig. 3.5 Angular variables describing the emission cone of forward diffraction radiation in the ultrarelativistic case

$$[1-\beta\cos(\psi_0+\psi)][1-\beta\cos(\psi_0-\psi)] = \sin^2\psi_0\left(\gamma^{-2}+\theta_x^2\right). \tag{3.99}$$

In the ultrarelativistic approximation used here, the explicit dependence of the exponential on angle ψ_0 disappears:

$$\exp\left(-\frac{4\pi h}{\lambda\beta_\perp}\sqrt{(1-\beta_z\cos\psi)^2-\beta_\perp^2\sin^2\psi}\right) \sim \exp\left(-\frac{4\pi h}{\lambda\beta_\gamma}\sqrt{1+\gamma^2\theta_x^2}\right) = \\ = \exp\left(-z\sqrt{1+\gamma^2\theta_x^2}\right). \tag{3.100}$$

In this expression, the wavelength dependence is separated in the form of parameter $z = \frac{4\pi h}{\lambda\beta\gamma} = \frac{\omega}{\omega_c}$. Further, expanding the numerator and denominator in Eq. (3.76) and retaining the terms no higher than γ^{-2}, we arrive at the following expression exactly coinciding with expression (3.59) (for Z = 1):

$$\frac{d^2W}{\hbar d\omega d\Omega} = \frac{\alpha}{4\pi^2}\exp\left(-\frac{4\pi h}{\gamma\lambda}\sqrt{1+\gamma^2\theta_x^2}\right)\frac{\gamma^{-2}+2\theta_x^2}{\left(\gamma^{-2}+\theta_x^2\right)\left(\gamma^{-2}+\theta_x^2+\theta_y^2\right)} \tag{3.101}$$

In other words, the general expression for the diffraction radiation intensity in the ultrarelativistic approximation coincides with the formula for the diffraction radiation intensity for the perpendicular geometry, but angular variables θ_x, θ_y are determined from Eq. (3.98).

The authors of the recent works [28, 29] suggested new approaches to calculate characteristics of DR from an ideally conducting inclined semiplane. The detailed comparison of results obtained in [28, 29] with the Kazantsev-Surdutovich model [22] was performed in the paper [29].

As was shown there the expression describing a DR spectral-angular distribution (see formula (3.56)) becomes incorrect for grazing inclination angles (particularly, for a parallel particle flight above a semi-plane).

Following to the approach developed in [29] let us write the DR spectral-angular distribution using the same angular variables as in the formula (3.56):

$$\frac{d^2W}{\hbar d\omega d\Omega} = \frac{\alpha}{4\pi^2}\frac{\exp\left[-\frac{4\pi h}{\gamma\beta\lambda}\sqrt{1+(\beta\gamma\cos\varphi)^2}\right]}{\left[1+(\beta\gamma\cos\varphi)^2\right]\left[\left(\frac{\cos\varphi_0}{\beta}-\sin\psi\cos\varphi\right)^2+\sin^2\varphi_0\left(\frac{1}{\beta^2\gamma^2}+\cos^2\psi\right)\right]}\times \\ \times\left[\sin^2\psi\left(\frac{\cos^2\varphi_0}{\beta^2\gamma^2}+\frac{\sin^2\varphi_0}{\beta^2}\left(1+(\beta\gamma\cos\psi)^2\right)\right)+\gamma^2\cos^2\psi\left(\cos^2\psi+\sin^2\psi\sin^2\varphi\right)+ \right. \\ \left. +\frac{2}{\beta}\sin\varphi\sin\psi\cos^2\psi\cos\varphi_0\right]. \tag{3.102}$$

Results of calculations for different tilting angles φ_0 with using the formula (3.102) and the traditional one (3.56) one may compare in Fig. 3.6. Note that the significant discrepancy between two models appears for a region of tilting angles $\varphi_0 \sim \gamma^{-1}$. For a parallel geometry ($\varphi_0 = 0$) in the ultrarelativistic limit the expression (3.102) can be written approximately to within quantities in the order of γ^{-2}:

$$\frac{d^2 W_{\parallel}}{\hbar d\omega d\Omega} = \frac{\alpha}{4\pi^2}\gamma^2 \frac{1+2t_x^2}{\left(1+t_x^2\right)\left(1+t_x^2+t_y^2\right)^2} \exp\left[-\frac{4\pi h}{\beta\gamma\lambda}\sqrt{1+t_x^2}\right], \tag{3.103}$$

where $t_x = \gamma\theta_x$, $t_y = \gamma\theta_y$, θ_x, θ_y are the projection angles (3.58).

For the perpendicular geometry the analogous approximation is just the same as from the model [22] (see, formula (3.59)). The Fig. 3.6 illustrates the confluence of forward and backward diffraction radiation cones into a single one directed along particle velocity with a decreasing of target inclination angle to zero (see also Fig. 3.7).

Finally, it should be noted that radiation losses remain the same as in the Kazantev-Surdutovich model (see, Eq. (3.62)).

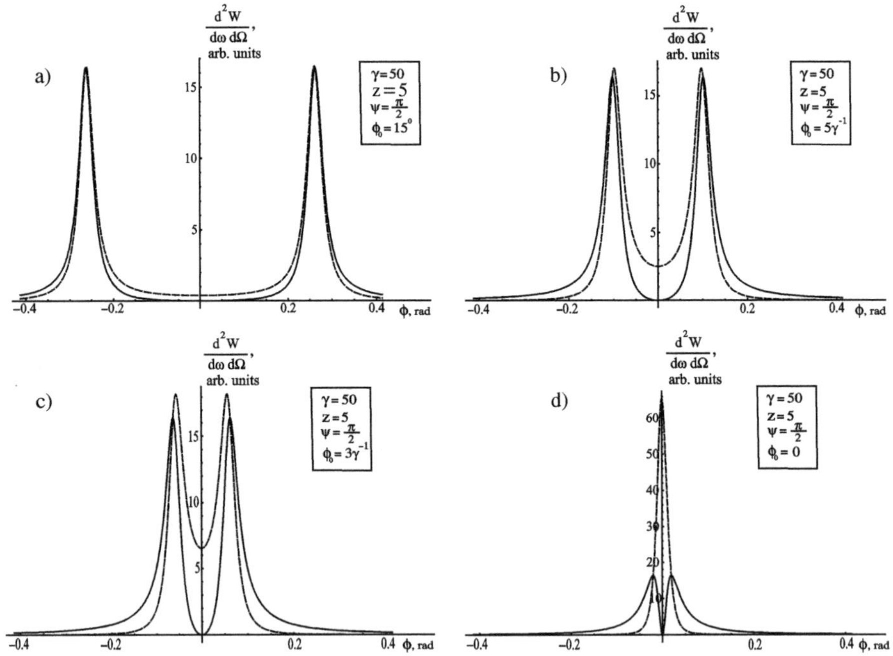

Fig. 3.6 Angular distribution of the diffraction radiation from inclined half-plane calculated using approach [22] (*low curves*) and approach [29] (*upper curves*) for different tilted angels

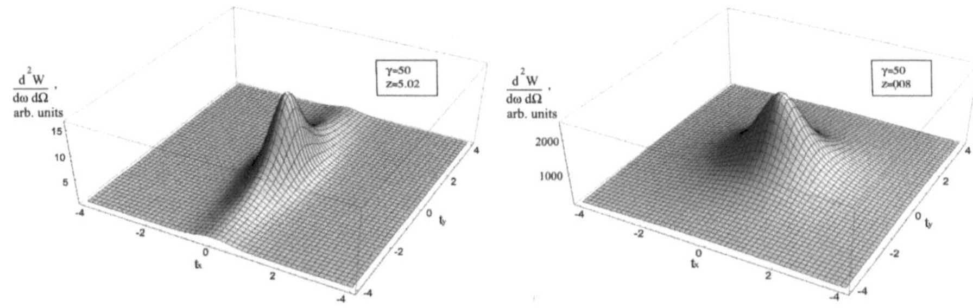

Fig. 3.7 Two dimensional angular distribution of the diffraction radiation from half-plane for the perpendicular geometry ($\psi_0 = \pi/2, \phi_0 = 0$)

3.3 Radiation Generated by a Charge Passing Through a Slit in a Perfectly Conducting Screen

The characteristics of diffraction radiation generated by a charged particle passing through a slit in a screen perpendicular to the particle trajectory were considered in [9, 30]. The author of the latter paper pointed to the possibility of using diffraction radiation from the slit for the non-invasive diagnostics of beams. However, for the experimental implementation of the proposed method, it is necessary to use backward diffraction radiation from the slit in the inclined target.

The diffraction radiation field generated when the charge passes through the slit with width a in the inclined perfectly conducting screen (see Fig. 3.8) can be considered as a superposition of the diffraction radiation fields from the upper and lower semi-planes [31]. The particle generally passes not through the center of the slit, but at a distance of h_1 from the upper edge of the slit and, correspondingly, at a distance of h_2 from the lower edge.

Interference effects in the resulting radiation are obviously manifested under the condition

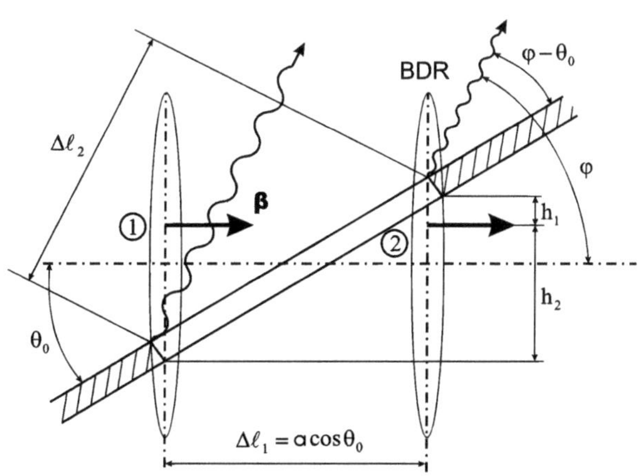

Fig. 3.8 Scheme of the determination of phase shifts for backward diffraction radiation at the slit

$$h_1 + h_2 = a \sin\theta_0 \leq \frac{\beta\gamma\lambda}{2\pi}, \tag{3.104}$$

i.e., mainly for ultrarelativistic particles. This case will be discussed below.

In the general case, the diffraction radiation field from the slit is written in the form

$$\mathbf{E}_{\text{slit}} = \mathbf{E}_U \exp(i\varphi_U) + \mathbf{E}_L \exp(-i\varphi_L), \tag{3.105}$$

where $\mathbf{E}_U$ ($\mathbf{E}_L$) is the diffraction radiation field from the upper (lower) semi-plane. Expressions for phase shifts φ_U, φ_L can be obtained on the basis of the description of the generation of diffraction radiation at the slit as being formed only at the edge of both semi-planes. Independently, as shown in [32], at distances exceeding $\gamma^2\lambda$, a source can be considered as a point (see Chap. 8). Thus, for optical diffraction radiation generated when an electron passes through a macroscopic slit wider than the wavelength, a simple relation for the phase shifts between radiations at the lower and upper edges of the slit can be obtained from a simple geometric representation (see Fig. 3.8). Diffraction radiation at the edge of the lower and upper semi-planes is formed when the charge passes points 1 and 2, respectively. The time interval between these waves, Δt_2, depends on slit width a, inclination θ_0, and radiation angle φ:

$$\Delta t_2 = \frac{a}{c}\cos(\varphi - \theta_0). \tag{3.106}$$

Taking into account the time necessary for the motion of the charge from point 1 to point 2 with velocity β, $\Delta t_1 = \frac{\Delta l_1}{\beta c} = \frac{a\cos\theta_0}{\beta c}$, we arrive at the following expression for the phase difference between diffraction radiations with wavelength λ that are emitted from both half-planes:

$$\varphi_{\text{slit}} = \frac{2\pi a}{\lambda}\left(\cos(\varphi - \theta_0) - \frac{\cos\theta_0}{\beta}\right). \tag{3.107}$$

With the use of the expression, $a = \dfrac{h_1 + h_2}{\sin\theta_0}$, for slit width a in terms of impact parameters h_1 and h_2, expression (3.107) for phase shift φ_{slit} reduces to the sum of two phase shifts:

$$\varphi_{\text{slit}} = \varphi_U + \varphi_L, \tag{3.108}$$

where phase shifts φ_U (φ_L) for radiation from the upper (lower) edge of the target with respect to the particle trajectory are given by the expressions

$$
\begin{aligned}
\varphi_U &= \frac{2\pi h_1}{\lambda \sin\theta_0}\left(\cos(\varphi-\theta_0) - \frac{\cos\theta_0}{\beta}\right),\\
\varphi_L &= \frac{2\pi h_2}{\lambda \sin\theta_0}\left(\cos(\varphi-\theta_0) - \frac{\cos\theta_0}{\beta}\right).
\end{aligned}
\tag{3.109}
$$

Introducing angular variable $\theta_y = \varphi - 2\theta_0$ and disregarding the terms $\sim \gamma^{-2}$ as compared to unit, we arrive at the simpler expressions

$$
\varphi_U \sim -\frac{2\pi h_1}{\lambda}\theta_y,\ \varphi_L \sim -\frac{2\pi h_2}{\lambda}\theta_y. \tag{3.110}
$$

As shown in the preceding section, the expressions for the diffraction radiation fields in the ultrarelativistic approximation are the same functions of variables θ_x, θ_y for any geometry. For this reason, we consider the perpendicular geometry ($\psi_0 = \pi/2$) for the sake of simplicity.

For the passage of the ultrarelativistic particle near the screen perpendicular to the particle trajectory in the geometry shown in Fig. 3.1 ($\varphi_0 = \pi/2$ corresponds to the lower half-plane), expressions (3.51) for backward diffraction radiation, where terms $\sim \gamma^{-2}$ are neglected, are written in the form

$$
\begin{aligned}
E_1 &= \frac{ie}{4\pi^2}\frac{\theta_x}{\sqrt{\gamma^{-2}+\theta_x^2}}\frac{\exp\left[\frac{-2\pi h_2}{\gamma\lambda}\sqrt{1+\gamma^2\theta_x^2}\right]}{\sqrt{\gamma^{-2}+\theta_x^2}+i\theta_y},\\
E_2 &= \frac{e}{4\pi^2}\frac{\exp\left[-\frac{2\pi h_2}{\gamma\lambda}\sqrt{1+\gamma^2\theta_x^2}\right]}{\sqrt{\gamma^{-2}+\theta_x^2}+i\theta_y}.
\end{aligned}
\tag{3.111}
$$

Note that the field components of backward diffraction radiation for an inclined semi-plane in the ultrarelativistic approximation coincide with the resulting formulas, but angles θ_x, θ_y are measured from the mirror-reflection direction.

The diffraction radiation field in the upper semi-plane is obtained from the field in the lower semi-plane after the complex conjugation with simultaneous change in sign and change $h_2 \to h_1$. Then, taking into account expressions (3.110) for the phase shifts, we arrive at the final expressions

$$
\begin{aligned}
E_{1,\text{slit}} = \frac{ie}{4\pi^2}\frac{\theta_x}{\sqrt{\gamma^{-2}+\theta_x^2}}\Bigg(&\frac{\exp\left[-\frac{2\pi h_1}{\lambda}\left(\sqrt{\gamma^{-2}+\theta_x^2}-i\theta_y\right)\right]}{\sqrt{\gamma^{-2}+\theta_x^2}-i\theta_y}+\\
&+\frac{\exp\left[-\frac{2\pi h_2}{\lambda}\left(\sqrt{\gamma^{-2}+\theta_x^2}+i\theta_y\right)\right]}{\sqrt{\gamma^{-2}+\theta_x^2}+i\theta_y}\Bigg),
\end{aligned}
$$

$$E_{2,\text{slit}} = \frac{e}{4\pi^2}\left(-\frac{\exp\left[-\frac{2\pi h_1}{\lambda}\left(\sqrt{\gamma^{-2}+\theta_x^2}-i\theta_y\right)\right]}{\sqrt{\gamma^{-2}+\theta_x^2}-i\theta_y}+\right.$$
$$\left.+\frac{\exp\left[-\frac{2\pi h_2}{\lambda}\left(\sqrt{\gamma^{-2}+\theta_x^2}+i\theta_y\right)\right]}{\sqrt{\gamma^{-2}+\theta_x^2}+i\theta_y}\right). \tag{3.112}$$

These expressions for backward diffraction radiation from the inclined slit completely coincide with the result obtained by Ter-Mikaelyan [9] for the passage of the relativistic particle through the slit in the screen perpendicular to the particle velocity. This is not surprising, because an explicit dependence on the screen inclination φ_0 is absent in the approximation under consideration.

Using the model of the diffraction of a packet of real waves that approximates the almost transverse field of the relativistic charge, Fiorito and Rule [33] considered the problem of the generation of diffraction radiation at the slit in the inclined imperfect screen. When the imperfection of the screen is neglected (i.e., when the Fresnel coefficients in the formulas presented in [33] are omitted), the results obtained in [33] coincide with expression (3.112).

Note that formulas (3.112) in the limit $h_1 \to 0$, $h_2 \to 0$ (i.e., when the slit width vanishes) reduce to the known expressions for the transition radiation fields at the infinite (perfect conductor—vacuum) interface, as should be expected:

$$E_1 = \frac{ie}{2\pi^2}\frac{\theta_x}{\gamma^{-2}+\theta_x^2+\theta_y^2},$$
$$E_2 = \frac{ie}{2\pi^2}\frac{\theta_y}{\gamma^{-2}+\theta_x^2+\theta_y^2}. \tag{3.113}$$

In order to calculate the spectral–angular density of diffraction radiation at the slit, we use the dimensionless angular variables, $t_x = \gamma\theta_x, \quad t_y = \gamma\theta_y$. As above, the diffraction radiation spectrum is described in terms of variable $z = \omega/\omega_c$, where the characteristic frequency

$$\omega_c = \frac{c\gamma}{2a\sin\theta_0} \tag{3.114}$$

depends on the width of the slit projection on the plane perpendicular to the particle velocity ($a\sin\theta_0 = h_1 + h_2$). In contrast to diffraction radiation generated on the semi-plane, the characteristic frequency of diffraction radiation for the slit is determined by the width of the slit projection on the plane perpendicular to the particle velocity rather than by the impact parameter. The asymmetric passage of the particle with respect to the center of the slit is characterized by the parameter

$$l = \frac{h_1}{a\sin\theta_0} - \frac{1}{2} = \frac{1}{2} - \frac{h_2}{a\sin\theta_0}. \tag{3.115}$$

For the passage through the center of the slit, $l = 0$, whereas $l = \pm 0.5$ for the trajectories touching the slit edges; i.e., parameter l varies in the interval

$$-0.5 \leq l \leq 0.5. \tag{3.116}$$

The spectral–angular distribution of diffraction radiation is easily obtained from Eq. (3.113) in the form

$$\frac{d^2W}{\hbar d\omega d\Omega} = 4\pi^2\left(|E_1|^2 + |E_2|^2\right) = \frac{\alpha}{2\pi^2}\gamma^2 \frac{\exp\left(-z\sqrt{1+t_x^2}\right)}{(1+t_x^2)(1+t_x^2+t_y^2)} \times$$
$$\left\{(1+2t_x^2)\cosh\left(2zl\sqrt{1+t_x^2}\right) - \frac{(1+t_x^2-t_y^2)\cos\left(zt_y\right) - 2t_y\sqrt{1+t_x^2}\sin\left(zt_y\right)}{(1+t_x^2+t_y^2)}\right\}. \tag{3.117}$$

Here, $t_{x,y} = \gamma\theta_{x,y}$.

Figures 3.9 and 3.10 show the spectral–angular distributions of diffraction radiation at the slit for various z values. It is seen that, as in the case of transition radiation, the radiation intensity in the case of the symmetric passage of the particle beam

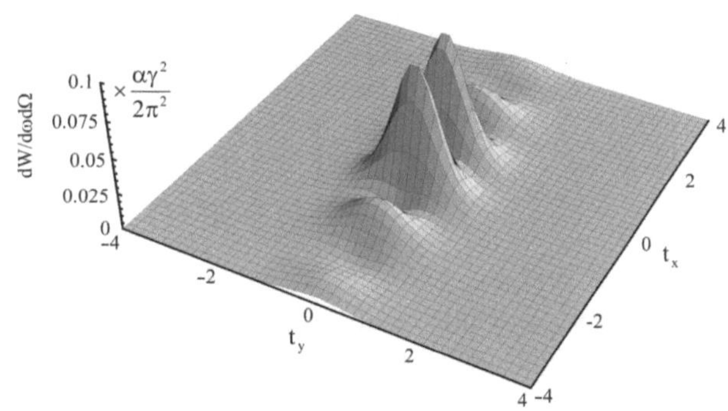

Fig. 3.9 Two-dimensional distribution of diffraction radiation for the case of the symmetric passage through the center of the slit at $z = 2.5$

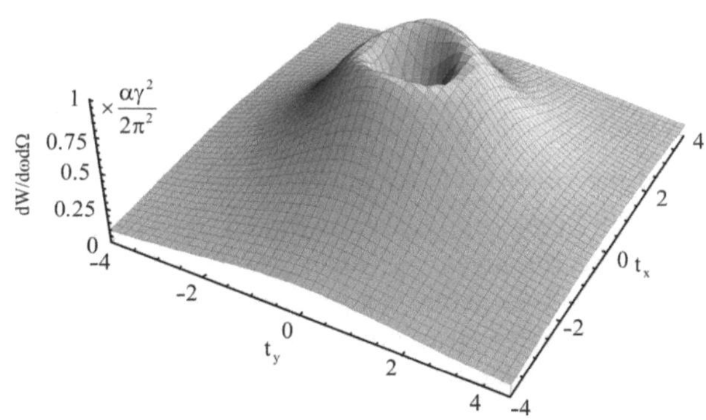

Fig. 3.10 Two-dimensional distribution of diffraction radiation for the case of the symmetric passage through the center of the slit at $z = 0.2$

through the slit is minimal at $\theta_x = \theta_y = 0$ and maximal at $\theta_x \approx \theta_y \approx \gamma^{-1}$. As follows from the above distributions for $z \geq 1$, a sharp interference pattern is observed in the plane perpendicular to the slit plane (see Fig. 3.9) and this pattern is smeared with decreasing spectral variable z (see Fig. 3.10). It is interesting that the distributions of diffraction radiation in the perpendicular planes (at $t_x = 0$ and $t_y = 0$) are asymmetric for $z \geq 1$, whereas the distribution of diffraction radiation in the limit $z \to 0$ approaches a similar distribution of transition radiation characterized by the azimuthal symmetry.

Indeed, as shown above, the characteristics of diffraction radiation are identical to those of transition radiation at $z \equiv 0$ ($h_1 = h_2 = 0$, i.e., the slit width is equal to zero). At the same time, the characteristics of diffraction radiation again become identical to those of transition radiation at $\lambda\gamma \gg a$, $a \neq 0$, and $z = \dfrac{2\pi a \sin\theta_0}{\lambda\gamma} \to 0$.

For $\lambda\gamma \leq a$, the strength of the diffraction radiation field decreases exponentially with increasing a, in contrast to the case $\lambda\gamma \geq a$, where the field strength is almost constant over the entire a interval; i.e., the pattern is the same for the passage of the particle through the screen (transition radiation) and slit (diffraction radiation).

The radiation losses are easily calculated from expression (3.117) preliminarily integrated with respect to the frequency:

$$\frac{dW}{d\Omega} = \hbar \int_0^\infty d\omega \frac{d^2W}{\hbar d\omega d\Omega} = \frac{\alpha}{2\pi^2}\gamma^2 \hbar\omega_c \frac{1}{\sqrt{\left(1+t_x^2\right)\left(1+t_x^2+t_y^2\right)}} \left\{ \frac{1+2t_x^2}{\left(1-4l^2\right)\left(1+t_x^2\right)} - \frac{1+t_x^2-3t_y^2}{\left(1+t_x^2+t_y^2\right)^2} \right\}. \tag{3.118}$$

The integration of this result with respect to angular variables $d\Omega = d\theta_x d\theta_y = \gamma^{-2} dt_x dt_y$ yields the total diffraction radiation intensity (radiation losses):

$$W = \int_{-\infty}^{\infty} dt_x \int_{-\infty}^{\infty} dt_y \frac{dW}{d\Omega} = \frac{\alpha\hbar\omega_c}{2\pi} \int_{-\infty}^{\infty} dt_x \frac{1+2t_x^2}{\left(1+t_x^2\right)^2\left(1-4l^2\right)} = \frac{3}{4}\frac{\alpha\hbar\omega_c}{1-4l^2}. \tag{3.119}$$

According to this formula, radiation losses of the particle passing through the center of the slit ($l = 0$, $h_1 = h_2 = \dfrac{a\sin\Theta_0}{2}$) correspond to the double losses for one edge of the semi-plane, as should be expected. For $l \to \pm 0.5$ (passage along one of the slit edges, i.e., for $h_1 \to 0$ or $h_2 \to 0$), expression (3.119) diverges, formally because terms independent of ω appear in the spectral radiation density at $h_1 \to 0$ (or $h_2 \to 0$) according to Eq. (3.113). For this reason, the formal integration with respect to the frequency from 0 to ∞ gives a divergent result. A similar situation appears in calculating the characteristics of transition radiation from a perfect conductor. Note that the perfect conductor approximation is inapplicable for real targets (e.g., metal foils) for $\omega \geq \omega_p$ (ω_p is the plasma frequency of the target). The inclusion of the dispersion of the relativistic particle at high frequencies leads to the finite contribution from the high-frequency part of the spectrum to the total losses (for more details, see Chap. 4 concerning diffraction radiation at frequencies above the plasma frequency).

Figure 3.11 shows the diffraction radiation intensity as a function of parameter l for various wavelengths. As well as for the total radiation losses, a minimum is observed in the case of the passage through the center of the slit ($\ell = 0$), whereas the diffraction radiation intensity increases with approaching the slit edge. This property is also illustrated in Fig. 3.12, where the frequency dependence of the spectral–angular density is shown for $\ell = 0$ and $\ell = 0.4$. Finally, Fig. 3.13 shows the diffraction radiation spectrum obtained by integrating the spectral–angular density (see expression (3.117)) over a fixed solid angle ($-t_{x\,\max} \leq t_x \leq t_{x\,\max}$, $-t_{y\,\max} \leq t_y \leq t_{y\,\max}$).

In contrast to diffraction radiation on the semi-plane, interference effects for the superposition of the fields from the both edges of the slit lead to nontrivial dependences of the characteristics of diffraction radiation on the distance between the particle trajectory and center of the slit. The dependence of the angular distributions of diffraction radiation at the slit on this parameter makes it possible to successfully

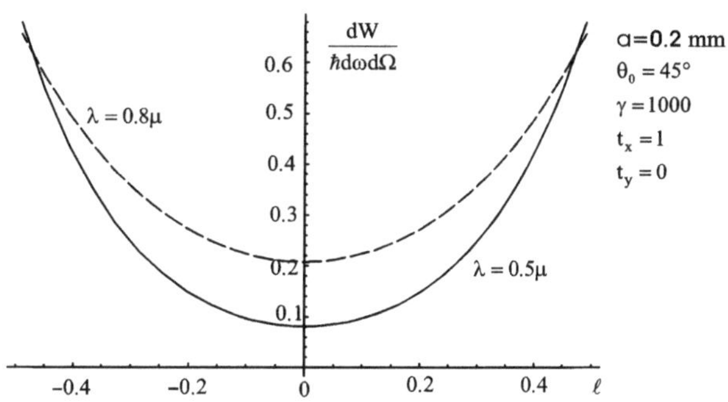

Fig. 3.11 Diffraction radiation yield with various wavelengths vs. the transverse coordinate of the electron passage through the slit

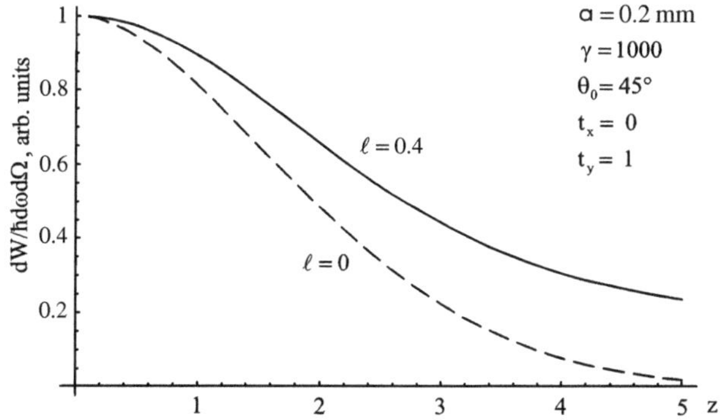

Fig. 3.12 Diffraction radiation spectrum for various ℓ values at fixed observation angles

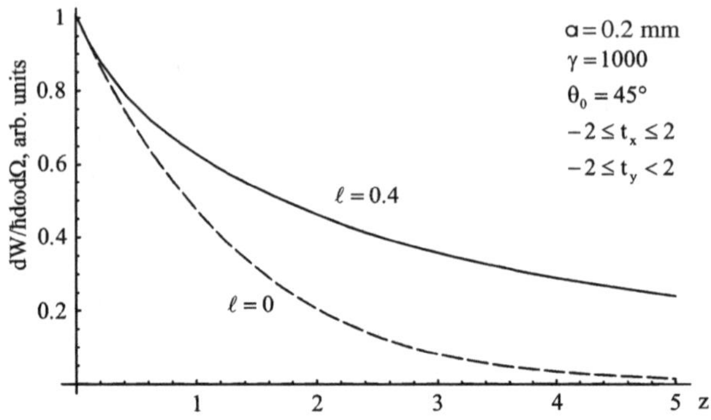

Fig. 3.13 Same as in Fig. 3.12, but for a finite angular aperture

use diffraction radiation to determine the transverse size of the beam passing through the slit (for more details, see Chap. 9).

In order to expand the diagnostic capabilities, it was proposed in [34] to use the interference pattern of diffraction radiation (interferogram) formed by two beams of diffraction radiation, one of which is generated at the fixed half of the slit through which the electron beam passes and the other beam, at the movable half (see Fig. 3.14).

In this case, phase shifts φ_L, φ_U (see expression (3.105) can be obtained on the basis of the obvious description of radiation from the slit as being formed only at the edge of the semi-planes, because radiation is considered in the far-field zone. We use the following notation (see Figs. 3.8 and 3.14): a is the width of the slit formed by two halves of the target, which are located in one plane; d is the shift of the lower (movable) half of the target with respect of this position; θ_0 is the inclination of the target with respect to the electron momentum; and h_1, h_2 is the distance between the electron trajectory and upper (lower) edge of the slit ($h_1 + h_2 = a \sin \theta_0$).

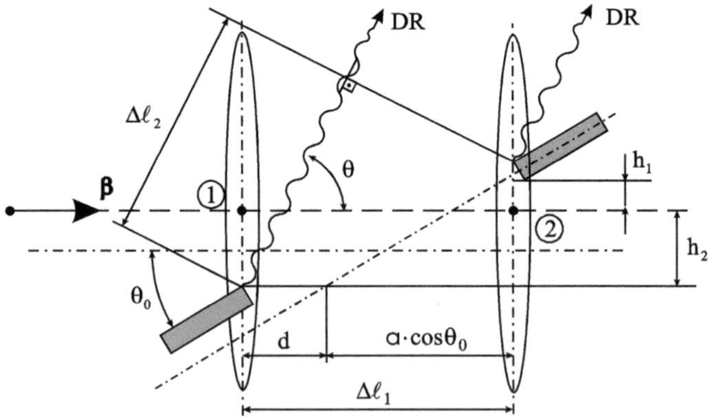

Fig. 3.14 Scheme of the generation of backward diffraction radiation in the case of the shift of the halves of the slit target at distance d

At the edge of the lower half of the target, diffraction radiation is formed when the electron passes point 1, whereas at the upper edge of the target, diffraction radiation is formed when the electron is at point 2. A simple geometric representation provides the result

$$\Delta\varphi = \varphi_U + \varphi_L = \frac{2\pi}{\lambda}\left[a\cos(\theta-\theta_0) + d\cos\theta - \frac{a\cos\theta_0 + d}{\beta}\right]. \tag{3.120}$$

Here, θ is the emission angle of a photon with wavelength λ and $\beta = v/c$ is the electron velocity divided by the speed of light. Let us introduce angle $\theta_y = \theta - 2\theta_0$, which is measured from the mirror-reflection direction, such that $\theta_y \sim \gamma^{-1}$. Then, neglecting the terms $\sim \gamma^{-2}$, we reduce Eq. (3.120) to the form

$$\Delta\varphi = \varphi_U + \varphi_L = -\frac{2\pi\,(h_1+h_2)}{\lambda}\left[\theta_y + 2\frac{d}{a}\left(\sin\theta_0 + \theta_y\cos\theta_0\right)\right], \tag{3.121}$$

or

$$\varphi_{U(L)} = -\frac{2\pi\,h_{1(2)}}{\lambda}\left[\theta_y + 2\frac{d}{a}\left(\sin\theta_0 + \theta_y\cos\theta_0\right)\right]. \tag{3.122}$$

In the same approximation, using Eqs. (3.112) and (3.122), we arrive at the following formulas for the diffraction radiation field components from the entire target:

$$\begin{aligned}
E_1 = {} & \frac{ie}{4\pi^2}\frac{\theta_x}{\sqrt{\gamma^{-2}+\theta_x^2}}\left\{\frac{\exp\left[-\frac{2\pi h_1}{\lambda}\left(\sqrt{\gamma^{-2}+\theta_x^2} + i\theta_y\left(1+\frac{2d\cos\theta_0}{a}\right) + i\frac{2d\sin\theta_0}{a}\right)\right]}{\sqrt{\gamma^{-2}+\theta_x^2}+i\theta_y} + \right.\\
& \left. + \frac{\exp\left[-\frac{2\pi h_2}{\lambda}\left(\sqrt{\gamma^{-2}+\theta_x^2} - i\theta_y\left(1+\frac{2d\cos\theta_0}{a}\right) - i\frac{2d\sin\theta_0}{a}\right)\right]}{\sqrt{\gamma^{-2}+\theta_x^2}-i\theta_y}\right\},\\
E_2 = {} & \frac{e}{4\pi^2}\left\{-\frac{\exp\left[-\frac{2\pi h_1}{\lambda}\left(\sqrt{\gamma^{-2}+\theta_x^2} + i\theta_y\right)\left(1+\frac{2d\cos\theta_0}{a}\right) + i\frac{2d\sin\theta_0}{a}\right]}{\sqrt{\gamma^{-2}+\theta_x^2}+i\theta_y} + \right.\\
& \left. + \frac{\exp\left[-\frac{2\pi h_2}{\lambda}\sqrt{\gamma^{-2}+\theta_x^2} - i\theta_y\left(1+\frac{2d\cos\theta_0}{a}\right) - i\frac{2d\sin\theta_0}{a}\right]}{\sqrt{\gamma^{-2}+\theta_x^2}-i\theta_y}\right\}.
\end{aligned} \tag{3.123}$$

With the use of formulas (3.123) for the standard procedure of the calculation of the spectral–angular density, the diffraction radiation intensity can be obtained as a function of shift d. For the case of the passage of the electron through the center of the slit ($h_1 = h_2 = a \sin \Theta_0/2$), this dependence can be represented by the formula [34]

$$\frac{d^2W}{\hbar d\omega d\Omega} = \frac{\alpha\gamma^2}{2\pi^2} \frac{1 + 2t_x^2 - \cos(2\beta + \chi)}{\left(1 + t_x^2\right)\left(1 + t_x^2 + t_y^2\right)} \exp\left(-\frac{2\pi a \sin\theta_0 \sqrt{1 + t_x^2}}{\gamma\lambda}\right),$$

$$2\beta = \frac{2\pi a \sin\theta_0}{\gamma\lambda}\left[t_y\left(1 + \frac{2d}{a}\cos\theta_0\right) + \frac{2\gamma d}{a}\sin\theta_0\right], \tag{3.124}$$

$$\chi = \arctan\frac{2t_y\sqrt{1 + t_x^2}}{1 + t_x^2 - t_y^2}.$$

The yield of diffraction radiation photons in the mirror-reflection direction ($t_x = t_y = 0$) is described by the simple formula

$$\frac{d^2W}{d\omega d\Omega}\left(t_x = t_y = 0\right) = \frac{\alpha\gamma^2}{\pi^2}\exp\left(-\frac{2\pi a \sin\theta_0}{\lambda\gamma}\right)\sin^2\left(\frac{2\pi d \sin^2\theta_0}{\lambda}\right). \tag{3.125}$$

In agreement with expectations, the yield of diffraction radiation for $d = 0$ (slit in the flat target) strictly vanishes in this case. However, the yield for $d = \dfrac{\lambda}{4\sin^2\theta_0}$ reaches the maximum value

$$\frac{d^2W_{\max}}{d\omega d\Omega} = \frac{\alpha\gamma^2}{\pi^2}\exp\left(-\frac{2\pi a \sin\theta_0}{\lambda\gamma}\right) = \frac{\alpha\gamma^2}{\pi^2}\exp(-z). \tag{3.126}$$

Figure 3.15 shows the angular distribution of diffraction radiation for the parameters $\gamma = 100$, $\theta_0 = 45°$, $d = 0$, and (a) $z = 0.04$ and (b) $z = 1$. The distribution close to the distribution of transition radiation (with a weak breaking of the

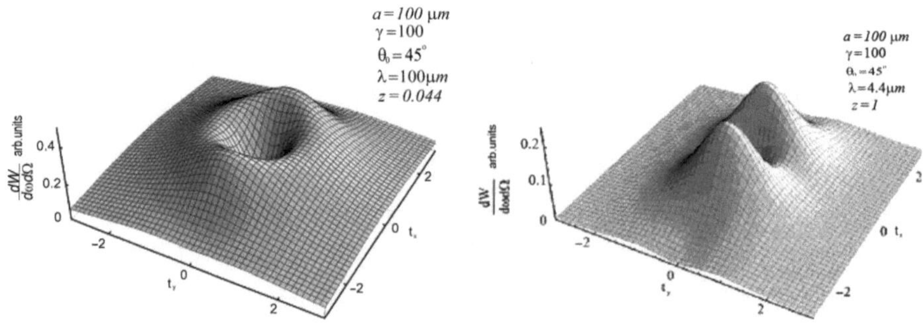

Fig. 3.15 Two-dimensional angular distribution of backward diffraction radiation from the flat slit target

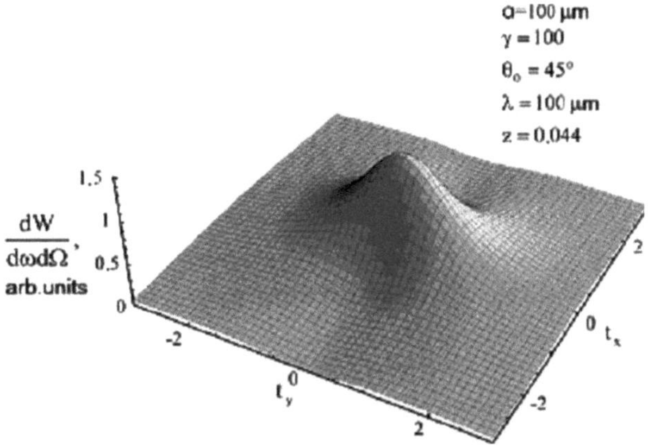

Fig. 3.16 Two-dimensional angular distribution of backward diffraction radiation for the case of the shift of the halves of the slit target by a distance of half the wavelength for $z = 0.044$

azimuthal symmetry) is seen in Fig. 3.15a, whereas the distribution in Fig. 3.15b exhibits two pronounced peaks in the $t_y = 0$ line.

Figure 3.16 shows the angular distributions of diffraction radiation for a nonzero value of $d = 50\,\mu$m at the parameters $\gamma = 100$, $a = 100\,\mu$m, $\theta_0 = 45°$, and $\lambda = 100\,\mu$m. As mentioned above, a single peak at $t_x = t_y = 0$ is observed in the distribution for $d = \dfrac{\lambda}{4\sin^2\theta_0} = \dfrac{\lambda}{2}$.

Figure 3.17 shows the evolution of the angular distribution of diffraction radiation for the same parameter values at four distances d from +50 to −25 µm. Note

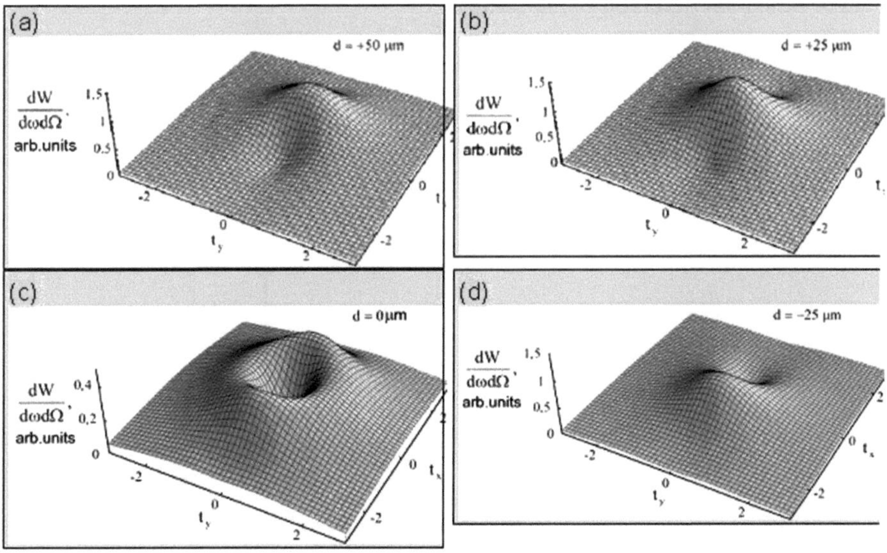

Fig. 3.17 Evolution of the angular distribution of backward diffraction radiation for various shifts d. The other parameters are the same as in Fig. 3.16

that an asymmetric distribution in angle θ_y is observed at $d = 25\,\mu$m and the shape of the distribution does not change under the change $d \to -d$ and $t_y \to -t_y$.

According to the last figures, an additional degree of freedom (shift of the semi-planes forming the slit) leads to additional interference effects in diffraction radiation from the slit, which can be used for the diagnostics of the length of an electron bunch.

3.4 Polarization Characteristics of Diffraction Radiation

In contrast to transition radiation characterized by a high degree of the azimuthal symmetry (particularly for ultrarelativistic particles), the azimuthal symmetry is absent in the case of diffraction radiation generated by the particle moving near the semi-plane. In accordance with this reason, it should be expected that the polarization characteristics of diffraction radiation differ significantly from similar characteristics of transition radiation.

Before calculating the polarization characteristics of diffraction radiation, we consider the polarization of transition radiation.

If an ultrarelativistic charge intersects the interface between vacuum and a perfectly conducting medium, the self field of the charge is reflected from the interface in the cone near the mirror-reflection direction. The transition radiation fields from which the polarization characteristics of transition radiation are easily determined were found in [35] in the framework of classical electrodynamics.

Figure 3.18 shows the geometry of the problem and angular variables used in this problem. Following [35], we write the expressions for the component of the electric field of transition radiation in the form

$$\begin{aligned}
E_x &= \omega^2 \left[\Pi_z \cos\theta_x \cos\theta_z \Pi_y \cos\theta_x \cos\theta_y\right], \\
E_y &= \omega^2 \left[\Pi_z \cos\theta_y \cos\theta_z \Pi_y \sin^2\theta_y\right], \\
E_z &= \omega^2 \left[\Pi_z \sin^2\theta_x \Pi_y \cos\theta_y \cos\theta_x\right].
\end{aligned} \tag{3.127}$$

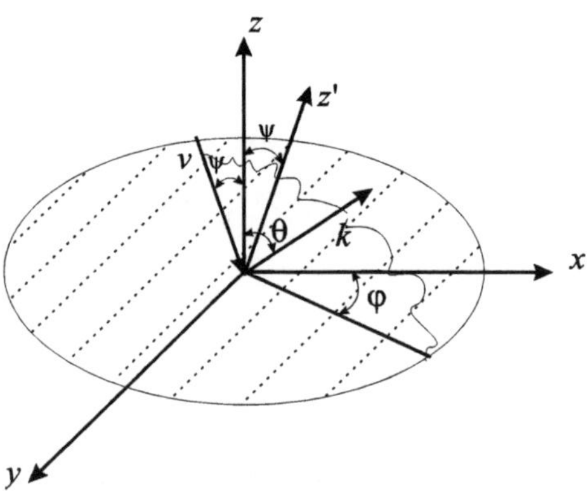

Fig. 3.18 Geometry of the problem for calculating the polarization characteristics of backward transition radiation

Here, ω is the frequency of transition radiation and $\cos\theta_{x,y,z}$ are the cosines of the angles between wave vector **k** and the coordinate axes:

$$\mathbf{k} = \omega\left\{\cos\theta_x,\ \cos\theta_y,\ \cos\theta_z\right\} = \omega\left\{\sin\theta\cos\varphi,\ \sin\theta\sin\varphi,\ \cos\theta\right\}. \qquad (3.128)$$

The components of the Hertz vector, Π_y, Π_x, for the interface between vacuum and the perfectly conducting medium are written as follows:

$$\begin{aligned} \Pi_y &= -C_0\left(1-\beta_y\cos\theta_y\right), \\ \Pi_z &= -C_0\beta_y\cos\theta_z, \\ C_0 &= \frac{\exp(i\omega R)}{R}\frac{e\beta_z}{\pi\omega^2}\frac{1}{\left(1-\beta_y\cos\theta_y\right)^2-\beta_z^2\cos^2\theta_z}. \end{aligned} \qquad (3.129)$$

Here, $\beta_{y,z}$ are the components of the particle velocity (see Fig. 3.18)

$$\boldsymbol{\beta} = \beta\left\{0,\ \sin\psi_0,\ \cos\psi_0\right\}. \qquad (3.130)$$

For the case under consideration,

$$\begin{aligned} E_x &= C_0\omega^2\cos\theta_x\cos\theta_z, \\ E_y &= C_0\omega^2\left(\cos\theta_y\cos\theta_z-\beta_y\cos\theta_z\right), \\ E_z &= C_0\omega^2\left(-\sin^2\theta_z+\beta_y\cos\theta_y\right). \end{aligned} \qquad (3.131)$$

Further calculations will be performed in a more convenient coordinate system (primed) whose z' axis is directed along the mirror-reflection direction. The primed system is obtained from the initial one by rotation around the x axis by angle $\pi-\psi_0$. The components of any vector a_i are transformed with the standard rotation matrix:

$$a_{x'} = a_x, \quad a_{y'} = -a_y\cos\psi - a_z\sin\psi, \quad a_{z'} = a_y\sin\psi - a_z\cos\psi. \qquad (3.132)$$

This system is chosen because backward transition radiation for the perfectly conducting medium is concentrated in a cone of $\sim\gamma^{-1}$ near the mirror-reflection direction. In the new system, we use the variables $\theta_{x'}$, $\theta_{y'} \sim \gamma^{-1}$ related to the angles θ, φ in the initial system as

$$\theta_{x'} = \left(\frac{\pi}{2}-\varphi\right)\sin\psi_0, \quad \theta_{y'} = \theta-(\pi-\psi_0). \qquad (3.133)$$

Thus, in the primed system with an accuracy of $\sim\gamma^{-2}$,

$$\mathbf{k} = \left\{k_{x'}, k_{y'}, k_{z'}\right\} = \omega \left\{\theta_{x'},\ \theta_{y'},\ 1 - \frac{\theta_{x'}^2 + \theta_{y'}^2}{2}\right\},$$
$$\mathbf{E} = \left\{E_{x'},\ E_{y'},\ E_{z'}\right\} = -C_0\omega^2 \cos\psi_0 \left\{\theta_{x'},\ \theta_{y'}, -\theta_{x'}^2 - \theta_{y'}^2\right\}. \tag{3.134}$$

In the ultrarelativistic approximation, quantity C_0 is expressed in terms of Eqs. (3.133) as

$$C_0 = \frac{\exp(i\omega R)}{R}\frac{e\beta_z}{\pi\omega^2}\frac{1}{\left(1 - \beta_y\cos\theta_y - \beta_z\cos\theta_z\right)\left(1 - \beta_y\cos\theta_y + \beta_z\cos\theta_z\right)} =$$
$$= \frac{\exp(i\omega R)}{R}\frac{e\beta\cos\psi_0}{\pi\omega^2}\frac{1}{\cos^2\psi_0\left(\gamma^{-2} + \theta_{x'}^2 + \theta_{y'}^2\right)}. \tag{3.135}$$

With the use of this expression, the spectral–angular density of transition radiation is easily expressed from Eq. (3.134) as

$$\frac{d^2W}{\hbar d\omega d\Omega} = R^2\left|\mathbf{E}\right|^2 = R^2\left|C_0\right|^2\omega^4\cos^2\psi_0\left(\theta_{x'}^2 + \theta_{y'}^2\right) =$$
$$= \frac{\alpha}{\pi^2}\frac{\theta_{x'}^2 + \theta_{y'}^2}{\left(\gamma^{-2} + \theta_{x'}^2 + \theta_{y'}^2\right)^2}. \tag{3.136}$$

Polarization characteristics of any radiation (including transition radiation) should be calculated in the system whose one axis coincides with the photon momentum. However, the primed (fixed) system, where the angle between the photon momentum and the z' axis is $\sim \gamma^{-1}$, can be used to this aim with an accuracy of $\sim \gamma^{-1}$. According to Eq. (3.134), $E_{z'}/E_{x',y'} \sim \gamma^{-1}$, $\mathbf{Ek} \sim \gamma^{-4}$ in this system.

The polarization of radiation is most completely characterized by the three Stokes parameters

$$\xi_1 = \frac{E_x^* E_y + E_x E_y^*}{\left|E_x\right|^2 + \left|E_y\right|^2},\quad \xi_2 = i\frac{E_x^* E_y - E_x E_y^*}{\left|E_x\right|^2 + \left|E_y\right|^2},\quad \xi_3 = \frac{\left|E_x\right|^2 - \left|E_y\right|^2}{\left|E_x\right|^2 + \left|E_y\right|^2}. \tag{3.137}$$

The substitution of the expressions for the components of the transition radiation field from Eq. (3.134) into Eq. (3.137) yields

$$\begin{aligned}
\xi_1 &= \frac{2\theta_x\theta_y}{\theta_x^2+\theta_y^2} = \sin 2\chi\,,\\
\xi_2 &= 0,\\
\xi_3 &= \frac{\theta_x^2-\theta_y^2}{\theta_x^2+\theta_y^2} = \cos 2\chi\,,\\
\chi &= \arctan\frac{\theta_y}{\theta_x}.
\end{aligned} \tag{3.138}$$

In other words, the polarization of transition radiation in the case under consideration (infinite vacuum—perfect conductor interface) is linear and independent of the target inclination ψ_0. The linear-polarization degree $P_\ell = \sqrt{\xi_1^2+\xi_3^2} = 1$ is independent of the emission angle of a transition radiation photon [36]. The polarization plane passes through the mirror-reflection direction and photon momentum; i.e., the linear polarization is "radial" similarly to the polarization of Cherenkov radiation.

In order to calculate the Stokes parameters of diffraction radiation generated by the particle moving near the inclined plane, we choose the polarization unit vectors in form Eq. (3.50). Note that unit vectors $\mathbf{e}_1$ and $\mathbf{e}_2$ are defined in the plane perpendicular to unit vector $\mathbf{n} = \mathbf{k}\omega$. In the general case, the polarization characteristics of diffraction radiation are calculated in this coordinate system, which is not fixed and is independent of the xyz system specified with respect to the screen.

However, in the ultrarelativistic case, where the emission angles of diffraction radiation photons are concentrated in two narrow cones with an opening of $\sim\gamma^{-1}$ (forward diffraction radiation and backward diffraction radiation), the spread of vectors $\mathbf{e}_1$ and $\mathbf{e}_2$ with respect to their mean values $\langle\mathbf{e}_1\rangle$ and $\langle\mathbf{e}_2\rangle$, respectively, is also on the order of γ^{-1}. For the perpendicular geometry, vector $\langle\mathbf{e}_1\rangle$ is directed along the z axis, vector $\langle\mathbf{n}\rangle$ is directed along the particle velocity (forward diffraction radiation) or along the mirror-reflection direction (backward diffraction radiation), and vector $\langle\mathbf{e}_2\rangle$ is perpendicular to the first two vectors. In this case, with an accuracy of $\sim\gamma^{-1}$, we can consider the rotation angle of the linear-polarization plane of diffraction radiation with respect to the plane passing through $\langle\mathbf{n}\rangle$ and the target edge.

With the use of expression (3.51) for the diffraction radiation field components, it is easy to calculate the Stokes parameters [37]

$$\begin{aligned}
\xi_1 &= \frac{\alpha_0^2\sin\phi\cos\psi_0}{D},\\
\xi_2 &= -\frac{\beta\alpha_0\sin\phi\sin\theta_0\,\sin\psi_0\cos\psi_0}{D},\\
\xi_3 &= \frac{\cos^2\psi_0\cos^2(\phi/2)\,[1-\beta\sin\psi_0\cos\theta_0]-\alpha_0^2\sin^2(\phi/2)\,[1+\beta\sin\psi_0\cos\theta_0]}{D},\\
D &= \cos^2\psi_0\cos^2\frac{\phi}{2}\,(1-\beta\sin\psi_0\cos\theta_0)+\alpha_0^2\sin^2\frac{\phi}{2}\,(1+\beta\sin\psi_0\cos\theta_0)\,,\\
\alpha_0 &= \sqrt{1-\beta^2\sin^2\psi}
\end{aligned} \tag{3.139}$$

According to these expressions, the polarization characteristics of diffraction radiation are independent of the impact parameter h and photon frequency.

Stokes parameters (3.139) satisfy the relation

$$\xi_1^2 + \xi_2^2 + \xi_3^2 = 1, \tag{3.140}$$

which is not surprising, because radiation generated by a single charge is considered.

Let us consider particular cases. For example, according to Eq. (3.139), the circular polarization is absent for the parallel motion ($\varphi_0 = 0$), whereas all three Stokes parameters are nonzero for the other orientations of the target with respect to the beam (except for the particular photon emission directions). Thus, in contrast to transition radiation, diffraction radiation is elliptically polarized.

Note also that circular polarization changes sign in the plane perpendicular to the target ($\cos\psi = 0$); i.e., the sign of circular polarization remains unchanged to the right and to the left of the vertical plane (right- or left-polarized radiation, respectively).

Let us consider the case where $\varphi_0 = \pi/2$ (particle trajectory is perpendicular to the target surface). For the angles $|\psi - \pi/2| \gg \gamma^{-1}$ ($\cos\psi_0 \gg \gamma^{-1}$), the following expressions simpler than expressions (3.139) are obtained:

$$\xi_1' = -\cos\psi\sin\varphi, \quad \xi_2' = -\sin\psi\sin\varphi, \quad \xi_3' = \cos\varphi, \tag{3.141}$$

therefore, radiation at angles $\varphi \sim \pm\pi/2$ (forward diffraction radiation and backward diffraction radiation) and $\psi \sim \pi/2$ is almost completely circularly polarized.

Figure 3.19 shows all three Stokes parameters for the geometry under consideration. In the ultrarelativistic approximation and for target inclinations $\varphi_0 \gg \gamma^{-1}$, $|\pi/2 - \varphi_0| \gg \gamma^{-1}$, the diffraction radiation fields are written in form (3.111). In this approximation, the Stokes parameters are expressed as

$$\xi_1 = 0, \quad \xi_2 = \pm\frac{2\theta_x\sqrt{\gamma^{-2} + \theta_x^2}}{\gamma^{-2} + 2\theta_x^2}, \quad \xi_3 = -\frac{\gamma^{-2}}{\gamma^{-2} + 2\theta_x^2}. \tag{3.142}$$

The upper and lower signs in the expression for ξ_2 correspond to forward diffraction radiation and backward diffraction radiation, respectively. Note that Stokes parameters (3.142) are independent of the photon emission angle θ_y up to the γ^{-1} terms. Further, $\xi_3 < 0$ for any θ_x angle; i.e., the linear polarization plane is perpendicular to the target plane. The universal plots for ξ_2 and ξ_3 are shown in Fig. 3.20.

In the ultrarelativistic case, the cone of forward diffraction radiation has an opening of γ^{-1} with respect to the electron momentum; hence, with an accuracy of γ^{-2}, we can consider the average polarization of the forward diffraction radiation (or backward diffraction radiation) beam passed through a given aperture. In this case, to calculate the Stokes parameters by formulas (3.137), the following bilinear field combinations $\langle E_i^* E_k\rangle$ averaged with respect to all the variables over a given region should be used:

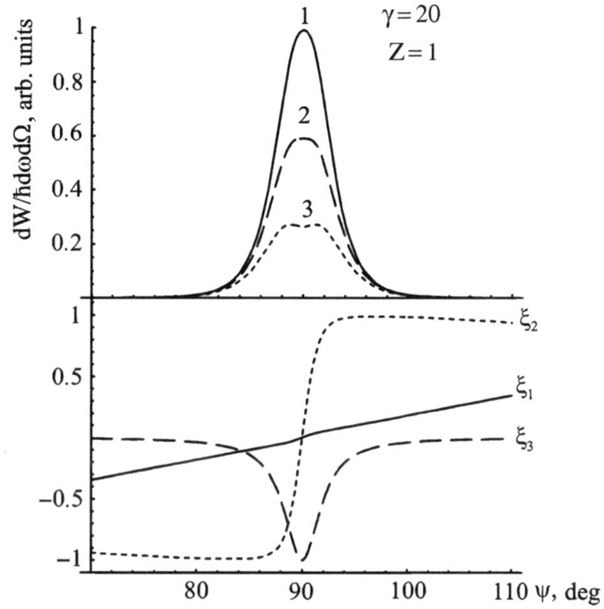

Fig. 3.19 Diffraction radiation intensity for (line (1))$\varphi = 90°$, (line (2)) $\varphi = 88.55°$, and (line (3)) $\varphi = 87.1$ and three Stokes parameters (for $\varphi = 90°$)

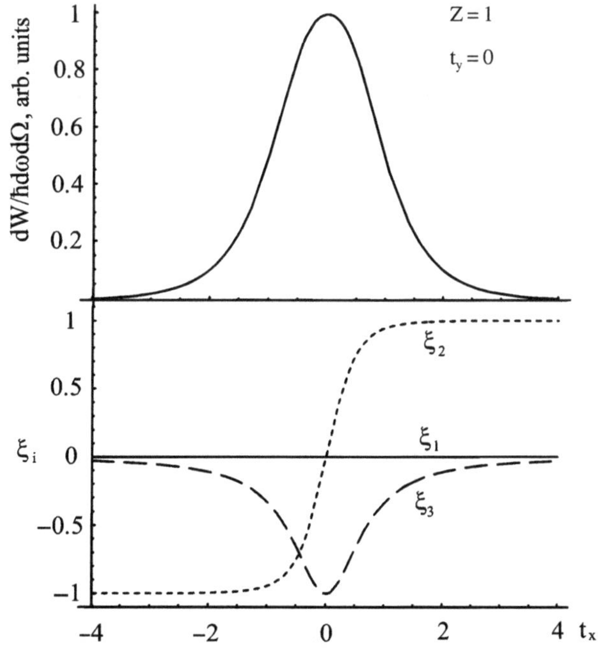

Fig. 3.20 (*Upper plot*) Diffraction radiation intensity for $z = 1$ and (*lower plot*) universal dependences for the Stokes parameters in the ultrarelativistic case

$$\langle E_i^* E_k \rangle = \int_{\Delta\Omega} d\Omega \int_{\Delta\omega} d\omega E_i^* (\varphi, \psi, \omega) \, E_k (\varphi, \psi, \omega) \, , \quad i, k = 1, 2. \tag{3.143}$$

Let us calculate the Stokes parameters of diffraction radiation in the case of the passage of the charge through the slit with the use of fields (3.112). As shown above, unit vector $\mathbf{e}_1$ coincides with the target edge with an accuracy of $\sim \gamma^{-1}$. As above, projection angle θ_x is defined in the plane containing the particle velocity and the direction along the target edge, whereas angle θ_y is defined in the perpendicular plane. Using the same dimensionless variables as in formula (3.117), we arrive at the following expressions for the Stokes parameters, where the Lorentz factor does not appear in the explicit form:

$$\begin{aligned}
\xi_1 &= \tfrac{1}{D} \left\{ 2t_x \sqrt{1+t_x^2} \left[\left(1+t_x^2-t_y^2\right) \sin\left(zt_y\right) - 2t_y\sqrt{1+t_x^2} \cos\left(zt_y\right) \right] \right\}, \\
\xi_2 &= \frac{1}{D} 2t_x \sqrt{1+t_x^2} \left(1+t_x^2+t_y^2\right) \sinh\left(2lz\sqrt{1+t_x^2}\right), \\
\xi_3 &= \frac{1}{D} \left\{ \left(1+2t_x^2\right) \left[\left(1+t_x^2-t_y^2\right) \cos\left(zt_y\right) - 2t_y\sqrt{1+t_x^2} \sin\left(zt_y\right) \right] - \right. \\
&\quad \left. - \left(1+t_x^2+t_y^2\right) \cosh\left(2lz\sqrt{1+t_x^2}\right) \right\}, \\
D &= \left(1+2t_x^2\right) \left(1+t_x^2+t_y^2\right) \cosh\left(2lz\sqrt{1+t_x^2}\right) - \left(1+t_x^2-t_y^2\right) \cos\left(zt_y\right) + \\
&\quad +2t_y\sqrt{1+t_x^2} \sin\left(zt_y\right).
\end{aligned} \tag{3.144}$$

It is easy to verify that the Stokes parameters of diffraction radiation in this case also satisfy the condition $\xi_1^2 + \xi_2^2 + \xi_3^2 = 1$, as should be expected.

Let us consider some particular cases. When the slit width vanishes, Stokes parameters (3.144) must describe the characteristics of transition radiation. The substitution of $z = \omega a \gamma^{-1} \sin\theta_0 = 0$ into Eqs. (3.144) yields the expressions

$$\xi_1 = \frac{2t_x t_y}{t_x^2 + t_y^2}, \quad \xi_2 = 0, \quad \xi_3 = \frac{t_x^2 - t_y^2}{t_x^2 + t_y^2}, \tag{3.145}$$

which coincide with expressions (3.138) as should be expected. For the case of the passage of the particle near the slit edge ($l \to \pm 0.5$), i.e., for $z \gg 1$,

$$\begin{aligned}
&\cosh\left(2lz\sqrt{1+t_x^2}\right) \to \frac{1}{2} \exp\left(z\sqrt{1+t_x^2}\right), \\
&\sinh\left(2lz\sqrt{1+t_x^2}\right) \to \frac{1}{2} \exp\left(z\sqrt{1+t_x^2}\right), \\
&\exp\left(z\sqrt{1+t_x^2}\right) \gg \cos\left(zt_y\right), \ \sin\left(zt_y\right)
\end{aligned} \tag{3.146}$$

and, therefore, expressions (3.145) change to

$$\xi_1 \approx 0, \quad \xi_2 \approx -\frac{2t_x\sqrt{1+t_x^2}}{1+2t_x^2}, \quad \xi_3 = -\frac{1}{1+2t_x^2}. \tag{3.147}$$

The Stokes parameters in this case are specified by the same formulas (3.139) as in the case of diffraction radiation on the semi-plane. Thus, for sufficiently short wavelengths

$$\frac{2\pi c}{\omega_p} \ll \lambda \ll \frac{2\pi a}{\gamma}\sin\theta_0, \tag{3.148}$$

the slit edge does not affect the polarization of diffraction radiation when the charge passes near the other slit edge.

For the case of the passage through the center of the slit ($l = 0$), circular polarization disappears and diffraction radiation has almost 100% linear polarization in the plane perpendicular to the slit edge, $\xi_1 \simeq 0$, $\xi_2 \simeq 0$, $\xi_3 \simeq -1$ (if $z \gg 1$). According to Eq. (3.147), the sign of circular polarization is determined by the photon emission angle θ_x. As seen in Fig. 3.21, diffraction radiation can be almost completely circularly polarized with some intensity loss.

Figure 3.22 shows parameters ξ_1 and ξ_3 as functions of the photon emission angle in the vertical plane for the case of the passage of the particle through the center of

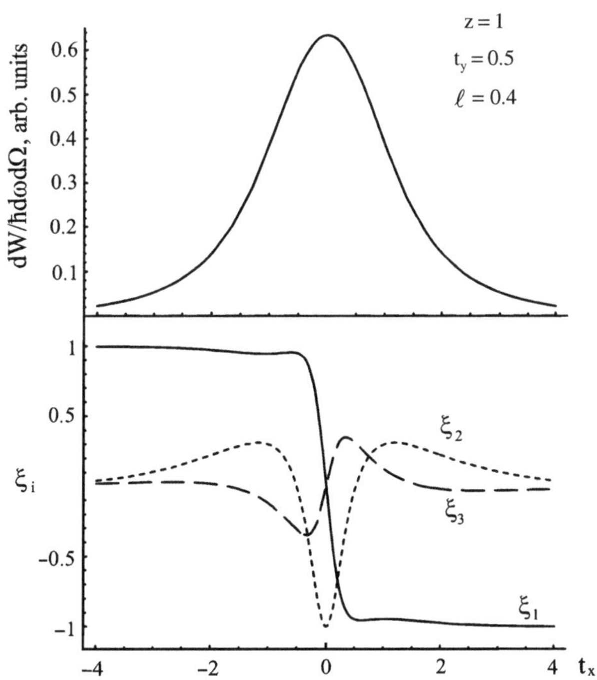

Fig. 3.21 (*Upper plot*) Diffraction radiation intensity and (*lower plot*) three Stokes parameters vs. emission angle $t_x = \gamma\theta_x$ for the parameters $t_y = 0.5$, $z = 1$, $\ell = 0.4$

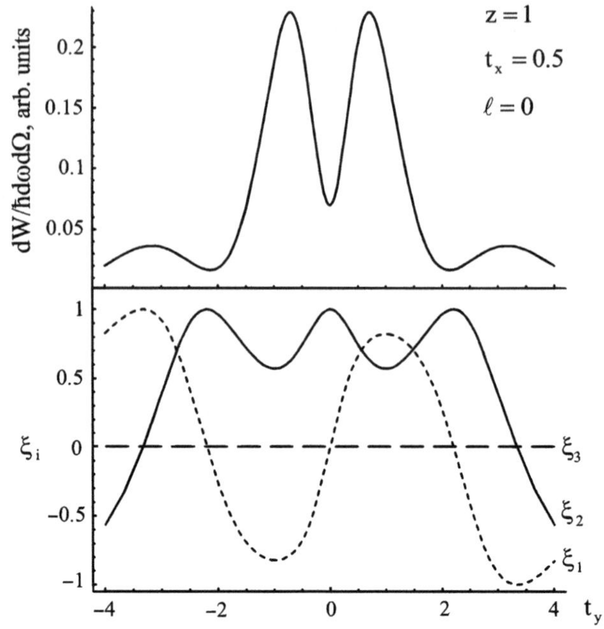

Fig. 3.22 ((*Upper plot*) Diffraction radiation intensity and (*lower plot*) three Stokes parameters vs. emission angle $t_y = \gamma\theta_y$ for the case of the passage through the center of the slit at the parameters $t_x = 0.5, z = 1, \ell = 0$

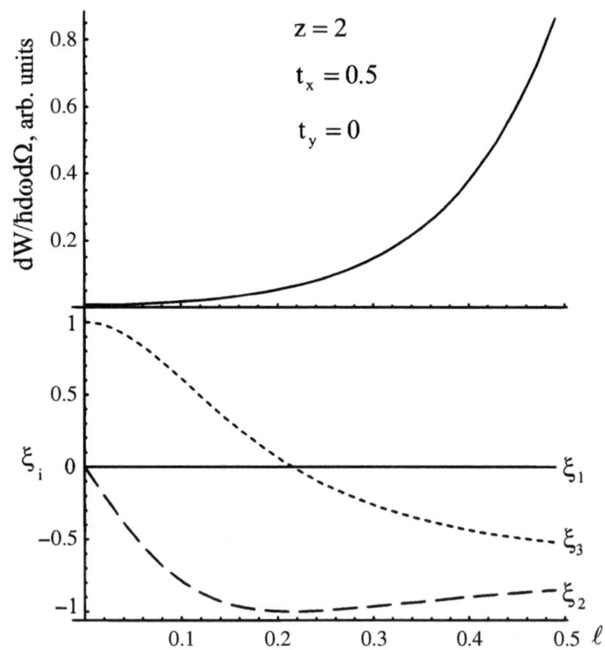

Fig. 3.23 (*Upper plot*) Diffraction radiation intensity and (*lower plot*) three Stokes parameters vs. eccentricity ℓ for the parameters $t_x = 0.5$, $t_y = 0, z = 2$

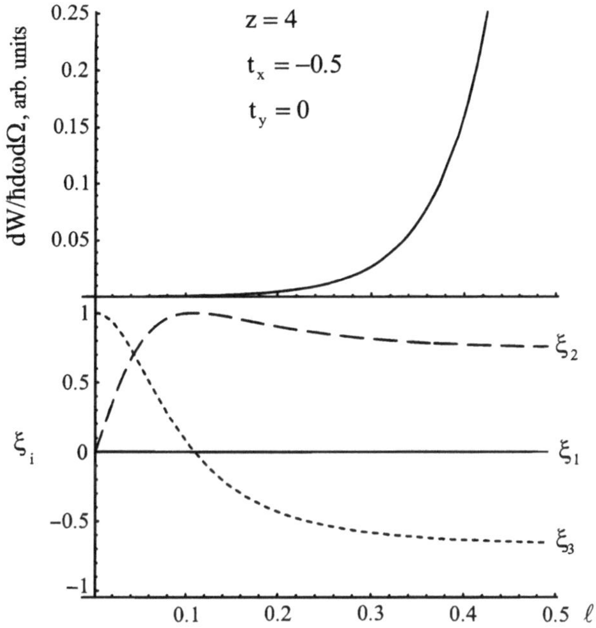

Fig. 3.24 (*Upper plot*) Diffraction radiation intensity and (*lower plot*) three Stokes parameters vs. eccentricity ℓ for the parameters $t_x = 0.5$, $t_y = 0$, $z = 5$

the slit. In this case, as follows from formula (3.142), circular polarization is equal to zero.

Figures 3.23 and 3.24 show the evolution of the Stokes parameters with varying eccentricity (offset) l for photons with various wavelengths. Note that parameter ξ_3 is an oscillating function of eccentricity and the eccentricity values at which ξ_3 vanishes are determined by the energy of the diffraction radiation photon.

In contrast to transition radiation, diffraction radiation has a wide spectrum of the characteristics depending on the geometry of the problem (in addition to the angular distributions, which are similar to those of transition radiation, the diffraction radiation spectrum is characterized by energy ω_c and radiation is elliptically polarized). It can be assumed that the polarization characteristics of diffraction radiation from the slit can be used to develop new tools for non-invasive diagnostics of beams.

References

1. Bobrinev, V., Braginskiy, V.: Radiation of the point charge moving along axis of round aperture in infinite perfect conducting surface. Dokl. Akad. Nauk. USSR **123**(4):634 (1958)
2. Dnestrovskiy, Yu.N., Kostomarov, D.P.: Radiation of modulated beam of charged particles moving across round aperture in flat screen. Sov. Phys. Dokl. **4**, 132 (1959)
3. Dnestrovskiy, Yu.N., Kostomarov D.P.: Radiation of ultra relativistic charges moving across round aperture in flat screen. Sov. Phys. Dokl. **4**, 158 (1959)
4. Khachatryan, B.V.: Mathematical proof of formulae of diffraction radiation. Izv. Arm. Akad. Nauk, **18** 133 (1965)
5. Jackson, J.D.: The Classical Electrodynamics. New York, NY (1998)

6. Stratton, J.A.: Theory of Electromagnetism. McGraw-Hill, New York, NY (1941)
7. Stratton, J.A., Chu, L.J.: Diffraction theory of electromagnetic waves. Phys. Rev. **56**, 99 (1939)
8. Ter-Mikaelyan, M.L., Khachatryan, B.V.: Diffraction radiation of fast-moving particles. Dokl. Akad. Nauk. ASSR **40**(XL), 13 (1965)
9. Ter-Mikaelyan, M.L.: High-Energy Electromagnetic Processes in Condensed Media. Wiley-Interscience, New York, NY (1972)
10. Xiang, D., Huang, W.-H.: Properties of diffraction radiation in practical conditions: Finite size target effect, surface roughness and pre-wave zone. Nucl. Instrum. Methods Phys. Res. B **248**, 163 (2006)
11. Potylitsyn, A.P.: Scattering of coherent diffraction radiation by a short electron bunch. Nucl. Instrum. Methods Phys. Res. A **455**, 213 (2000)
12. Masullo, M.R., Panariello, G., Schettino, F., Vaccaro, V.G.: Longitudinal coupling impedance of a plane conducting ring. Phys. Rev. ST-AB **2**, 124402 (1999)
13. Gianfelice, E., Palumbo, L., Vaccaro, V.G., Verolino, L.: A canonical problem for the understanding of the energy diffraction losses in high-energy accelerators. Nuovo Cimeto A. **104**, 885 (1991)
14. Bolotovskiy, B.M., Galst'yan, E.A.: Diffraction and diffraction radiation. Phys.-Uspekhi **43**, 755 (2000)
15. Weinstein, L.A.: Electromagnetic Waves. Radio i Svyaz, Moskow (in Russian) (1988)
16. Shulga, N.F., Dobrovolsky, S.N.: Theory of relativistic-electron transition radiation in a thin metal target. JETP **90**(4), 579 (2000)
17. Dobrovolsky, S.N., Shul'ga, N.F.: Transversal spatial distribution of transition radiation by relativistic electron in the formation zone by the dotted detector. Nucl. Instrum. Methods Phys. Res. B **201**, 123 (2002)
18. Dobrovolsky, S.N., Shul'ga, N.F.: Transition and diffraction radiation by relativistic electrons in a pre-wave zone. Proceedings of EPAC 2002, Paris, France, p. 1867 (2002)
19. Landau, L.D., Lifshitz, E.M.: The Classical Theory of Fields. Pergamon Press, (1987)
20. Weinstein, L.A.: Theory of Diffraction and the Factorization Method. The Golden Press Boulder, Colorado (1969)
21. Bolotovskiy, B.M., Voskresenskiy, G.V.: Diffraction radiation. Phys.-Uspekhi **9**, 73 (1966)
22. Kazantsev, A.P., Surdutovich, G.I.: Radiation of a charged particle passing close to a metal screen. Sov. Phys. Dokl. **7**, 990 (1963)
23. Potylitsyn, A.P.: Transition radiation and diffraction radiation. Similarities and differences. Nucl. Instrum. Methods Phys. Res. B **145**, 169 (1998)
24. Potylitsyna-Kube, N.A., Artru, X.: Diffraction radiation from ultrarelativictic particles passing through a slit. Determination of the electron beam divergence. Nucl. Instrum. Methods Phys. Res. B, **201**, 172 (2003)
25. Akhiezer, A.I., Berestetzkiy, V.B.: Quantum Electrodynamics. Moskow, Nauka (in Russian) (1989)
26. Sedrakyan, D.M.: Diffraction radiation of one-dimensional source moving near edge of perfect conducting semi-surface. Izv. Arm. Akad. Nauk **16**, 115 (1963)
27. Sedrakyan, D.M.: Radiation of charged particle moving across metal screen. Izv. Arm. Akad. Nauk **17**, 103 (1964)
28. Xiang, D., Huang, W-H., Lin, Y-Z. et al.: Wake of a beam passing through a diffraction radiation target. Phys. Rev. ST Accel. Beams **11**(2), 024001 (2008)
29. Karlovets, D.V., Potylitsyn, A.P.: On the theory of diffraction radiation. JETP **107**, 755 (2008)
30. Castellano. M.: A new non-intercepting beam size diagnostics using diffraction radiation from a slit. Nucl. Instrum. Methods Phys. Res. A **34**, 275 (1997)
31. Potylitsyn, A.P., Potylitsyna, N.A.: Diffraction radiation of ultra relativistic particles at moving across sloping slit. Russ. Phys. J. **43**(4), 303 (2000)
32. Verzilov, V.A.: Transition radiation in the pre-wave zone. Phys. Lett. A **273**, 135 (2000)
33. Fiorito, R.B., Rule, D.W.: Diffraction radiation diagnostics for moderate to high energy beams. Nucl. Instrum. Methods Phys. Res. B **173**, 67 (2001)

34. Potylitsyn, A.P.: Coherent diffraction radiation interferometry and short bunch length measurements. Nucl. Instrum. Methods Phys. Res. B **227**, 191 (2005)
35. Pafomov, V.E.: Radiation of charged particle at boundary. Trudy FIAN **44**, 28 (in Russian) (1969)
36. Korkhmazyan, N.A.: Transition radiation at grazing incidence of charge. Izv. Arm. Akad. Nauk. Fizika **11**, 7 (1958)
37. Potylitsyn, A.P.: Linear polarization of diffraction radiation from slit and beam size determination. Nucl. Instrum. Methods Phys. Res. B **201**, 161 (2003)

Chapter 4
Diffraction Radiation in the Ultraviolet and Soft X-Ray Regions

4.1 Polarization Current and the Radiation Field

The majority of the problems on diffraction radiation that have been already studied theoretically were solved for ideally conducting targets. The perfect-conductivity model is applicable at large values of the imaginary part of the relative permittivity, i.e., for most metals at optical, infrared, and radio frequencies. However, the relative permittivity decreases sharply near the frequency

$$\omega_L = \sqrt{4\pi n_e e^2/m} = A\sqrt{2\pi\alpha\bar{\lambda}_e n_e}, \tag{4.1}$$

where $\bar{\lambda}_e$ is the electron Compton wavelength, n_e is the number of the conduction electrons per unit volume, m is the electron mass, and e is the elementary charge. This frequency is usually called the Langmuir frequency. For the frequencies$\omega > \omega_L$, the relative permittivity of metals is close to unit and the perfect-conductivity model becomes inapplicable. For most metals, frequency ω_L is located in the near ultraviolet region and even in the visible spectrum for some metals such as copper or gold, which is responsible for their red color: only the harmonics with $\omega < \omega_L$ from the incident-wave field are reflected (see [1], page 525).

At frequencies higher than the eigenfrequencies of the electrons in an atom, an external field interacts with the electrons bound in atoms so as with conduction electrons. For this reason, beginning with the frequency

$$\omega_p = \sqrt{4\pi NZe^2/m} = A\sqrt{2\pi\alpha\bar{\lambda}_e NZ}, \tag{4.2}$$

the responses of dielectrics and conductors to the external field become the same and are determined by only the total number of electrons per unit volume of the medium, NZ, where N is the atomic number density and Z is the atomic number of the element. Note that, in metal optics and plasma physics, frequency (4.1) is called the Langmuir frequency and plasma frequency, respectively, whereas frequency (4.2) is not considered. The characteristic ω_p values for typical metals of the target are several tens of electron volts [2].

A.P. Potylitsyn et al., *Diffraction Radiation from Relativistic Particles*, STMP 239, 105–136, DOI 10.1007/978-3-642-12513-3_4,

It is convenient to represent the relative permittivity in the form

$$\varepsilon(\omega) = 1 + \chi'(\omega) + i\chi''(\omega). \tag{4.3}$$

For the frequencies

$$\omega \gg \omega_p \tag{4.4}$$

beyond the absorption region, the real part of the electric susceptibility $\chi'(\omega)$ has the form

$$\chi'(\omega) = -\frac{\omega_p^2}{\omega^2} < 0 \tag{4.5}$$

and $\chi'(\omega) \gg \chi''(\omega)$. Near the bands of the characteristic X-ray radiation from atoms or narrow lines of Mössbauer nuclear transitions, the condition $\chi'(\omega) > 0$ can be satisfied and $\chi'(\omega)$ and $\chi''(\omega)$ can be on the same order of magnitude [3].

Note that expression (4.3) is independent of the coordinates. For this reason, the results obtained with the use of formula (4.3) are applicable only for the wavelengths

$$a_B < \lambda < \frac{2\pi c}{\omega_p}. \tag{4.6}$$

where a_B is the atomic Bohr radius. This interval corresponds to the energy interval from tens of electron volts to tens of keVs. For radiation with shorter wavelengths, fluctuations of the relative permittivity that are caused by the inhomogeneity of the electron distribution in the medium should be taken into account so as scattering is taken into account in macroscopic electrodynamics [4, 5]. Note that the inclusion of the coordinate dependence of the relative permittivity can also be required for radiation in wavelength range (4.6), e.g., for a crystalline medium, where the coherence effects leading to new phenomena in radiation, e.g., to the existence of the parametric X-ray radiation are possible. However, according to Eq. (3.72), the consideration of diffraction radiation in the hard X-ray region is of insignificant interest, because it requires either too high energies of the particles or too small distances between the particle trajectory and target surface.

Diffraction radiation is generated due to the dynamic polarization of the target medium by the field of the passing particle. In order to determine the diffraction radiation field, let us reveal a quantity that serves as the polarization current density. To this end, we first consider an infinite medium. Then, finding the induced current density, we consider a finite target.

In the infinity homogeneous medium with relative permittivity $\varepsilon(\omega)$, the magnetic field is determined from the equation

$$\Delta\mathbf{H}(\mathbf{r},\omega) + k^2\varepsilon(\omega)\mathbf{H}(\mathbf{r},\omega) = -\frac{4\pi}{c}\mathrm{rot}\mathbf{j}_0(\mathbf{r},\omega), \tag{4.7}$$

where $k^2 = \omega^2/c^2$ and $\mathbf{j}_0$ is the current density corresponding to the motion of the charged particle. The field of the charge in vacuum, $\mathbf{H}_0(\mathbf{r}, \omega)$, is determined by the current density $\mathbf{j}_0(\mathbf{r}, \omega)$ from the equation

$$\Delta \mathbf{H}_0(\mathbf{r}, \omega) + k^2 \mathbf{H}_0(\mathbf{r}, \omega) = -\frac{4\pi}{c}\mathrm{rot}\mathbf{j}_0(\mathbf{r}, \omega). \tag{4.8}$$

Let us introduce the notation

$$\mathbf{H}' = \mathbf{H} - \mathbf{H}_0. \tag{4.9}$$

The self field of the charge in vacuum, $\mathbf{H}_0$, does not make contribution to the radiation field at long distances. Therefore, field $\mathbf{H}'$ contains all possible kinds of radiation at uniform motion. Substituting Eq. (4.8) from Eq. (4.7), we arrive at the following equation for $\mathbf{H}'$:

$$\Delta \mathbf{H}'(\mathbf{r}, \omega) + k^2 \varepsilon(\omega) \mathbf{H}'(\mathbf{r}, \omega) = k^2 [1 - \varepsilon(\omega)] \mathbf{H}_0(\mathbf{r}, \omega). \tag{4.10}$$

Note that Eq. (4.10) is derived without any approximations. Taking into account the equation $\mathbf{H}_0(\mathbf{r}, \omega) = -i(c/\omega)\,\mathrm{rot}\mathbf{E}_0(\mathbf{r}, \omega)$ in a transparent medium, i.e., at $\mathrm{Im}\varepsilon(\omega) \ll \mathrm{Re}\varepsilon(\omega)$, at large distances $r \gg k^{-1}$, we obtain the solution of Eq. (4.10) in the form of the divergent spherical wave

$$\mathbf{H}^r(\mathbf{r}, \omega) = -\frac{\omega}{4\pi c}[\varepsilon(\omega) - 1]\frac{e^{ik'r}}{r}\left[\mathbf{k}' \int d^3r' \exp\left(-i\mathbf{k}'\mathbf{r}'\right) \mathbf{E}_0\left(\mathbf{r}', \omega\right)\right], \tag{4.11}$$

where $\mathbf{k}' = \sqrt{\varepsilon}(\omega/c)\,\mathbf{n}$ is the wave vector of the radiation field in the medium and $\mathbf{n} = \mathbf{r}/r$ is the unit vector in the observation direction. Taking into account that the relation between the Fourier images of the electric and magnetic components of the radiation field is the same as that in the plane wave, i.e.,

$$\mathbf{E}^r(\mathbf{r}, \omega) = -\frac{1}{\sqrt{\varepsilon(\omega)}}\left[\mathbf{n}\mathbf{H}^r(\mathbf{r}, \omega)\right], \tag{4.12}$$

we can represent the radiation field in the form

$$\mathbf{E}^r(r, \omega) = \sqrt{\varepsilon(\omega)}\frac{\omega^2}{4\pi c^2}[\varepsilon(\omega) - 1]\frac{e^{ik'r}}{r}\left[\mathbf{n}\left[\mathbf{n}\int d^3r' \exp\left(-i\mathbf{k}'\mathbf{r}'\right)\mathbf{E}_0\left(\mathbf{r}', \omega\right)\right]\right]. \tag{4.13}$$

The Fourier image of the self field of the charged particle can be represented in the form

$$\mathbf{E}_0(\mathbf{q},\omega) = \int \frac{d^3r'}{(2\pi)^3}\exp\left(-i\mathbf{q}\mathbf{r}'\right)\mathbf{E}_0\left(\mathbf{r}',\omega\right) = -\frac{ie}{2\pi^2}\frac{\mathbf{q}-\mathbf{v}\omega/c^2}{q^2-\omega^2/c^2}\delta\left(\omega-\mathbf{q}\mathbf{v}\right). \tag{4.14}$$

Note that expression (4.13) provides the ordinary Cherenkov radiation in the infinite homogeneous medium. Indeed, the substitution of Eq. (4.14) into Eq. (4.13) yields

$$\mathbf{E}^r(\mathbf{r},\omega) = \frac{ie\omega}{c^2r}\left[\mathbf{n}\left[\mathbf{n}\mathbf{v}\right]\right]\exp\left(ik'r\right)\delta\left(\omega-\mathbf{v}\mathbf{k}'\right). \tag{4.15}$$

According to Eq. (4.13), the quantity

$$\mathbf{j}(\mathbf{r},\omega) = \frac{\omega}{4\pi i}\left(\varepsilon(\omega)-1\right)\mathbf{E}_0(\mathbf{r},\omega) \tag{4.16}$$

serves as the current density induced in the infinite homogeneous medium by the field of the passing charged particle.

Let us consider a finite target. Further, we consider the case where $\left|\chi'(\omega)\right|$, $\chi''(\omega) \ll 1$, so that there is the small parameter

$$|\varepsilon(\omega)-1| \ll 1. \tag{4.17}$$

We use this fact for solving the problem of diffraction radiation in the ultraviolet and X-ray regions of the spectrum [6, 7].

The self field of the passing particle polarizes the target. As a result of dynamic polarization, emission from electrons occurs in each elementary volume of the target. The electromagnetic wave of diffraction radiation propagates in the target medium, generally undergoes refraction and reflection at the edges of the target, and then propagates from the target surface to vacuum. At frequencies satisfying condition (4.17), the difference of the refractive index from one is much smaller than one. At the same time, the polarization current density is proportional to $|\varepsilon(\omega)-1| \ll 1$. Therefore, the inclusion of refraction and reflection gives the second-order terms in the parameter specified by Eq. (4.17), which can be disregarded. Thus, in the frequency range given by Eq. (4.4), the reflection and refraction of radiation at the plate surface can be neglected. This conclusion can be invalid for radiation emitted through the corner of the target, i.e., through the surface part whose shape and, possibly, optical properties are far from the optical properties of the plane, which are described by the Fresnel formulas. However, this relatively small contribution to radiation can be disregarded.

Thus, we assume the rectilinear propagation of radiation from the emission points inside the plate to its surface in the form of a plane wave with the real wave vector $\sqrt{\varepsilon}\mathbf{k}$ and subsequent rectilinear propagation from the surface to the observation point in the form of a plane wave with wave vector $\mathbf{k}$. Such an approach was successfully used by Durand [8] to analyze transition radiation at high frequencies. A similar method was also used in the framework of the eikonal approximation in transition radiation theory [9]. In this case, the radiation field is determined by the integral of the polarization current density given by Eq. (4.16) over the target volume V and can be calculated by the formula

$$\mathbf{E}^r(\mathbf{r},\omega) = \sqrt{\varepsilon(\omega)}\frac{\omega^2}{4\pi c^2}\left[\varepsilon(\omega)-1\right]\frac{e^{ikr}}{r}\left[\mathbf{n}\left[\mathbf{n}\int_V d^3r' \exp\left(-i\mathbf{k}'\mathbf{r}'\right)\mathbf{E}_0\left(\mathbf{r}',\omega\right)\right]\right]. \tag{4.18}$$

We emphasize that this expression describes the radiation field in the wave zone, i.e., sufficiently far from the target.

4.2 Forward Diffraction Radiation

Let us consider an $a\times\infty\times b$ plate as a target (Fig. 4.1). Let a particle with charge e uniformly move with velocity $\mathbf{v}=\left(v_x,v_y,0\right)=(v\cos\alpha, v\sin\alpha, 0)$ over the plate. At $t=0$, the particle is located at point $(0,0,h)$.

By integrating with respect to coordinates x, y, and z over intervals $(-a,0)$, $(-\infty,\infty)$, and $(-b,0)$, respectively, in formula (4.18), we obtain the radiation field in the form

$$\begin{gathered}\mathbf{E}^r_{\mathrm{DR}}(\mathbf{r},\omega) = \frac{e^{ikr}}{r}\frac{ie}{2\pi}\frac{\chi(\omega)}{v_x}\exp(-h\rho)\times \\ \frac{\exp(-ia\varphi)-1}{\varphi}\left[\mathbf{k}\left[\mathbf{k}\left(\frac{\mathbf{L}}{\rho}-i\mathbf{e}_z\right)\right]\right]\frac{1-\exp(-b\rho)\exp\left(ib\sqrt{\varepsilon}k_z\right)}{\rho-i\sqrt{\varepsilon}k_z}\end{gathered} \tag{4.19}$$

Here,

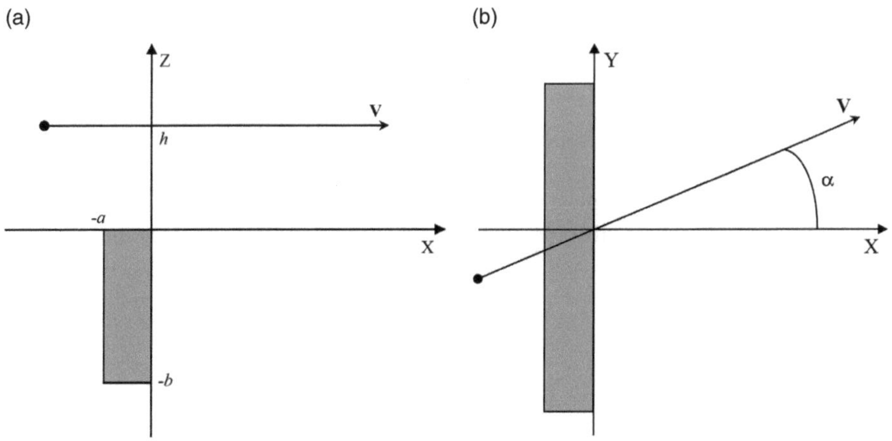

Fig. 4.1 Geometry of the passage of the charged particle near the target: (**a**) side view and (**b**) plan view

$$\begin{aligned}\varphi &= \frac{\omega - \sqrt{\varepsilon}\mathbf{v}\mathbf{k}}{v_x},\\ \rho^2 &= \left(\frac{\omega - \sqrt{\varepsilon}v_y k_y}{v_x}\right)^2 + \varepsilon k_y^2 - \frac{\omega^2}{c^2} \quad ,\\ \mathbf{L} &= \frac{\omega - \sqrt{\varepsilon}v_y k_y}{v_x}\mathbf{e}_x + \sqrt{\varepsilon}k_y\mathbf{e}_y - \mathbf{v}\frac{\omega}{v^2}\end{aligned} \tag{4.20}$$

where $v_x \neq 0$, h is the impact parameter, i.e., is the shortest distance between the particle trajectory and target surface, and b and a are the width and thickness of the plate, respectively. The physical meaning of φ is the phase difference between radiations arriving from different edges of the plate.

For the sake of simplicity, let us consider radiation far from the absorption lines, i.e., at

$$\chi(\omega) = -\omega_p^2/\omega^2 << 1. \tag{4.21}$$

The spectral–angular distribution of the emitted energy density can be calculated by the formula

$$d^2W(\mathbf{n}, \omega) = cr^2 d\Omega d\omega \left|\mathbf{E}^r(\mathbf{r}, \omega)\right|^2, \tag{4.22}$$

which, in view of Eq. (4.19), provides

$$\frac{d^2W(\mathbf{n}, \omega)}{d\Omega d(\hbar\omega)} = \alpha\left(\frac{\omega_p^2}{2\pi\omega^2}\right)^2 F_b \frac{\sin^2(\varphi a/2)\exp\{-2h\rho\}}{(c/\omega)^2\left(\rho^2 + \varepsilon k_z^2\right)} \frac{1 - n_z^2 + \frac{\left(L^2 - (\mathbf{L}\mathbf{n})^2\right)}{\rho^2}}{\left(1 - \mathbf{n}\mathbf{v}\sqrt{\varepsilon}/c\right)^2}, \tag{4.23}$$

where α is the fine structure constant and

$$F_b(b, \omega) = 1 - 2\exp\{-b\rho\}\cos\left(\sqrt{\varepsilon}bk_z\right) + \exp\{-2b\rho\} \tag{4.24}$$

is the factor determining the dependence of diffraction radiation on the target thickness b.

Let us consider the simplest case of the normal passage of the particle near semi-infinite plate, where absorption is first disregarded. In this case, the velocity is directed along the x axis, i.e., $\alpha = 0$ and $\mathbf{v} = (v, 0, 0)$. From Eq. (4.23), we obtain

$$\begin{aligned}\frac{d^2W(\mathbf{n}, \omega)}{d\Omega d(\hbar\omega)} &= \alpha\left(\frac{\omega_p^2}{2\pi\omega^2}\right)^2 \frac{\sin^2\left(\frac{a\omega}{2v}\left(1 - \sqrt{\varepsilon}\beta n_x\right)\right)}{\left(1 - \sqrt{\varepsilon}\beta n_x\right)^2} \times\\ &\times F_b \frac{1 - n_z^2 + n_y^2\left(1 - n_y^2\right)/\left(\gamma^{-2} + n_y^2\right)}{\gamma^{-2} + 1 - n_x^2}\exp\{-2h\rho\}\end{aligned} \tag{4.25}$$

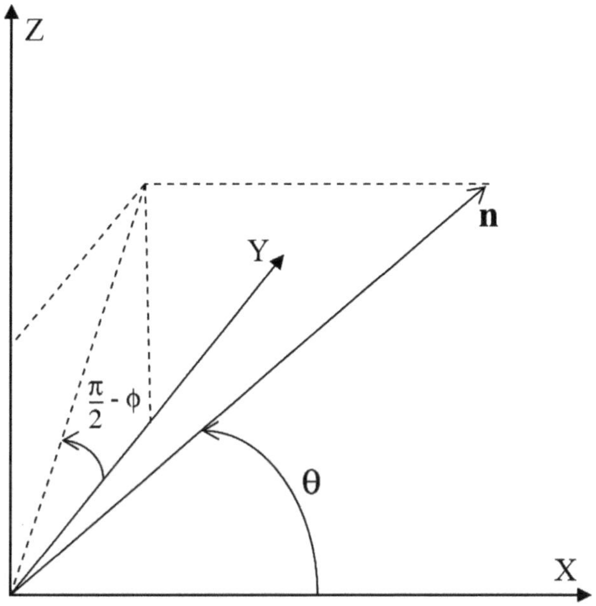

Fig. 4.2 Observation angles

where ρ is given by the expression

$$\rho = \frac{\omega}{c\gamma}\sqrt{1+\gamma^2 n_y^2}. \tag{4.26}$$

We specify a spherical coordinate system (see Fig. 4.2) as

$$n_x = \cos\theta, \quad n_y = \sin\theta\sin\phi, \quad n_z = \sin\theta\cos\phi. \tag{4.27}$$

It is easy to see that diffraction radiation has a sharp peak at $\phi \simeq 0$ and is located in a narrow cone of $\theta \sim \gamma^{-1}$. At $\alpha = 0$, $b \gg c\gamma/\omega$, and $\theta \ll 1$, the spectral–angular distribution of the emitted energy is obtained from Eq. (4.23) in the form

$$\frac{d^2W(\mathbf{n},\omega)}{d\Omega d(\hbar\omega)} = \frac{\alpha}{\pi^2}\left(\frac{\omega_p}{\omega}\right)^4 \frac{\sin^2\left[\frac{a\omega}{4c}\left(\theta^2+\gamma^{-2}+\omega_p^2/\omega^2\right)\right]}{\left(\theta^2+\gamma^{-2}+\omega_p^2/\omega^2\right)^2\left(\theta^2+\gamma^{-2}\right)}\frac{1+2\gamma^2n_y^2}{1+\gamma^2n_y^2}\times$$
$$\times\exp\left\{-\frac{2h\omega}{c\gamma}\sqrt{1+\gamma^2n_y^2}\right\}, \tag{4.28}$$

where the terms on the order of γ^{-2} and ω_p^2/ω^2 are neglected as compared to one.

Let us compare the expression describing radiation under the condition that the particle trajectory near the target surface is characterized by impact parameter h with the expression describing transition radiation generated when the particle passes through the infinite plate of the same width. From formula (4.18), it is easy to obtain the spectral—angular distribution density of transition radiation from the infinite plate of thickness a:

$$\frac{d^2W_{\mathrm{TR}}(\mathbf{n},\omega)}{d\Omega d(\hbar\omega)}=\frac{4}{\hbar}\frac{e^2}{c}\frac{1}{\pi^2}\left(\frac{\omega_p^2}{\omega^2}\right)^2\frac{\theta^2\sin^2\left[\frac{a\omega}{4c}\left(\theta^2+\gamma^{-2}+\omega_p^2/\omega^2\right)\right]}{\left(\gamma^{-2}+\theta^2+\omega_p^2/\omega^2\right)^2\left(\gamma^{-2}+\theta^2\right)^2}. \quad (4.29)$$

This expression is well known in the theory of X-ray transition radiation [2, 3]). The ratio of expressions (4.28) and (4.29) is given by the formula

$$\frac{d^2W_{\mathrm{DR}}(\mathbf{n},\omega)}{d^2W_{\mathrm{TR}}(\mathbf{n},\omega)}=\frac{1+\gamma^2\theta^2}{4\gamma^2\theta^2}\frac{1+2\gamma^2n_y^2}{1+\gamma^2n_y^2}\exp\left\{-\frac{2h\omega}{c\gamma}\sqrt{1+\gamma^2n_y^2}\right\}. \quad (4.30)$$

It is seen that, at $\theta\geq\gamma^{-1}$ and under condition (4.33), the maximum of the spectral density of transition radiation is larger than the maximum of diffraction radiation by a factor of $2\exp\{2\omega/\omega_c\}$. Note that a similar situation is observed for frequencies lower than the plasma frequency [10]. Figure 4.3 shows the angular distributions of transition radiation and diffraction radiation calculated by formulas (4.29) and (4.30), respectively.

It is interesting to compare the spectral–angular distributions of diffraction radiation and transition radiation. The main features of radiation generated when the particle trajectory does not intersect the target as compared to the case of the intersection of the target are as follows.

First, the spectrum of diffraction radiation is cut by the exponential factor $\exp\left(-\frac{\omega}{\omega_c}\right)$. When

$$h<c/\omega_p, \quad (4.31)$$

the limiting frequency of the spectrum,

$$\omega_c^{\mathrm{TR}}\simeq\gamma\omega_p, \quad (4.32)$$

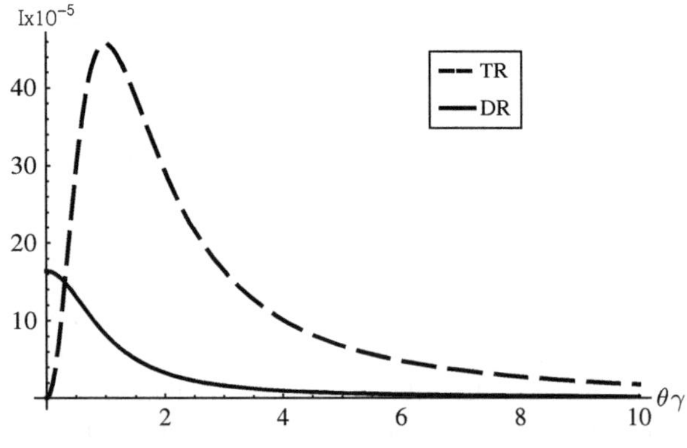

Fig. 4.3 Angular distributions of the intensities of forward transition and diffraction radiations for beryllium target ($\omega_p=4.1\times10^{16}\,\mathrm{c}^{-1}$, $\hbar\omega_p=26.1\,\mathrm{eV}$, $\omega=10\omega_p$, $h=30\,\mu\mathrm{m}$; $\gamma=8\times10^4$, and the target thickness $a=5\,\mu\mathrm{m}$)

coincides with the frequency known from the theory of X-ray transition radiation. Condition (4.31) is rather exotic. It means that the particle trajectory almost touches the plate surface. When

$$h > c/\omega_p, \tag{4.33}$$

the cutoff frequency is determined by the decreasing exponential and is independent of the dielectric properties of the target:

$$\omega_c \simeq \frac{c\gamma}{2h}. \tag{4.34}$$

Further, we assume that inequality (4.33) is satisfied and the cutoff frequency is determined by quantity (4.34). This value is usually much lower than $\gamma\omega_p$. For example, at $h = 30\,\mu\text{m}$, $\gamma = 10^5$ and the limiting energy of the forward diffraction radiation spectrum is $\hbar\omega_c \approx 0.65\,\text{keV}$.

Second, diffraction radiation is distributed in a cone of $\theta \sim \gamma^{-1}$, but has a peak at zero angle in contrast to transition radiation, which vanishes strictly along the particle velocity.

Note that the sine in Eq. (4.28) at $\omega_c \ll \gamma\omega_p$ is a rapidly oscillating function of frequency ω and thereby its square can be replaced by a mean value of $1/2$ under the condition

$$\pi^2 ah \gg (2\pi c/\omega_p)^2 \gamma. \tag{4.35}$$

In this case, formula (4.28) is almost independent of ω_p and we arrive at an expression similar to that for optical diffraction radiation [10] for two perfectly conducting, infinity thin screens located at $x = 0$ and $x = a$:

$$\frac{d^2W(\mathbf{n},\omega)}{d\Omega d(\hbar\omega)} \simeq \frac{\alpha}{2\pi^2}\frac{\gamma^{-2}+2n_y^2}{\left(\theta^2+\gamma^{-2}\right)\left(\gamma^{-2}+n_y^2\right)}\exp\left\{-2\frac{\omega}{\omega_c}\sqrt{1+\gamma^2 n_y^2}\right\}. \tag{4.36}$$

Condition (4.35) is satisfied, e.g., if $\omega_p = 4\times 10^{16}\,c^{-1}$ ($\hbar\omega_p = 26.1\,\text{eV}$ for beryllium), $h \geq 50\,\mu\text{m}$, $a \geq 10\,\mu\text{m}$, and $\gamma \leq 10^5$.

Let us consider the dependence of the diffraction radiation intensity on the plate width b. This dependence is determined by the factor $F_b^0(b,\omega)$ (see Eq. (4.24)). For $b \gg \rho^{-1}$ ($\rho^{-1} = c\gamma/\omega$ at $n_y = 0$),

$$F_b \approx 1; \tag{4.37}$$

i.e., the dependence on the plate width disappears for sufficiently large b values, as should be expected. For $b \ll \rho^{-1}$,

$$F_b \approx 2(1-\cos bk_z); \tag{4.38}$$

this condition means the appearance of oscillations with a period of λ/n_z, where λ is the radiation wavelength. Thus, the dependence of the diffraction radiation intensity on the target width b is determined by the ratio of b to a certain value b_{eff} given by the expression

$$b_{\text{eff}} \sim \rho^{-1} = \frac{c\gamma}{\omega\sqrt{1+\gamma^2 n_y^2}}. \tag{4.39}$$

The quantity b_{eff} is the characteristic distance at which the self field of the ultrarelativistic charged particle decreases. Under the condition $n_y \leq \gamma^{-1}$, $b_{\text{eff}}\ \gamma\lambda$. For example, at $n_y = 0$, $\gamma\tilde{}6 \times 10^4$, and $\omega \sim 10^{17}\ c^{-1}$, we obtain $b_{\text{eff}} \approx 180\,\mu\text{m}$.

The maximum of the diffraction radiation intensity is located between the frequencies ω_p and $\omega_c \simeq c\gamma/h$ and depends on the target width b. This is formally caused by the fact that the behavior of the function $F_b\,(b,\omega)$ at certain b values determined by the ratio of frequency ω to a certain effective frequency

$$\omega_m \sim \gamma c/b. \tag{4.40}$$

Figure 4.4 shows the factor $F_b\,(b,\omega)$ as a function of the energy of emitted photons in a range from $\hbar\omega_p$ to $20\hbar\omega_p$ for $\alpha = 0$ at $\theta = \gamma^{-1}$, $n_y = 0$, $\omega_p = 4.1 \times 10^{16}\ c^{-1}$, $\hbar\omega_p = 26.1$ eV (beryllium), $h = 50\,\mu\text{m}$; and $\gamma = 6 \times 10^4$. Thickness a satisfies condition (4.35). For these parameters, the cutoff frequency is $\hbar\omega_c = 219$ eV. For these resonances, $\omega_m \sim c\gamma/b$ is equal to $3.48 \times 10^{16}\ c^{-1}$.

Figure 4.5 shows the forward diffraction radiation spectrum at various b values. With a further increase in b, the plate changes insignificantly. It is seen that the soft part of the radiation spectrum ($\omega \sim \omega_p$) is somewhat suppressed at $b \leq b_{\mathit{eff}}$. Such a behavior is caused by the function $F_b\,(b,\omega)$, see Fig. 4.4.

Using formula (4.28), we can estimate the total losses on diffraction radiation in the frequency range $\omega > \omega_p$. Losses can be roughly estimated by multiplying the maximum of $d^2W\,(\mathbf{n},\omega)\,/d\Omega d\,(\hbar\omega)$ by the width of the range, $\Delta\Omega\Delta\,(\hbar\omega)$, in which radiation is maximal. On the order of magnitude, $(\Delta\omega)_{\text{eff}} \sim \omega \sim \omega_c \sim c\gamma/h$ and $(\Delta\Omega)_{\text{eff}} \sim \gamma^{-2}$. Under condition (4.35), we obtain

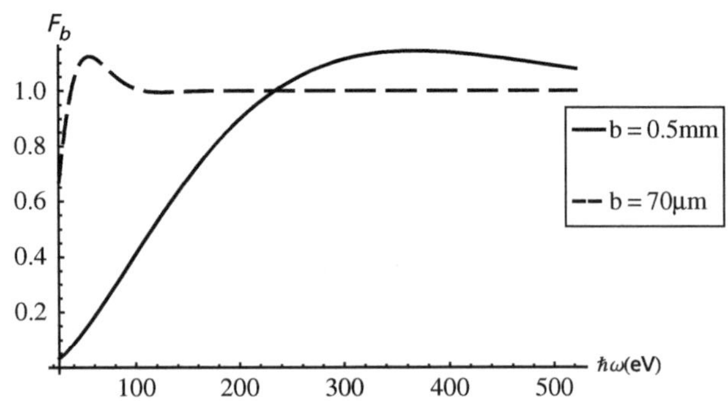

Fig. 4.4 Factor $F_b\,(b,\omega)$ for beryllium at $h = 50\,\mu\text{m}$ and $\gamma = 6 \times 10^4$

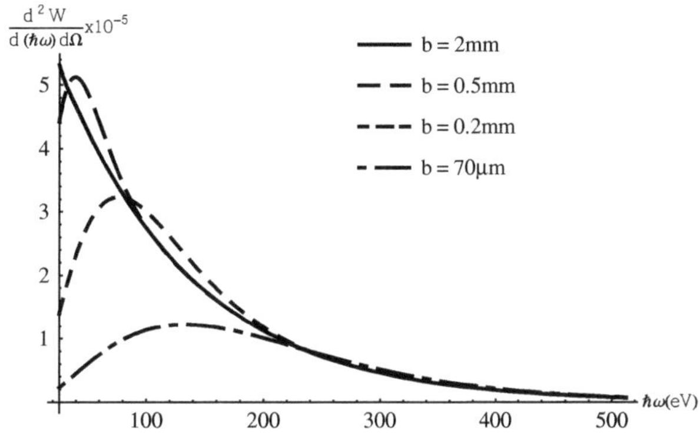

Fig. 4.5 Spectrum of diffraction radiation for various target thicknesses b. The parameters are the same as those in Fig. 4.4

$$W \sim \alpha\hbar\omega_c \left(1 - 2\exp\left\{-\frac{b}{h}\right\}\cos\left(\frac{b}{h}\right) + \exp\left\{-\frac{2b}{h}\right\}\right). \tag{4.41}$$

For $b \gg h$,

$$W \sim \alpha\hbar\omega_c = e^2\frac{\gamma}{h}. \tag{4.42}$$

Thus, the total losses on diffraction radiation are proportional to the Lorentz factor γ and are inversely proportional to the impact parameter h. Note that a similar estimate for transition radiation is $W_{\text{TR}} \sim \frac{1}{3}\frac{e^2}{c}\gamma\omega_p$, which is much larger than expression (4.42). This is mainly caused by the difference between the spectrum cutoff frequencies for transition radiation and diffraction radiation: $c\gamma/h \ll \gamma\omega_p$. As mentioned above, under condition (4.31), the cutoff frequencies for diffraction radiation and transition radiation are the same and, therefore, losses are of the same order of magnitude. The above estimate of losses on radiation is rather rough and is applicable only for qualitative analysis.

4.3 Backward Diffraction Radiation

Let us discuss the total form of expression (4.23). First, we analyze radiation maxima characteristics. It is easy to see that the diffraction radiation intensity is maximal when the argument of the exponential $\exp(-2h\rho)$ is minimal, whereas the pre-exponential factor is maximal. Analysis of ρ at fixed h, ω and γ values provides the following condition at which the exponential reaches its maximum

$$n_y = \frac{\beta_y}{\beta^2\sqrt{\varepsilon}} = \frac{\sin\alpha}{\beta\sqrt{\varepsilon}}. \tag{4.43}$$

The particle velocity components are determined in terms of the angle α as $\beta_x = \beta\cos\alpha$ and $\beta_y = \beta\sin\alpha$ (see Fig. 4.1). Substituting Eq. (4.43) into formulas (4.20), which is conveniently represented in the form

$$\varphi = \frac{\omega}{v_x}\left(1 - \sqrt{\varepsilon}\boldsymbol{\beta}\mathbf{n}\right),$$

$$\rho^2 = \frac{\omega^2}{c^2}\left[\left(\frac{1 - \sqrt{\varepsilon}\beta_y n_y}{\beta_x}\right)^2 + \varepsilon n_y^2 - 1\right], \tag{4.44}$$

$$\mathbf{L} = \frac{\omega}{c}\left\{\frac{1 - \sqrt{\varepsilon}\beta_y n_y}{\beta_x} - \frac{\beta_x}{\beta^2}, \sqrt{\varepsilon}n_y - \frac{\beta_y}{\beta^2}\right\},$$

we obtain

$$\varphi = \frac{\omega}{v_x}\left(\beta^{-2}\beta_x - \sqrt{\varepsilon}n_x\right), \quad \mathbf{L} = 0, \quad \rho = \frac{\omega}{c\gamma}. \tag{4.45}$$

In view of Eq. (4.43), according to formula (4.23), the radiation intensity is proportional to the factor

$$\frac{1 - n_z^2 + \left(L^2 - (\mathbf{Ln})^2\right)/\rho^2}{\left(c^2/\omega^2\right)\rho^2 + n_z^2} = \frac{1 - n_z^2}{\gamma^{-2} + n_z^2}. \tag{4.46}$$

This factor is maximal at

$$n_z = 0. \tag{4.47}$$

In other words, radiation intensity is maximal in the xy plane. With the use of Eqs. (4.47) and (4.43), we can show that there are two maximum radiation directions

$$n_x = \pm\sqrt{1 - \frac{\sin^2\alpha}{\beta^2\varepsilon}} = \pm\frac{\sqrt{\cos^2\alpha - \left(1 - \varepsilon\beta^2\right)}}{\beta\sqrt{\varepsilon}}. \tag{4.48}$$

Here, the sign + corresponds to the maximum of radiation along the particle trajectory and the sign—corresponds to radiation at the mirror reflection angle from the target surface (see Eq. 4.6). Further, we assume that

$$\omega \leq \omega_c = c\gamma/h \ll \gamma\omega_p. \tag{4.49}$$

The case of grazing, $\cos\alpha \leq \sqrt{1 - \varepsilon\beta^2}$, requires a separate consideration. For sufficiently large angles α satisfying the condition

$$\cos^2\alpha \gg 1 - \varepsilon\beta^2 \approx \omega_p^2/\omega^2, \tag{4.50}$$

formula (4.48) takes the form

$$n_x \approx \pm\frac{\cos\alpha}{\beta\sqrt{\varepsilon}}\left(1 - \frac{\omega_p^2/\omega^2}{2\cos^2\alpha}\right). \tag{4.51}$$

Disregarding the second term in Eq. (4.51), we can conclude that there are two radiation directions in the xy plane:

$$\begin{aligned} &\varphi = 0, \theta = \alpha - \max \mathrm{FDR} \\ &\varphi = 0, \theta = \pi - \alpha - \max \mathrm{BDR} \end{aligned} \tag{4.52}$$

The first of these conditions provides the maximum of forward diffraction radiation.

The second condition corresponds to radiation at the mirror reflection angle (see Fig. 4.6). This backward diffraction radiation is a sufficiently convenient tool for non-invasive diagnostics of charged particles [11], because the radiation observation angle with respect to the beam direction can be arbitrary. Backward diffraction radiation in the optical range was analyzed in detail in [10, 12]. In particular, it was shown that backward diffraction radiation is emitted in a narrow cone similarly to forward diffraction radiation. Figure 4.7 shows the angular distribution of backward diffraction radiation emitted at the angles $\theta = \pi - \alpha$ and $\alpha = 70°$. The observation frequency is $\omega = 10\omega_p$. A plate width is equal to $b = 0.5$ mm. The other parameters are the same as those in Fig. 4.4. The plot is obtained by Eq. (4.23) with the inclusion of condition (4.35).

Using Eqs. (4.43) and (4.51), we can obtain

$$1 - \mathbf{nv}\frac{\sqrt{\varepsilon}}{c} \approx \frac{\omega_p^2}{2\omega^2} \tag{4.53}$$

for forward diffraction radiation (sign + in Eq. (4.51)) and

$$1 - \mathbf{nv}\frac{\sqrt{\varepsilon}}{c} \approx 2\cos^2\alpha \tag{4.54}$$

for backward diffraction radiation (sign—in Eq. (4.51)). The same expressions (4.53) and (4.54) can be obtained with the use of Eq. (4.52). The maxima in forward diffraction radiation and backward diffraction radiation are obtained from

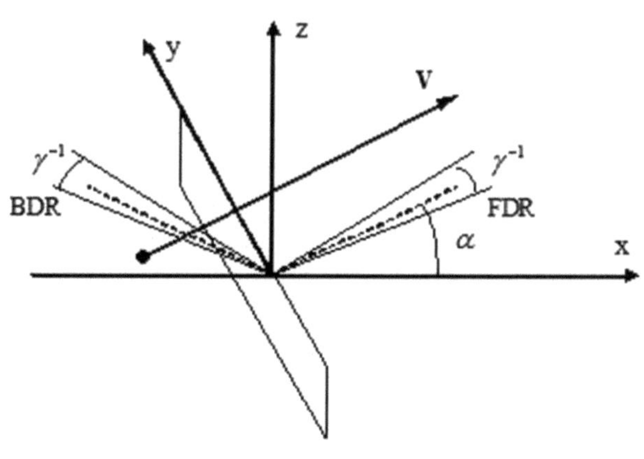

Fig. 4.6 Forward- and backward-diffraction radiation cones (FDR and BDR, respectively)

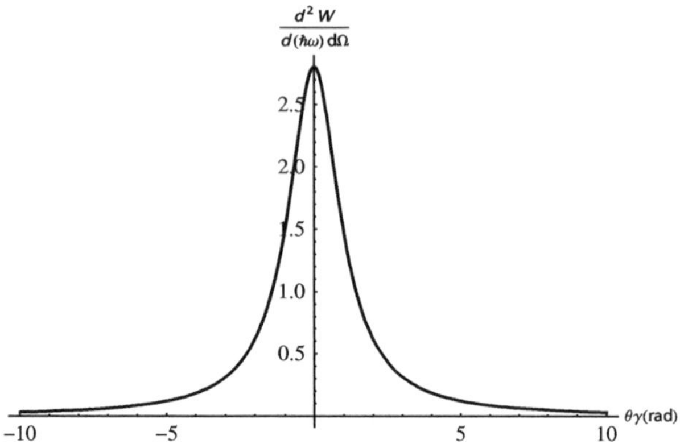

Fig. 4.7 Azimuth distribution of backward diffraction radiation

Eq. (4.23) with the use of formulas (4.45), (4.53), and (4.54). For the maximum forward diffraction radiation,

$$\frac{d^2W_{\max}^{\mathrm{FDR}}}{d\Omega\, d\,(\hbar\omega)} = \alpha\frac{\gamma^2}{2\pi^2}\left[1-\exp\left\{-\frac{b\omega}{c\gamma}\right\}\right]^2 \exp\left\{-\frac{2h\omega}{c\gamma}\right\} \tag{4.55}$$

and, for backward diffraction radiation,

$$\frac{d^2W_{\max}^{\mathrm{BDR}}}{d\Omega\, d\,(\hbar\omega)} = \alpha\frac{\gamma^2}{2\pi^2}\frac{(\omega_p/\omega)^4}{16\cos^4\alpha}\left[1-\exp\left\{-\frac{b\omega}{c\gamma}\right\}\right]^2 \exp\left\{-\frac{2h\omega}{c\gamma}\right\}. \tag{4.56}$$

In formulas (4.55) and (4.56), the sine squared is changed to $1/2$ according to inequality (4.35). The comparison of Eqs. (4.55) and (4.56) shows that backward diffraction radiation is much lower than forward diffraction radiation under condition (4.50). However, the maximum backward diffraction radiation intensity increases with angle α, as seen in Fig. 4.8, where this dependence is shown for radiation at angles $\phi = 0$ and $\theta = \pi - \alpha$ and frequency $\omega = 10\omega_p$ ($\hbar\omega_p = 26.1$ eV).

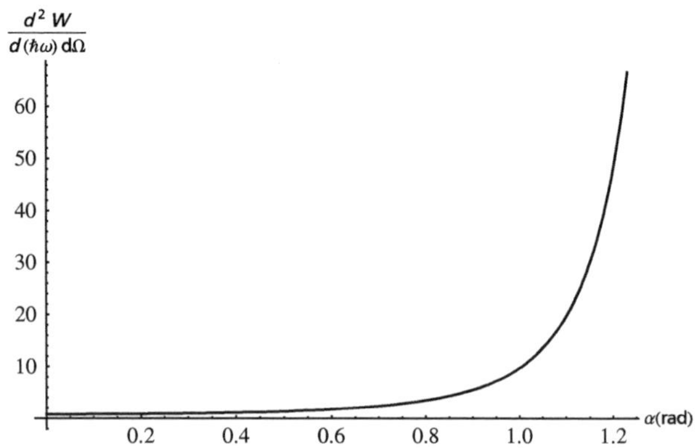

Fig. 4.8 Intensity at the peak of backward diffraction radiation vs. angle α

The impact parameter is equal to $h = 50\,\mu\text{m}$, the Lorentz factor is $\gamma = 6 \times 10^4$, and the plate width is $b = 0.5\,\text{mm}$. The plate thickness a is assumed to satisfy relation (4.35). Angle α varies from 0 to 70°. With a further increase in α, the radiation intensity continues to increase. However, the theory becomes inapplicable for grazing when inequality (4.50) is not satisfied.

Owing to these two features, x-ray diffraction radiation differs from optical diffraction radiation, for which the spectral characteristics of backward diffraction radiation are almost the same as those of forward diffraction radiation (see Sect. 3.2).

The factor $\left(\frac{\omega_p}{\omega}\right)^4 \left[1 - \exp\left\{-\frac{b\omega}{c\gamma}\right\}\right]^2$ determines the spectrum of backward diffraction radiation in the frequency range from ω_p to ω_c. The decreasing law is close to ω^{-4} if

$$b\omega \gg c\gamma \tag{4.57}$$

and is close to ω^{-2} if

$$b\omega \ll c\gamma. \tag{4.58}$$

Conditions (4.49) and (4.58) are satisfied simultaneously if $b \ll h$. In applications, this condition is satisfied for very thin plate ($a \gg b$) or thin filament ($a \sim b$).

4.4 X-ray Diffraction Radiation Under Conditions of the Cherenkov Effect

As known [3, 13], conditions of the Cherenkov effect, $v\sqrt{Re\varepsilon} > c$, can be satisfied in the ultraviolet and X-ray regions of the spectrum in the absorption region, where $\chi''(\omega) \neq 0$. In the soft ultraviolet region (several electron volts), the condition

$$\chi'(\omega) > 0 \tag{4.59}$$

can be satisfied when the binding energy of valence electrons is higher than the energy of emitted photons. In a harder part of the spectrum, condition (4.59) is satisfied near the edges of photoabsorption by electrons of various atomic shells (K, L, M edges, usually, tens and hundreds of electron volts). Data on the atomic properties in this spectral range can be found in [14–17]. Moreover, condition (4.59) can be satisfied in the gamma range near the Mössbauer nuclear transition line [3, 13, 16, 18]. The theoretical and experimental results obtained to 1987 are most completely presented in [3]. After that, the theory was developed to the case of grazing [19] (see also [20, 21]) in order to explain a sharp increase in the Cherenkov radiation in this case [22]. The theory of radiation in crystals under the Cherenkov condition was developed in [23–26]. The latest experiments were devoted to the development of the compact sources of intense narrow-band X-ray radiation [27, 28] on the basis of accelerators with an energy of several MeVs.

Expression (4.19) remains valid with the inclusion of the imaginary part of the relative permittivity. In obtaining the characteristic diffraction radiation with the inclusion of absorption and analyzing the spectrum of generated Cherenkov radiation, we follow [29].

For the sake of simplicity, we consider the case $\alpha = 0$. In this case, the radiation field can be written in the form

$$\mathbf{E}^r(r,\omega) = -\frac{e^{ikr}}{r}\frac{e\chi}{4\pi\beta c}\frac{\exp(-h\rho)}{\rho - ik_z\sqrt{\varepsilon}}\frac{\exp(-ia\varphi)-1}{\varphi}\left[\mathbf{k}\left[\mathbf{k}\left(\frac{\mathbf{L}}{\rho} - ie_z\right)\right]\right], \tag{4.60}$$

where

$$\varphi = \frac{k}{\beta}\left(1 - n_x\beta\sqrt{\varepsilon}\right),\ \rho = \frac{\omega}{c\beta\gamma}\sqrt{1+\varepsilon\gamma^2\beta^2 n_y^2},\ \mathbf{L} = \left\{0, \frac{\omega}{c}\sqrt{\varepsilon}n_y, 0\right\}. \tag{4.61}$$

Taking into account that the real part of the relative permittivity is close to one, we can write

$$\sqrt{\varepsilon} = 1 + \frac{\chi' + i\chi''}{2}. \tag{4.62}$$

In this case,

$$\varphi = \frac{\omega}{2\beta c}\left(\gamma^{-2} + \theta^2 - \chi'\right) - i\frac{\omega}{2c}\chi'', \tag{4.63}$$

$$\left|\rho - ik_z\sqrt{\varepsilon}\right|^2 \simeq \frac{\omega^2}{c^2}\left(\gamma^{-2} + \theta^2\right). \tag{4.64}$$

It is easy to see that radiation has a peak at $\theta \ll 1$. With the use of Eqs. (4.63) and (4.64), the spectral–angular distribution of radiation can be written in the form

$$\frac{d^2W(\mathbf{n},\omega)}{d\Omega d(\hbar\omega)} = \alpha\frac{|\chi|^2}{\pi^2}\frac{1+2\theta^2\gamma^2\sin^2\phi}{1+\theta^2\gamma^2\sin^2\phi}\exp\left(-\frac{2h\omega}{c\gamma}\sqrt{1+\gamma^2\theta^2\sin^2\phi}\right)\times \\ \times\frac{\left|1-\exp\left(-i\frac{a\omega}{2c}\left(\gamma^{-2}+\theta^2-\chi'\right)\right)\exp\left(-\frac{a\omega}{2c}\chi''\right)\right|^2}{\left(\gamma^{-2}+\theta^2\right)\left[\left(\gamma^{-2}+\theta^2-\chi'\right)^2+(\chi'')^2\right]} \tag{4.65}$$

where $n_x = \cos\theta$, $n_y = \sin\theta\sin\phi$, $n_z = \sin\theta\cos\phi$ according to Eq. (4.27). Expression (4.65) at $\chi'' = 0$ is transformed to Eq. (4.28).

Let the plate width be much larger than the photon absorption length in the medium:

$$a \gg \frac{2c}{\omega\chi''}. \tag{4.66}$$

In this case, the exponential in expression (4.65) can be disregarded. Let us compare the resulting expression describing radiation under the condition that the particle trajectory near the target surface is characterized by the impact parameter h with the following expression for transition radiation generated when the particle passes through the infinite plate of the same width [3]:

$$\frac{d^2 W_{\mathrm{TR}}(\mathbf{n},\omega)}{d\Omega d(\hbar\omega)} = \alpha \frac{\chi^2}{\pi^2} \frac{\theta^2 \left|1 - \exp\left\{-\frac{a\omega}{2c}\left(\gamma^{-2}+\theta^2+1-\varepsilon\right)\right\}\right|^2}{\left(\gamma^{-2}+\theta^2\right)^2 \left|\gamma^{-2}+\theta^2+1-\varepsilon\right|^2}. \tag{4.67}$$

The comparison of expressions (4.65) and (4.67) with Eq. (4.30) shows that the ratio of the yields of X-ray transition radiation and diffraction radiation does not change after the inclusion of absorption and is expressed by the same formula (4.30).

Let us consider the denominator of expression (4.65)

$$\left(\gamma^{-2}+\theta^2\right)\left[\left(\gamma^{-2}+\theta^2-\chi'\right)^2+\left(\chi''\right)^2\right]. \tag{4.68}$$

The first factor is minimal at $\theta = 0$, which provides a radiation peak with a width of $\Delta\theta \sim \gamma^{-1}$. The second factor is minimal at $\theta = \sqrt{\chi' - \gamma^{-2}}$ and the width of the corresponding peak is $\Delta\theta = \chi''/\sqrt{\chi' - \gamma^{-2}}$. This peak exists only in the frequency range where $\chi'(\omega) > \gamma^{-2}$ and $\chi''(\omega) \neq 0$ and describes X-ray Cherenkov radiation or, in other words, diffraction radiation under conditions of the Cherenkov effect. Thus, the radiation intensity is maximal at

$$\begin{array}{ll} \theta_{\mathrm{DR}} = 0, & \Delta\theta = \gamma^{-1} \\ \theta_C = \sqrt{\chi' - \gamma^{-2}}, & \Delta\theta = \chi''/\sqrt{\chi' - \gamma^{-2}} \end{array}. \tag{4.69}$$

We assume that condition (4.33) is satisfied; i.e., $h > c/\omega_p$. In this case, according to expression (4.65), for the radiation intensity to be noticeable, in addition to conditions (4.69), the following condition should be satisfied:

$$\gamma\theta\sin\phi \leq 1. \tag{4.70}$$

Typical χ' values near the absorption line are no more than 10^{-2}. Thus, for ultrarelativistic particles with $\gamma \gg \left(\chi'\right)^{-1/2} \sim 10$, the position and width of the Cherenkov peak are determined only by the properties of the target medium: $\theta_C \simeq \sqrt{\chi'}$, $\Delta\theta \simeq \chi''/\sqrt{\chi'}$. Figure 4.9 shows the two radiation peaks generated by a 5-GeV electron. The peak at $\theta = 85.6$ mrad with a width of $\Delta\theta = 3.2$ mrad corresponds to the Cherenkov effect and the peak at $\theta = 0$ with a width of $\Delta\theta = \gamma^{-1}$

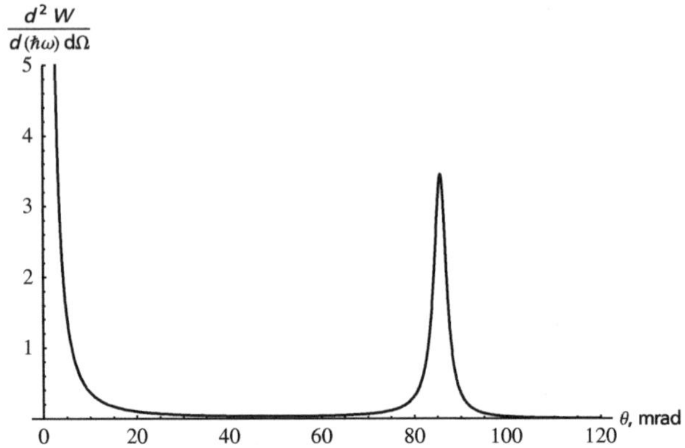

Fig. 4.9 Angular distribution of diffraction radiation generated by 5-GeV electrons from a carbon target (parameters are $\hbar\omega = 284\,\text{eV}$, $\chi' = 7.34 \times 10^{-3}$, $\chi'' = 2.76 \times 10^{-4}$, and $\rho_C = 2.20\,\text{g/cm}^3$). The impact parameter is $h = 10\,\mu\text{m}$, $\phi = 0$ thickness is $a >> \frac{2c}{\omega\chi''}\,5\,\mu\text{m}$

corresponds to forward diffraction radiation. The peak at zero angle, i.e., the forward diffraction radiation peak, is two orders of magnitude higher than the Cherenkov diffraction radiation peaks.

As seen in Fig. 4.9, Cherenkov radiation is emitted at quite large angles θ_C. Therefore, the constraint on angles, $\gamma\theta\sin\phi \leq 1$, following from the minimization of the argument of the exponential under conditions of the Cherenkov effect provides the condition

$$\varphi \simeq (\gamma\theta_C)^{-1} \simeq \gamma^{-1}/\sqrt{\chi'} \ll 1. \tag{4.71}$$

In other words, the Cherenkov radiation cone is degenerated into two rays along the directions specified by the angles $\theta = \theta_C$ and $\phi = 0, \pi$ with the widths $\Delta\theta \simeq \chi''/\sqrt{\chi'}$ and $\Delta\phi \simeq \gamma^{-1}/\sqrt{\chi'}$. Figure 19.2 schematically shows the forward radiation peak and two rays remaining from the Cherenkov radiation cone. The upper and lower rays correspond to angles $\theta = \theta_C$ and $\phi = 0, \pi$, respectively (Fig. 4.10). They are attributed to the Cherenkov radiation mechanism and their width is equal to $\Delta\theta \simeq \chi''/\sqrt{\chi'}$ and $\Delta\phi \simeq \gamma^{-1}/\sqrt{\chi'}$. The central ray corresponds to forward diffraction radiation and its maximum is located at $\theta = 0$ and the angular widths are equal to $\Delta\theta = \gamma^{-1}$ and $\Delta\phi \ll 1$.

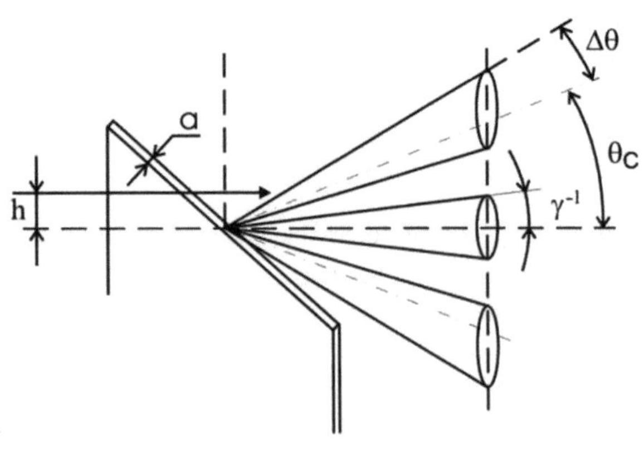

Fig. 4.10 Forward diffraction radiation cone and two Cherenkov radiation rays

4.5 Diffraction Radiation from a Crystal Target

In this section, we use the method applied in Sect. 4.1. Let us consider the generation of radiation in an infinite crystal, determine the density of polarization currents, and then consider a finite crystal target.

Let us consider an inhomogeneous medium. Maxwell's equations can be written in the form

$$\begin{aligned}
&\mathrm{rot}\mathbf{H}(\mathbf{r},\omega) = \frac{4\pi}{c}\left(\mathbf{j}_0(\mathbf{r},\omega) + \mathbf{j}(\mathbf{r},\omega)\right) - i\frac{\omega}{c}\varepsilon(\omega)\mathbf{E}(\mathbf{r},\omega)\\
&\varepsilon(\omega)\,\mathrm{div}\mathbf{E}(\mathbf{r},\omega) = 4\pi\left(\rho_0(\mathbf{r},\omega) + \rho(\mathbf{r},\omega)\right)\\
&\mathrm{rot}\mathbf{E}(\mathbf{r},\omega) = i\frac{\omega}{c}\mathbf{H}(\mathbf{r},\omega)\\
&\mathrm{div}\mathbf{H}(\mathbf{r},\omega) = 0.
\end{aligned} \tag{4.72}$$

Here, the coordinate dependent part of the response function is included in the current density $\mathbf{j}(\mathbf{r},\omega)$, the coordinate independent part is determined by the function $\varepsilon(\omega)$, and $\mathbf{j}_0$ is the current density describing the motion of the charged particle. The equation for the field $\mathbf{H}' = \mathbf{H} - \mathbf{H}_0$ is derived similarly to Eq. (4.10) and has the form

$$\Delta\mathbf{H}'(\mathbf{r},\omega) + k^2\varepsilon(\omega)\mathbf{H}'(\mathbf{r},\omega) = -\frac{4\pi}{c}\mathrm{rot}\mathbf{j}(\mathbf{r},\omega) + k^2\left(1-\varepsilon(\omega)\right)\mathbf{H}_0(\mathbf{r},\omega)\,. \tag{4.73}$$

The solution of this equation at large distances is a sum of two terms the first of which determined by the first term on the right-hand side of the equation provides transition radiation and diffraction radiation, depending on the geometry of the problem, and was discussed above.

For the sake of simplicity, let us consider the case $\varepsilon'' = 0$. In the frequency range given by Eq. (4.4), the average relative permittivity of the crystal has the form

$$\varepsilon(\omega) = 1 - \omega_p^2/\omega^2. \tag{4.74}$$

Under condition (4.4), $\mathbf{j}(\mathbf{r},\omega)$ in the crystal can be written as

$$\mathbf{j}(\mathbf{r},\omega) = \frac{i\omega}{4\pi}\sum_{\mathbf{g}\neq 0}\chi_g e^{i\mathbf{g}\mathbf{r}}\mathbf{E}_{\mathrm{act}}(\mathbf{r},\omega)\,, \tag{4.75}$$

where the summation is performed over all reciprocal lattice vectors $g \neq 0$. Following [30], we use the notation $\chi_g = -\frac{\omega_p^2}{\omega^2}f(\mathbf{g})$ and

$$f(\mathbf{g}) = \frac{F(\mathbf{g})}{Z}\frac{S(\mathbf{g})}{N_{\mathrm{cell}}}e^{-W(\mathbf{g})}, \tag{4.76}$$

where $F(\mathbf{g})$ is the atomic form factor ($F(0) = Z$), Z is the number of electrons in the atom, $W(\mathbf{g})$ is the Debye—Waller factor, and $S(\mathbf{g})$ is the geometric structure factor characterizing the crystal with an unit cell consisting of N_{cell} atoms. In the Debye approximation, $W(\mathbf{g}) = g^2u^2/2$, where u^2 is the average amplitude squared of atomic thermal vibrations.

The field $\mathbf{E}_{\text{act}}$ acting on each atom of the medium is composed of the self field of the charge given by Eq. (4.14) and the sum of the secondary fields from the other atoms of the medium. Note that contribution from the secondary fields leads to the "renormalization" of the self field of the charge in the medium. Solving the system of equations (4.72) for a homogeneous medium, we obtain

$$\mathbf{E}_0^m(\mathbf{q}, \omega) = -\frac{ie}{2\pi^2\varepsilon(\omega)} \frac{\mathbf{q} - \mathbf{v}\varepsilon(\omega)\,\omega/c^2}{q^2 - k^2\varepsilon(\omega)} \delta(\omega - \mathbf{q}\mathbf{v}). \tag{4.77}$$

Equation (4.73) at $\mathbf{j}(\mathbf{r}, \omega) \neq 0$ is solved approximately. To this end, in the first approximation, we change the field $\mathbf{E}_{\text{act}}$ to the field given by expression (4.77). In this case, the polarization current density determining radiation has the form

$$\mathbf{j}(\mathbf{r}, \omega) = \frac{i\omega}{4\pi} \int d^3q \mathbf{E}_0^m(\mathbf{q}, \omega) \exp\{i\mathbf{q}\mathbf{r}\} \sum_{\mathbf{g}\neq 0} \chi_g e^{i\mathbf{g}\mathbf{r}}. \tag{4.78}$$

In addition to the current density given by expression (4.78), the current density associated with the coordinate independent part of relative permittivity $\varepsilon(\omega)$ should be taken into account. The corresponding polarization current density is given by formula (4.16) and the contribution to the radiation field is specified by expression (4.15). Cherenkov radiation condition $c = \sqrt{\varepsilon}\mathbf{n}\mathbf{v}$ in frequency range (4.4) is satisfied only near the absorption lines. Beyond these narrow spectral ranges in region (4.4), $\sqrt{\varepsilon} < 1$, so that solution (4.15) provides zero and the radiation field obtained from solution (4.73) with the use of Eq. (4.78) has the form

$$\mathbf{E}^r(\mathbf{r}, \omega) = -\frac{\left[\mathbf{n}\mathbf{H}^r(\mathbf{r}, \omega)\right]}{\sqrt{\varepsilon(\omega)}} = ie\frac{e^{ik'r}}{r} \sum_{\mathbf{g}\neq 0} \chi_g \frac{\left[\mathbf{k}'\left[\mathbf{k}'\left(\mathbf{g}/\varepsilon(\omega) + \mathbf{v}\omega/c^2\right)\right]\right]}{(\mathbf{k}' - \mathbf{g})^2 - k^2\varepsilon(\omega)} \tag{4.79}$$

$$\delta\left(\omega - \left(\mathbf{k}' - \mathbf{g}\right)\mathbf{v}\right).$$

Formula (4.79) coincides with formula (3) from review [31]. The presence of the delta function determines the dispersion relation

$$\omega = \mathbf{v}\mathbf{k}' - \mathbf{g}\mathbf{v}. \tag{4.80}$$

Formulas (4.79) and (4.80) are well known in the theory of parametric X-ray radiation [30, 31]. They are obtained in the framework of an approach similar to the kinematical theory of X-ray scattering in crystals. The existence of parametric X-ray radiation was theoretically indicated in [32–34] and the first experiments on the observation of parametric X-ray radiation were carried out at Tomsk synchrotron in 1985 [35–37]. The kinematical theory is formally applicable only for sufficiently thin crystals with thickness $a \leq l_e \sim 1/\left|k\chi_g\right|$ [3, 4, 38]. At $\omega \sim 10\,\text{keV}$, this condition for "light" crystals such as silicon corresponds to the thickness $l_e \sim 10\,\mu\text{m}$. However, the kinematical theory of parametric X-ray radiation is also applicable for thick crystals, because the dynamic effects in radiation are almost compensated by the effects of the multiple scattering of a charged particle in the target medium [39]. The kinematical theory of parametric radiation is repeatedly confirmed by the experimental data [40–42]. The recent results (both theoretical and experimental) for parametric x-ray radiation were described in detail in [30].

Thus, the sum of the currents given by Eqs. (4.16) and (4.78) can be considered as the polarization current in the kinematical approximation. In this case, the radiation field is specified by the formula

$$\mathbf{E}^r(\mathbf{r},\omega) = \frac{1}{4\pi}\frac{e^{ikr}}{r}\left[\mathbf{k}\left[\mathbf{k}\int_V d^3r' \exp\left\{-i\mathbf{k}'\mathbf{r}'\right\}\left(\frac{\omega_p^2}{\omega^2}\mathbf{E}_{0\perp}(\mathbf{r}',\omega) + \mathbf{E}^m_{0\perp}(\mathbf{r}',\omega)\sum_{\mathbf{g}\neq 0}\chi_{\mathbf{g}}e^{i\mathbf{g}\mathbf{r}'}\right)\right]\right]. \tag{4.81}$$

Here, as compared to Eqs. (4.13) and (4.79), fields $\mathbf{E}^m_0$, $\mathbf{E}_0$ are changed to their transverse components $\mathbf{E}^m_{0\perp}$, $\mathbf{E}_{0\perp}$, because the field of an ultrarelativistic particle with the Lorentz factor

$$\gamma \gg 1 \tag{4.82}$$

is almost transverse to the velocity of this particle.

The transverse component of the self field of a charge moving in a medium, $\mathbf{E}^m_0$, is determined by the formula $\left(\mathbf{E}^m_{0\perp}\right)_i = \left(\delta_{ij} - v_i v_j/v^2\right)\left(\mathbf{E}^m_0\right)_j$. Let the charge move with a velocity of $\mathbf{v} = \{v\cos\alpha, v\sin\alpha, 0\}$, so that it passes through the coordinate origin at the time $t = 0$. The transverse part of the self field, $\mathbf{E}^m_{0\perp}$, at such initial conditions has the form

$$\mathbf{E}^m_{0\perp}(\mathbf{q},\omega) = -\frac{ie}{2\pi^2}\frac{\mathbf{q} - \mathbf{v}\omega/v^2}{q^2 - k^2\varepsilon(\omega)}\delta(\omega - \mathbf{q}\mathbf{v}), \tag{4.83}$$

where, in view of Eq. (4.10), $\varepsilon = 1$ is set everywhere except a pole factor of $\left(q^2 - k^2\varepsilon\right)^{-1}$. The quantity $E_{0\perp}$ is determined from Eq. (4.83) at $\varepsilon(\omega) = 1$.

To exemplify the application of formula (4.81) to a finite target and for convenience of the subsequent comparison of diffraction radiation with transition radiation and parametric X-ray radiation, let us consider radiation generated in the passage of the ultrarelativistic charged particle through the plate as shown in Fig. 4.11.

In the first approximation, we take field (4.83) as an active field. Calculating integrals in Eq. (4.81), we can easily obtain the radiation field as the sum of transition radiation and parametric radiation:

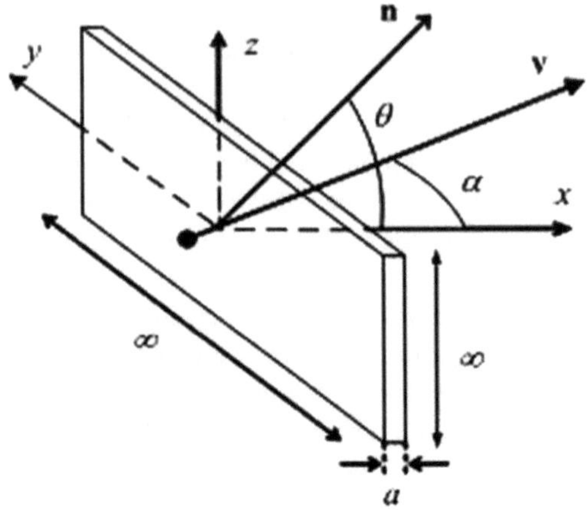

Fig. 4.11 Geometry of the passage of the charged particle in the process of the generation of transition radiation and parametric X-ray radiation

$$\mathbf{E}^{r}_{\mathrm{TR+PXR}}(\mathbf{r},\omega) = \mathbf{E}^{r}_{\mathrm{TR}}(\mathbf{r},\omega) + \mathbf{E}^{r}_{\mathrm{PXR}}(\mathbf{r},\omega)\,, \tag{4.84}$$

where

$$\mathbf{E}^{r}_{\mathrm{TR}}(\mathbf{r},\omega) = -\frac{e^{ikr}}{r}\frac{ie}{2\pi}\frac{\omega_p^2}{\omega^2}\frac{1}{v_x}\frac{1-e^{-ia\phi}}{\phi}\frac{\left[\mathbf{k}\left[\mathbf{k}\left(\mathbf{L}+k'_z\mathbf{e}_z\right)\right]\right]}{\rho^2+k'^2_z}, \tag{4.85}$$

$$\mathbf{E}^{r}_{\mathrm{PXR}}(\mathbf{r},\omega) = -\frac{e^{ikr}}{r}\frac{ie}{2\pi}\frac{1}{v_x}\sum_{\mathbf{g}\neq 0}\chi_{\mathbf{g}}\frac{1-e^{-ia\phi_{\mathbf{g}}}}{\phi_{\mathbf{g}}}\frac{\left[\mathbf{k}\left[\mathbf{k}\left(\mathbf{L_g}+\left(k'_z-g_z\right)\mathbf{e}_z\right)\right]\right]}{\rho_1^2+\left(k'_z-g_z\right)^2}. \tag{4.86}$$

Here,

$$\begin{aligned}
\varphi_g &= \frac{\omega-\mathbf{v}\mathbf{k}'+\mathbf{g}\mathbf{v}}{v_x},\\
\rho_1^2 &= \left(\frac{\omega-v_y\left(k'_y-g_y\right)}{v_x}\right)^2+\left(k'_y-g_y\right)^2-\varepsilon\omega^2/c^2 \\
\mathbf{L_g} &= \frac{\omega-v_y\left(k'_y-g_y\right)}{v_x}\mathbf{e}_x+\left(k'_y-g_y\right)\mathbf{e}_y-\mathbf{v}\omega/v^2
\end{aligned} \tag{4.87}$$

$$\begin{aligned}
\varphi &= \varphi_g\left(\mathbf{g}=0\right),\\
\mathbf{L} &= \mathbf{L_g}\left(g_y=0\right)\\
\rho^2 &= \left(\frac{\omega-v_yk'_y}{v_x}\right)^2+k'^2_y-\omega^2/c^2
\end{aligned} \tag{4.88}$$

Usually, in considering parametric X-ray radiation even in the framework of the kinematical approach [43–45] (the more so for the dynamic theory [44–47]), the effects of the diffraction of transition radiation are taken into account; i.e., the polarization radiation field is obtained in the form of the sum of parametric X-ray radiation and diffracted transition radiation. Artru and Rullhusen [44, 45] showed that the contributions from the effects of the diffraction of transition radiation can be on the same order of magnitude as compared to parametric X-ray radiation. However, in this work, we consider only the diffraction of the self field of the charge by inhomogeneities of the medium, i.e., transition radiation and parametric x-ray radiation, whereas the effects of the diffraction of real photons of transition radiation and parametric X-ray radiation are disregarded. The effects of the diffraction of real photons of transition radiation are lost when assuming that the radiation propagates through a medium with a certain average relative permittivity $\varepsilon(\omega) = 1 - \omega_p^2/\omega^2$ independent of the coordinates.

Note that the difference of the diffracted transition radiation from transition radiation insignificantly changes the qualitative pattern of polarization radiation for thin crystals, but such a difference can be decisive for thick crystals with the thickness $a \gg l_e \sim 1/\left|k\chi_{\mathbf{g}}\right|$. In this case, in the framework of the dynamic approximation, the fundamentally new dynamic effects can appear in radiation. In particular, the Borrmann effect is possible for diffracted transition radiation (anomalously small absorption of diffracted transition radiation propagating at the Bragg angle) [48, 49].

Let us consider the case of the passage of the charge in vacuum near the thin crystal plate of the finite width b (see Fig. 4.11). Here, in the first approximation, the acting field is changed to the field of the passing charge, but it should be taken into account that the charge moves in vacuum and the field is sought in the medium. Let the charge pass at a height of h above the medium located at $z < 0$. In the absence of radiation, the total field in vacuum is equal to the self field of the charge (only the transverse field component is again taken into account):

$$\mathbf{E}_{0\perp}(z, q_\perp, \omega) = -\frac{ie}{2\pi}\exp\left\{-(h-z)\sqrt{q_\perp^2 - k^2}\right\}\left(\frac{\mathbf{q}_\perp - \mathbf{v}\omega/v^2}{\sqrt{q_\perp^2 - k^2}} - i\mathbf{e}_z\right)\delta(\omega - \mathbf{q}_\perp\mathbf{v}). \tag{4.89}$$

The solution of Maxwell's equations in the case of an infinite homogeneous medium in the absence of external sources can be written in the form

$$\mathbf{E}_{0\perp}^m(z, \mathbf{q}_\perp, \omega) = \mathbf{B}(\mathbf{q}_\perp, \omega)\exp\left\{z\sqrt{q_\perp^2 - \varepsilon k^2}\right\} + \mathbf{C}(\mathbf{q}_\perp, \omega)\exp\left\{-z\sqrt{q_\perp^2 - \varepsilon k^2}\right\}. \tag{4.90}$$

Since $\omega = \mathbf{q}_\perp\mathbf{v}$, the second term in Eq. (4.90) at $z < 0$ describes the solution exponentially increasing inside the medium. Setting $\mathbf{C}(\mathbf{q}_\perp, \omega) = 0$ and disregarding, in view of Eq. (4.10), the difference of relative permittivity $\varepsilon(\omega)$ from unit, we obtain the coefficient $\mathbf{B}$ from the usual boundary conditions (equality of the tangential components of the electric field and the normal induction components at the boundary):

$$\mathbf{B}\left(\mathbf{q}_{\perp},\omega\right)=-\frac{ie}{2\pi}\exp\left\{-h\sqrt{q_{\perp}^{2}-k^{2}}\right\}\left(\frac{\mathbf{q}_{\perp}-\mathbf{v}\omega/v^{2}}{\sqrt{q_{\perp}^{2}-k^{2}}}-i\mathbf{e}_{z}\right)\delta\left(\omega-\mathbf{q}_{\perp}\mathbf{v}\right). \tag{4.91}$$

By calculating the integrals in Eq. (4.81) with the use of Eqs. (4.89), (4.90), (4.91), it is easy to obtain the radiation field in the form of the sum of the fields of the ordinary diffraction radiation and parametric diffraction radiation in the complete analogy with formulas (4.84), (4.85), (4.86):

$$\mathbf{E}^{r}_{\mathrm{DR+PDR}}\left(\mathbf{r},\omega\right)=\mathbf{E}^{r}_{\mathrm{DR}}\left(\mathbf{r},\omega\right)+\mathbf{E}^{r}_{\mathrm{PDR}}\left(\mathbf{r},\omega\right), \tag{4.92}$$

where

$$\mathbf{E}^{r}_{\mathrm{DR}}\left(\mathbf{r},\omega\right)=-\frac{e^{ikr}}{r}\frac{ie}{4\pi}\frac{1}{v_{x}}\frac{\omega_{p}^{2}}{\omega^{2}}\frac{1-e^{-ia\phi}}{\phi}\times \tag{4.93}$$

$$\times\exp\left\{-h\rho\right\}\left[\mathbf{k}\left[\mathbf{k}\left(\frac{\mathbf{L}}{\rho}-i\mathbf{e}_{z}\right)\right]\right]\frac{1-\exp\left\{-b\rho\right\}\exp\left\{ibk_{z}'\right\}}{\rho-ik_{z}'},$$

$$\mathbf{E}^{r}_{\mathrm{PDR}}\left(\mathbf{r},\omega\right)=-\frac{e^{ikr}}{r}\frac{ie}{4\pi}\frac{1}{v_{x}}\sum_{\mathbf{g}\neq 0}\chi_{\mathbf{g}}\frac{1-e^{-ia\varphi_{g}}}{\varphi_{g}}\times$$
$$\times\exp\left\{-h\rho_{2}\right\}\left[\mathbf{k}\left[\mathbf{k}\left(\frac{\mathbf{L}_{\mathbf{g}}}{\rho_{2}}-i\mathbf{e}_{z}\right)\right]\right]\frac{1-\exp\left\{-b\rho_{1}\right\}\exp\left\{ib\left(k_{z}'-g_{z}\right)\right\}}{\rho_{1}-i\left(k_{z}'-g_{z}\right)} \tag{4.94}$$

Here,

$$\rho_{2}^{2}=\left(\frac{\omega-v_{y}\left(k_{y}'-g_{y}\right)}{v_{x}}\right)^{2}+\left(k_{y}'-g_{y}\right)^{2}-\omega^{2}/c^{2}. \tag{4.95}$$

We point to the difference of the expression for ρ_2 from the expression for ρ_1 specified by the second of formulas (4.87).

Thus, in the case of the crystal target, diffraction radiation and transition radiation are supplemented by parametric diffraction radiation or by parametric X-ray radiation when the particle passes near the target or intersects it, respectively. Let us consider formulas (4.94) and (4.86). It is easy to see that the maximum contribution comes from **g** values satisfying the relation $\varphi_{\mathbf{g}}=0$, i.e., condition (4.80). For this reason, each of diffraction maxima can be considered separately and the total emitted energy is the sum over all possible **g** values.

The spectral–angular distribution of the radiation intensity near a maximum corresponding to a certain fixed reciprocal lattice vector **g** in the case of diffraction radiation can be represented in the form

$$\frac{d^2 W_{\mathbf{g}}^{\mathrm{PDR}}(\mathbf{n},\omega)}{d\Omega d(\hbar\omega)} = F_1 F_b(\mathbf{g}) \frac{\exp\{-2h\rho_2\}}{4} \frac{1-n_z^2+\left(L_g^2-(\mathbf{nL_g})^2\right)/\rho_2^2}{4(c/\omega)^2\left(\rho_1^2+(k_z'-g_z)^2\right)}, \quad (4.96)$$

where

$$F_1 = \alpha \frac{|\chi_g|^2}{\pi^2} \frac{\sin^2\left(\frac{\omega a}{2v_x}\left(1-\sqrt{\varepsilon}\mathbf{nv}/c+\mathbf{gv}/\omega\right)\right)}{\left(1-\sqrt{\varepsilon}\mathbf{nv}/c+\mathbf{gv}/\omega\right)^2} \quad (4.97)$$

and the factor F_b determines the dependence of the emitted energy on the plate width b:

$$F_b(b,\omega,\mathbf{g}) = 1 - 2\exp\{-b\rho_1\}\cos\left(b\left(k_z'-g_z\right)\right) + \exp\{-2b\rho_1\}. \quad (4.98)$$

Similarly, the expression for parametric x-ray radiation in the case of the passage of the charge through the infinite plate is obtained from Eq. (4.86) in the form

$$\frac{d^2 W_{\mathbf{g}}^{\mathrm{PXR}}(\mathbf{n},\omega)}{d\Omega d(\hbar\omega)} = F_1 \frac{\left(1-n_z^2\right)\left(k_z'-g_z\right)^2+\left(L_{\mathbf{g}}^2-(\mathbf{nL_g})^2\right)+2\left(k_z'-g_z\right)n_z(\mathbf{nL_g})}{(c/\omega)^2\left(\rho_1^2+(k_z'-g_z)^2\right)^2}. \quad (4.99)$$

For the real crystal target, the sum of the delta functions appears in formulas (4.96) and (4.99) in view of the limiting relation

$$\frac{\sin^2\left(\frac{\omega a}{2v_x}\left(1-\sqrt{\varepsilon}\mathbf{nv}/c+\mathbf{gv}/\omega\right)\right)}{\left(1-\sqrt{\varepsilon}\mathbf{nv}/c+\mathbf{gv}/\omega\right)^2} \to \pi N \frac{\omega d_x}{2v_x}\delta\left(1-\sqrt{\varepsilon}\mathbf{nv}/c+\mathbf{gv}/\omega\right), \quad (4.100)$$

which is valid under the condition

$$\frac{\omega a}{2v_x} \approx N\frac{\omega d_x}{2v_x} \approx N\frac{d_x}{\lambda}\frac{\pi}{\cos\alpha} \gg 1, \quad (4.101)$$

where N is the number of periods and d_x is the size of the unit cell along the x axis. Therefore, the spectral–angular distribution of parametric diffraction radiation includes a set of sharp peaks. Condition (4.80) determines the relation between the radiation frequency, particle energy, lattice parameters, and observation angles. The width of individual peaks is inversely proportional to the number of periods N and the height is proportional to N^2.

Integration of expression (4.99) with respect to the frequency with the use of the delta function provides

$$\frac{dN(\mathbf{n})}{d\Omega} = \frac{1}{\hbar\omega_B}\int_0^\infty d\omega \frac{d^2W(\mathbf{n},\omega)}{d\Omega d\omega}$$
$$= \alpha\frac{\omega_B\left|\chi_{\mathbf{g}}\right|^2}{4\pi v_x}\frac{a}{2\sin^2\theta_B}\frac{\left|\left[\mathbf{k}\left(\mathbf{g}+\mathbf{v}\omega/v^2\right)\right]\right|^2}{\left(\left(\mathbf{k}'-\mathbf{g}\right)_\perp^2+\frac{\omega_B^2}{v^2}\left(1-\beta^2\varepsilon\right)\right)^2}. \tag{4.102}$$

The introduction of angles θ_x and θ_y with respect to the quantum emission direction near the Bragg reflection direction makes it possible to transform the last formula to the form

$$\frac{dN(\mathbf{n})}{d\Omega} = \frac{e^2}{\hbar c}a\frac{\omega_B\left|\chi_{\mathbf{g}}\right|^2}{4\pi v_x}\frac{1}{2\sin^2\theta_B}\frac{\theta_x^2\cos^2 2\theta_B+\theta_y^2}{\left(\theta_x^2+\theta_y^2+\gamma^{-2}+\omega_p^2/\omega_B^2\right)^2}. \tag{4.103}$$

This formula is well known in the kinematical theory of parametric radiation [30, 31].

Note that, in contrast to diffraction radiation from the amorphous target (see Eq. (4.23)), parametric diffraction radiation has different decreasing ranges as a function of the impact parameter h and plate thickness b. In order to reveal this fact, we represent the coefficients ρ_1^2 and ρ_2^2 in the form

$$\rho_1^2 = \frac{\omega^2}{v_x^2}\left[1-\varepsilon(\omega)\beta^2-2\beta_y(c/\omega)\left(k_y'-g_y\right)+(\beta c/\omega)^2\left(k_y'-g_y\right)^2+\varepsilon(\omega)\beta_y^2\right]$$
$$\rho_2^2 = \frac{\omega^2}{v_x^2}\left[1-\beta^2-2\beta_y(c/\omega)\left(k_y'-g_y\right)+(\beta c/\omega)^2\left(k_y'-g_y\right)^2+\varepsilon(\omega)\beta_y^2\right] \tag{4.104}$$

The minimum condition at given h, ω, and γ values for ρ_1^2 and ρ_2^2 has the same form

$$k_y'-g_y = \frac{\omega}{c}\frac{\beta_y}{\beta^2} = \frac{\omega}{c}\frac{\sin\alpha}{\beta}. \tag{4.105}$$

In this case,

$$\left(\rho_1^2\right)_{\min} = \frac{\omega^2}{v^2}\left(1-\varepsilon\beta^2\right) = \frac{\omega^2}{v^2}\left(\gamma^{-2}+\omega_p^2/\omega^2\right)$$
$$\left(\rho_2^2\right)_{\min} = \frac{\omega^2}{v^2}\left(1-\beta^2\right) = \frac{\omega^2}{v^2}\gamma^{-2} \tag{4.106}$$

From these expressions, the characteristic impact parameter h is determined as

$$h_{\text{eff}} \sim (\rho_2)_{\min}^{-1} = \gamma\lambda. \tag{4.107}$$

At the same time, the characteristic size of the emitting layer of the crystal structure is determined from the condition

$$b_{\text{eff}} \sim (\rho_1)^{-1}_{\min} = \begin{cases} \gamma\lambda, \ \omega \geq \gamma\omega_p \\ c/\omega_p, \ \omega \ll \gamma\omega_p \end{cases} \tag{4.108}$$

The quantities given by Eq. (4.107) and (4.108) are the characteristic ranges at which the self field of the ultrarelativistic charged particle decreases in vacuum and in a medium, respectively. The difference between them is caused by a quite strong screening of the self field of the ultrarelativistic charged particle in the medium at high frequencies $\omega_p \ll \omega \ll \gamma\omega_p$ (whereas the medium is almost transparent for the radiation field). At the same time, for ordinary diffraction radiation (considered in the preceding sections of this chapter), the quantities h_{eff} and b_{eff} coincide and are equal to $\gamma\lambda$, because diffraction radiation in this case is determined only by the shape, sizes, and average relative permittivity $\varepsilon(\omega)$ of the target. Indeed, according to formula (4.81), usual diffraction radiation is determined by the scattering of the self field of the charged particle from the target in vacuum (the first term), whereas parametric diffraction radiation is a result of the charge field scattering from the crystal structure of the target in the medium.

Note that the spectrum cutoff frequency for the particle passing near the target surface at a distance of $h \leq A/\omega_p$ is $\omega_c \approx \gamma c/h \geq \gamma\omega_p$ and the parametric diffraction radiation spectrum is similar to the parametric x-ray radiation spectrum.

Condition (4.105) needs not to be satisfied. For example, in view of the requirement $n_y^2 \leq 1$, condition (4.105) can be satisfied only if $cg_y/\omega \simeq \lambda/d_y < 1$, where d_y is the lattice period along the y axis. At $d_y \sim 0.5$ nm, the condition $\lambda < d_y$ corresponds to the photon energy $\hbar\omega > 2.5$ keV. The omission of condition (4.105) means only that the functions ρ_1^2 and ρ_2^2 in this case have no minimum in the form of an extremum point.

Let us now consider the characteristic maxima of parametric diffraction radiation. When condition (4.105) is satisfied, the argument of the exponential $\exp\{-2h\rho_2\}$ is minimal and $A_{\mathbf{g}} = 0$. The substitution of Eq. (4.105) into Eqs. (4.96) and (4.99) yields

$$\frac{d^2 W_{\mathbf{g}}^{\text{PDR}}(\mathbf{n}, \omega)}{d\Omega d(\hbar\omega)} = \alpha \frac{|\chi_{\mathbf{g}}|^2}{\pi^2 \cos^2\alpha} \frac{\sin^2\left(\frac{\omega a}{2c}\Phi\right)}{\Phi^2} \frac{1 - n_z^2}{\gamma^{-2} + \frac{\omega_p^2}{\omega^2} + \beta^2\left(\sqrt{\varepsilon}n_z - \frac{cg_z}{\omega}\right)^2} \frac{F_b(\mathbf{g})}{4} \exp\left\{-\frac{2h\omega}{c\gamma}\right\} \tag{4.109}$$

$$\frac{d^2 W_{\mathbf{g}}^{\text{PDR}}(\mathbf{n}, \omega)}{d\Omega d(\hbar\omega)} = \alpha \frac{|\chi_{\mathbf{g}}|^2}{\pi^2 \cos^2\alpha} \frac{\sin^2\left(\frac{\omega a}{2c}\Phi\right)}{\Phi^2} \frac{1 - n_z^2}{\gamma^{-2} + \frac{\omega_p^2}{\omega^2} + \beta^2\left(\sqrt{\varepsilon}n_z - \frac{cg_z}{\omega}\right)^2} \left(\sqrt{\varepsilon}n_z - \frac{cg_z}{\omega}\right)^2. \tag{4.110}$$

Here,

$$\Phi = \beta^{-1}\cos\alpha + \frac{\lambda}{d_x}m - \sqrt{\varepsilon}n_x \tag{4.111}$$

and the component of the reciprocal lattice vector is taken in the form

$$g_x = \frac{2\pi m}{d_x}, \quad m\text{–integer}, \tag{4.112}$$

where d_x is the lattice period along the x axis. Note that the angles specified by n_z and n_x in Eqs. (4.109) and (4.110) are not arbitrary and are restricted by the constraint given by Eq. (4.105). Therefore, the peaks of both parametric diffraction radiation and parametric X-ray radiation are concentrated in narrow peaks determined by the equality $\Phi = 0$, i.e.,

$$-\beta \frac{\lambda}{d_x} m = \cos\alpha - \beta\sqrt{\varepsilon} n_x. \tag{4.113}$$

However, there is the additional condition

$$k_z' - g_z = \frac{\omega}{c}\sqrt{\varepsilon} n_z - g_z = 0 \tag{4.114}$$

under which the parametric diffraction radiation is maximal and parametric X-ray radiation vanishes.

In agreement with Eq. (4.27), $n_x = \cos\theta$. Let $\theta = \pi - \alpha$. In this case, from Eq. (4.113), we obtain the usual Bragg condition

$$-\frac{\lambda}{d_x} m = 2\cos\alpha. \tag{4.115}$$

In other words, at the Bragg frequency $\omega_B = \frac{\pi c}{d_x \cos\alpha}$, the radiation peak is concentrated near $\theta = \pi - \alpha$. Let us now consider radiation along the particle velocity, i.e., $\theta = \alpha$. In this case, condition (4.113) is modified to the form

$$-2\frac{\lambda}{d_x} m = \left(\gamma^{-2} + \frac{\omega_p^2}{\omega^2}\right)\cos\alpha. \tag{4.116}$$

Condition (4.116) leads to the suppression of parametric diffraction radiation along the particle velocity. Indeed, relation (4.116) at $\omega \ll \gamma\omega_p$ takes the form

$$-\frac{4\pi A}{d_x \omega_p} m = \frac{\omega_p}{\omega}\cos\alpha. \tag{4.117}$$

At typical values $\omega_p \simeq 4 \cdot 10^{16}\, c^{-1}$, $4\pi c/\omega_p \sim 10^{-5}$cm. The period of the crystal structure is $d_x \sim 10^{-7} \ll 4\pi c/\omega_p$; hence, the left-hand side of Eq. (4.117) is larger than unit. Since the right-hand side in this case is smaller than unit, condition (4.117) cannot be satisfied for the crystals.

At high frequencies $\omega \sim \gamma\omega_p$, condition (4.116) is expressed in the form

$$d_x \approx \gamma \frac{2\pi c}{\omega_p}\left(\frac{-m}{\cos\alpha}\right). \tag{4.118}$$

This condition also cannot be satisfied for crystals. The above discussion shows that parametric radiation from a crystal along the particle velocity at frequencies $\omega \leq \gamma\omega_p$ is suppressed in the kinematical approximation. Indeed, such radiation is investigated both theoretically [46, 50] and experimentally [51], but its calculation requires the use of the dynamic theory of parametric radiation. This radiation can physically be considered as a result of the secondary diffraction of a wave scattered at the Bragg angle.

Note that condition (4.116) at $\omega \gg \gamma\omega_p$ takes the form

$$\lambda = \gamma^{-2} d_x \left(\frac{\cos\alpha}{-2m}\right). \tag{4.119}$$

This condition can certainly be satisfied; i.e., parametric radiation along the velocity is possible even in the framework of the kinematical approximation.

Note that expression (4.116) follows from the conservation laws. For this reason, in the case of targets with the periodic structure with a period on the order of micron or larger, parametric X-ray radiation (both parametric X-ray radiation and parametric diffraction radiation) along the particle velocity is possible even in the framework of kinematical theory. Artificial layered media are examples of such structures.

Results concerning diffraction radiation from the crystal target can be formulated as follows. The spectrum of parametric diffraction radiation consists of the peaks satisfying the same condition as the peaks of parametric X-ray radiation (see Eq. (4.80)). The spectrum of parametric diffraction radiation has the cutoff frequency characteristic for any diffraction radiation: $\omega_c \approx c\gamma/h$. Moreover, we point to the qualitative difference between the thicknesses b_{eff} of the effectively emitting layer for diffraction radiation and parametric diffraction radiation: $b_{\mathrm{eff}}^{\mathrm{DR}} \sim \gamma\lambda$, whereas

$$b_{\mathrm{eff}}^{\mathrm{PDR}} \sim \begin{cases} \gamma\lambda, & \omega \geq \gamma\omega_p \\ c/\omega_p, & \omega \ll \gamma\omega_p \end{cases}. \tag{4.120}$$

As mentioned above, such a difference is associated with the screening of the self field of the charged particle in the medium at the distances $\geq c/\omega_p$ in the frequency range $\omega_p \ll \omega \ll \gamma\omega_p$.

It is interesting that the transition radiation and diffraction radiation fields are related to the polarization current density

$$\frac{i\omega}{4\pi}\left(\varepsilon(\omega) - 1\right)\mathbf{E}_0(\mathbf{r}, \omega) \tag{4.121}$$

(see Eq. (4.13)), where $\mathbf{E}_0$ is the field of a charge in vacuum. This result is exact (up to possible corrections associated with the effects of the local field) and leads to well known expressions for the Cherenkov radiation and transition radiation fields at $\omega \gg \omega_p$. At the same time, when considering parametric X-ray radiation, the known Ter-Mikaelyan formula (4.79) is obtained when the charge field in the medium (see Eq. (4.77)) rather than in vacuum is chosen as the first approximation in polarization current (4.75).

References

1. Ahmanov, S.A., Nikitin, S.Yu.: Physical Optics. Moscow State University, Moscow (in Russian) (1998)
2. Dolgoshein, B.: Transition radiation detectors. Nucl. Instrum. Methods Phys. Res. A **326**, 434 (1993)
3. Bazylev, V.A., Zhevago, N.K.: Radiation from Fast Particles in Medium and External Field. Moskow, Nauka (in Russian) (1987)
4. Landau, L.D., Lifshitz, E.M.: Electrodynamics of Continuous Media. Addison-Wesley, Reading, MA (1984)
5. Ryazanov, M.I.: Electrodynamics of Condensed Media. Moscow, Nauka (in Russian) (1984)
6. Tishchenko, A.A., Potylitsyn, A.P., Strikhanov, M.N.: Diffraction radiation from an ultrarelativistic charge in the plasma frequency limit. Phys. Rev. E **70**, 066501, (2004)
7. Tishchenko, A.A., Strikhanov, M.N., Potylitsyn, À.P.: X-ray transition radiation from an ultrarelativistic charge passing near the edge of a target or through a thin wire. Nucl. Instrum. Methods Phys. Res. B **227**, 63 (2005)
8. Durand, L.: Transition radiation from ultra-relativistic particles. Phys. Rev. D **11**, 89 (1975)
9. Alikhanian, A.I., Chechin, V.A.: Eikonal approximation in X-ray transition radiation theory. Phys. Rev. D. **19**, 1260 (1979)
10. Potylitsyn, A.P.: Transition radiation and diffraction radiation. Similarities and differences. Nucl. Instrum. Methods Phys. Res. B **145**, 169 (1998)
11. Urakawa, J., Hayano, H., Kubo, K. et al.: Feasibility of optical diffraction radiation for a non-invasive low-emittance beam diagnostics. Nucl. Instrum. Methods Phys. Res. A **472**, 309 (2001)
12. Potylitsyn, A.P., Potylitsyna, N.A.: Diffraction radiation of ultrarelativistic particles at passing of an inclined slit. Russ. Phys. J. **43**(4), 303 (2000)
13. Bazylev, V.A., Zhevago, N.K.: Intense electromagnetic radiation from relativistic particles. Phys. Uspekhi **25**, 565 (1982)
14. Piestrup, M.A., Pantell, R.H., Puthoff, H.E. et al.: Cerenkov radiation as a source of ultraviolet radiation. J. Appl. Phys. **44**, 5160, (1973)
15. Henke, B.L., Gullikson, E.M., Davis, J.C.: At. Data Nucl. Data Tables 54, 181 (1993)
16. Samsonov, V.M.: On Cherenkov and transition radiation in the frequency domain of γ-resonance. Sov. Zh. Eksp. Teor. Fiz. **75**(1), 88 (1978)
17. Samsonov, V.M.: On Cherenkov and transition radiations in the frequency domain with anomaly dispersion. Proceedings of the Second Symposium on Transition Radiation. Yerevan, Armenia (1983)
18. Fedorov, V.V., Smirnov, A.I.: On a possibility of Cherenkov radiation of γ-quanta by electrons. Sov. Phys. ZhETF Lett. **23**, 34 (1976)
19. Zhevago, M.K., Glebov, V.I.: X-ray Cerenkov radiation at grazing incidence of electrons. Phys. Lett. A. **160**, 564 (1991)
20. Gary, C., Kaplin, V., Kubankin, A. et al.: An investigation of the Cherenkov X-rays from relativistic electrons. Nucl. Instrum. Methods Phys. Res. B **227**, 95 (2005)

21. Kubankin, A., Nasonov, N., Kaplin, V. et al.: X-ray Cherenkov radiation under conditions of grazing incidence of relativistic electrons onto a target surface. Rad. Phys. Chem. **75**, 913, (2006)
22. Moran, M.J., Chang, B., Schneider, M.B., Maruyama, X.K.: Grazing-incidence Cherenkov X-ray generation. Nucl. Instrum. Methods Phys. Res. B **48**, 287 (1990)
23. Caticha, A. Transition-diffracted radiation and the Cerenkov emission of x rays. Phys. Rev. A. **40**, 4322 (1989)
24. Nasonov, N., Zhukova, P.: Anomalous photoabsorbtion in the parametric X-rays in conditions of Cherenkov effect. Phys. Lett. A. **346**, 367 (2005)
25. Nasonov, N., Zhukova, P., Hubbell, J.H.: Parametric X-rays along the velocity direction of an emitting particle under conditions of the Cherenkov effect. Rad. Phys. Chem. **75**, 923 (2006)
26. Caticha, A.: Quantum theory of the dynamical Cerenkov emission of x rays. Phys. Rev. B. **45**, 9541 (1992)
27. Knulst, W., Luiten, O.J., van der Wiel, M.J., Verhoeven, J.: Observation of narrow-band Si L-edge Cerenkov radiation generated by 5 MeV electrons. Appl. Phys. Lett. **79**, 2999 (2001)
28. Knulst, W., van der Wiel, M.J., Luiten, O.J., Verhoeven, J. High brightness, narrowband, and compact soft x-ray Cherenkov sorces in the water window. Appl. Phys. Lett. **83**, 4050 (2003)
29. Tishchenko, A.A., Potylitsyn, A.P., Strikhanov, M.N.: X-ray diffraction radiation in conditions of Cherenkov effect. Phys. Lett. A. **359**, 509, (2006)
30. Baryshevsky, V.G., Feranchuk, I.D., Ulyanenkov, A.P.: Parametric X-ray Radiation: Theory, Experiment and Applications. Springer, Heidelberg (2006)
31. Ter-Mikhaelyan, M.L.: Electromagnetic radiative processes in periodic media at high energies. Phys.-Uspekhi **44**, 571 (2001)
32. Ter-Mikaelyan, M.L.: High-Energy Electromagnetic Processes in Condensed Media. Wiley-Interscience, New York, NY (1972)
33. Garibyan, G.M., Yang, C.: X-ray transition radiation. Sov. Phys. JETP **61**, 930 (1971)
34. Baryshevskiy, V.G., Feranchuk, I.D.: On transition radiation of γ-quanta in a crystal. Sov. Phys. JETP. **61**, 944 (1971)
35. Vorobiev, S.A., Kalinin, B.N., Pak, S., Potylitsyn, A.P.: Detection of monochromatic x-ray radiation at interaction of ultrarelativistic electrons with diamond single crystal. Sov. JETP Lett. **41** (5), 3 (1985)
36. Adishev, Yu.N., Baryshevsky, V.G., Vorobiev, S.A. et al.: Experimental detection of parametric x-ray radiation. Sov. JETP Lett. **41**(7), 295 (1985)
37. Vorobiev, S.A., Kalinin, B.N., Pak, S., Potylitsyn, A.P. et al.: Spectra investigation of parametric quasi-Cherenkov radiation of ultrarelativistic electrons in diamond single crystal. Sov. Phys. JETP **90**, 829 (1986)
38. Garibyan, G.M., Yang, C.: X-ray Transition Radiation. Yerevan, Armenia (1983)
39. Nitta, H.: Theoretical notes on parametric X-ray radiation. Nucl. Instrum. Methods Phys. Res. B **115**, 401 (1996)
40. Asano, S., Endo, I., Harada, M., Ishii, S. et al.: How intense is parametric X radiation? Phys. Rev. Lett. **70**, 3247 (1993)
41. Fiorito, R.B., Rule, D.W., Maruyama, X.K. et al.: Observation of higher order parametric x-ray spectra in mosaic graphite and single silicon crystals. Phys. Rev. Lett. **71**, 704 (1993)
42. Shchagin, A.V., Khizhnyak, N.A.: Properties of parametric X-ray radiation from a thin crystal. Nucl. Instrum. Methods Phys. Res. B **119**, 115 (1996)
43. Potylitsyn, A.P., Verzilov, V.A.: Parametric X-rays and transition diffracted radiation in crystal stacks. Phys. Lett. A. **209**, 380 (1995)
44. Artru. X., Rullhusen. P.: Parametric x-rays and diffracted transition radiation in perfect and mosaic crystals. Nucl. Instrum. Methods Phys. Res. B **145**, 1 (1997)
45. Artru, X., Rullhusen, P.: Parametric x-rays and diffracted transition radiation in perfect and mosaic crystals. Nucl. Instrum. Methods Phys. Res. B **173**, 16 (2001)
46. Kubankin, A., Nasonov, N., Sergienko, V., Vnukov, I.: An investigation of the parametric x-rays along the velocity of emitting particle. Nucl. Instrum. Methods Phys. Res. B **201**, 97 (2003)

47. Imanishi, N., Nasonov, N., Yajima, K.: Dynamical diffraction effects in the transition radiation of a relativistic electron crossing a thin crystal. Nucl. Instrum. Methods Phys. Res. B **173**, 227 (2001)
48. Nasonov, N.N.: Borrmann effect in parametric x-ray radiation. Phys. Lett. A **260**, 391 (1999)
49. Nasonov, N.N.: On the effect of anomalous photoabsorption in the parametric X-rays. Phys. Lett. A **292**, 146 (2001)
50. Baryshevsky, V.G.: Parametric x-ray radiation at a small angle near the velocity direction of the relativistic particle. Nucl. Instrum. Methods Phys. Res. B **122**, 13 (1997)
51. Aleinik, A.N., Baldin, A.N., Bogomazova, E.A. et al.: Experimental observation of parametric X-ray radiation directed along the propagation velocity of relativistic electrons in a tungsten crystal. JETP Lett. **80**, 393 (2004)

Chapter 5
Diffraction Radiation at the Resonant Frequency

5.1 Diffraction Radiation at the Resonant Frequency from a Nonplanar Surface

Diffraction radiation is microscopically a result of the scattering of the self field of a uniformly moving charge from the atoms of a medium. The cross section for the scattering of an electromagnetic wave from an atom is maximal near resonance, so that the diffraction radiation intensity should increase at resonant frequencies. As known, the transverse field of a source at small distances is much lower than the longitudinal field. Therefore, the energy transfer from an exited atom to an unexcited atom in dense media occurs primarily through the transverse field by the dipole—dipole interaction rather than through the emission and absorption of resonant transverse waves. As a result, the interaction of a resonant photon with an atom most likely leads to disappearance of the photon and appearance of an electron excitation further migrating in the medium as an exciton. For this reason, the resonant photon, i.e., the photon whose energy is close to the exciton energy does not penetrate inside the medium. For the same reason, the emission of resonant transverse waves by an atom from the depth of a dense medium is impossible. Thus, the probability of the formation of diffraction radiation at the resonant frequency in the process of scattering from an atom in a dense medium is much lower than the probability of the formation of an exciton. Therefore, diffraction radiation is generated due to the scattering of the self field of the particle from the atoms of the surface layer. The thickness of this layer is determined by the absorption coefficient of transverse resonant waves. Since this layer is thin, we can use the approximation of the single scattering of the resonant component of the self field of the fast particle from the atoms of the medium. This approximation in the problem of the reflection of resonant electromagnetic waves from the surface of a medium was proposed by Fermi [1]. This makes it possible to solve the reflection problem without the usual macroscopic boundary conditions and provides a good agreement with experimental data.

Note that the field of the charge moving with constant velocity v and energy $E \equiv \gamma mc^2$ decreases with an increase in the distance h in the transverse direction as $\exp(-h\omega/\gamma v)$. Hence, polarization currents that are sources of diffraction radiation

A.P. Potylitsyn et al., *Diffraction Radiation from Relativistic Particles*, STMP 239, 137–147, DOI 10.1007/978-3-642-12513-3_5,

decrease exponentially with in increase in the distance from the surface. This circumstance increases the contribution to resonant radiation from the surface atoms and thereby increases the accuracy of the single-scattering approximation for the resonant component of the fast-particle field from the atoms of the medium. Diffraction radiation is usually considered in the framework of macroscopic electrodynamics including the boundary effects with the use of the boundary conditions [2–5]. These methods are convenient in the problems of diffraction radiation from surfaces of a quite simple profile. However, for media with a complex-profile surface, the accurate inclusion of the boundary conditions gives rise to significant difficulties and requires special approximate methods for each particular profile of the surface. A generalization of the method developed in [6] for considering resonant wave scattering to the problem of diffraction radiation at resonant frequencies [7] makes it possible to comparatively easy treat diffraction radiation at resonant frequencies for complex-profile surfaces.

Let us consider diffraction radiation at a frequency close to an eigenfrequency of the medium, when the motion of the particle is described by the law given by the equalities $x = a$, $y = 0$, and $z = vt$ near the homogeneous medium located in $X_a < \varsigma(Y_a, Z_a)$ with molecule number density n_0. As mentioned above, radiation is not generated when $\varsigma =$ const; i.e., when the charge uniformly moves in parallel to the surface of the homogeneous medium. Owing to an exponential decrease in the field of the uniformly moving charge, the polarization currents in the medium depth do not contribute to diffraction radiation, whereas radiation from the homogeneous medium is generated by the polarization currents near the irregularities of the surface.

Let us find a point with the minimum X coordinate on the medium surface $X = \varsigma(Y, Z)$ and take the coordinate axes so that the $X = 0$ plane passes through this point. As mentioned above, the polarization currents in the region $X < 0$ do not affect diffraction radiation. Owing to this circumstance, the polarization current density can be calculated with the inclusion of only the current density in the layer between the $X = \varsigma(Y, Z)$ surface and $X = 0$ plane. To avoid insignificant complications, we consider diffraction radiation from the medium whose surface specified by the function $X = \varsigma(Z)$. As the molecule is disposed in the medium, the amplitude of a transverse resonant wave emitted by a molecule located at the point $\mathbf{R}_a$ decreases exponentially due to inelastic scattering. The observation direction is specified by the unit vector n with the components n_x, n_y, and n_z. Let this direction intersect the medium surface at the point $\mathbf{R}_a + \mathbf{R}'(X', Y', Z')$ whose z coordinate $Z_a + Z' \equiv Z_a + Z'(\mathbf{R}_a)$ is determined from the equation

$$\varsigma\left(Z_a + Z'\right) = X_a + Z'\frac{n_x}{n_z}. \tag{5.1}$$

The distance between the emitting molecule and the intersection point with the surface is $\left|\mathbf{R}'\left(X', Y', Z'\right)\right| = \left|Z'(\mathbf{R}_a)/n_z\right|$. A decrease in the amplitude of the resonant wave in the path inside the medium can be taken into account by multiplying the intramolecular current density by $\exp\left\{-g\left|Z'(\mathbf{R}_a)/2n_z\right|\right\}$, where g is the absorption coefficient. Therefore, the absorption of the transverse resonance wave inside the medium can be taken into account by changing the usual expression for the Fourier transform of the polarization current density in the single scattering approximation,

$$\mathbf{j}(\mathbf{q},\omega) = -i\frac{\omega}{(2\pi)^3}\alpha(\omega)\sum_a \int d^3p\mathbf{E}_0(\mathbf{p},\omega)\exp\{-i(\mathbf{p}-\mathbf{q})\mathbf{R}_a\}, \tag{5.2}$$

to the expression

$$\mathbf{j}(\mathbf{q},\omega) = -i\frac{\omega}{(2\pi)^3}\alpha(\omega)\sum_a \int d^3p\mathbf{E}_0(\mathbf{p},\omega)\exp\left\{-i(\mathbf{p}-\mathbf{q})\mathbf{R}_a - g\left|Z'(\mathbf{R}_a)/2n_z\right|\right\}. \tag{5.3}$$

Here, $\alpha(\omega)$ is the polarizability of the molecule, $\mathbf{E}_0(\mathbf{p},\omega)$ is the Fourier transform of the electric field of the particle passing near the medium surface, $\mathbf{R}_a$ is the radius vector of the center of mass of the molecule, and summation is performed over all the molecules of the medium. The radiation wavelength corresponding to the resonant frequency is much larger than atomic sizes. In this case, it is possible to disregard the effect of fluctuations of the polarization current in the homogeneous medium and change expression (5.3) to its value averaged over the molecule coordinates

$$\begin{aligned}\mathbf{j}(\mathbf{q},\omega) = -i\frac{\omega}{(2\pi)^3}\alpha(\omega)\int d^3p\mathbf{E}_0(\mathbf{p},\omega)\times \\ \times\left\langle\sum_a \exp\left\{-i(\mathbf{p}-\mathbf{q})\mathbf{R}_a - g\left|Z'(\mathbf{R}_a)/2n_z\right|\right\}\right\rangle.\end{aligned} \tag{5.4}$$

Radiation energy emitted from the medium during the passage of the particle in the frequency range $d\omega$ near the resonant frequency within the solid angle element $d\Omega$ in the direction $\mathbf{k}$ is expressed in the form

$$\begin{aligned}\frac{d^2W(\mathbf{n},\omega)}{d\omega d\Omega} = \frac{1}{c}(2\pi)^6\left|[\mathbf{kj}(\mathbf{k},\omega)]\right|^2 = \\ = \frac{\omega^2}{c}\left|\alpha(\omega)\int d^3p\left[\mathbf{kE}_0(\mathbf{p},\omega)\right]\left\langle\sum_a \exp\left\{-i(\mathbf{p}-\mathbf{k})\mathbf{R}_a - g\left|Z'(\mathbf{R}_a)/2n_z\right|\right\}\right\rangle\right|^2.\end{aligned} \tag{5.5}$$

Introducing the notation

$$S(\mathbf{p}-\mathbf{k}) = (2\pi)^{-3}\int_{-\infty}^{\infty}dY\int_{-\infty}^{\infty}dZ\int_0^{\varsigma(Y,Z)}dX\exp\left\{-i(\mathbf{p}-\mathbf{k})\mathbf{R} - g\left|Z'(\mathbf{R})/2n_z\right|\right\}, \tag{5.6}$$

we can write

$$\left\langle \sum_a \exp\left\{-i\left(\mathbf{p}-\mathbf{k}\right)\mathbf{R}_a - g\left|Z'\left(\mathbf{R}_a\right)/2n_z\right|\right\}\right\rangle = (2\pi)^3 n_0 S\left(\mathbf{p}-\mathbf{k}\right). \tag{5.7}$$

According to Eq. (5.6), it is seen that the region of large X values does not noticeably contribute to the integral due, first, to the decreasing exponential $e^{-g\left|\frac{Z'(\mathbf{R})}{2n_z}\right|}$ and, second, to the upper limit. Therefore, two qualitatively different cases are possible. In the first case, the thickness of the effective emitting layer is smaller than the absorption length of the resonant wave. In this case, the upper limit is determining, whereas the decreasing exponential does not strongly vary in the integration interval, the decreasing exponential can be disregarded, and the current can be taken in the form of Eq. (5.2) rather than Eq. (5.3). In this case, $S\left(\mathbf{p}-\mathbf{k}\right)$ is given by the expression

$$S\left(\mathbf{p}-\mathbf{k}\right) = (2\pi)^{-3}\int_{-\infty}^{\infty} dY \int_{-\infty}^{\infty} dZ \int_0^{\varsigma(Y,Z)} dX \exp\left\{i\left(\mathbf{p}-\mathbf{k}\right)\mathbf{R}\right\} \tag{5.8}$$

rather than by expression (5.6). In the second case, the absorption length of the resonant wave is smaller than the thickness of the effective inhomogeneous layer. In this case, the upper integration limit with respect to X in Eq. (5.6) is insignificant and can be replaced by infinity and the $Z'\left(\mathbf{R}_a\right)$ dependence is obtained from Eq. (5.1), which is completely determined by the surface profile. Then, Eq. (5.6) is expressed as

$$S\left(\mathbf{p}-\mathbf{k}\right) = (2\pi)^{-3}\int_{-\infty}^{\infty} dY \int_{-\infty}^{\infty} dZ \int_0^{\varsigma(Y,Z)} dX \exp\left\{i\left(\mathbf{p}-\mathbf{k}\right)\mathbf{R} - g\left|Z'\left(R\right)/2n_z\right|\right\}. \tag{5.9}$$

Estimate by Eq. (5.9) can be performed with the use of the properties of the function $Z'\left(\mathbf{R}\right)$, i.e., with the use of a particular profile of the medium surface.

Let us consider the first case where the inhomogeneity layer thickness is smaller than the length at which a transverse resonant wave is absorbed. This means that the wave changes insignificantly when passing the inhomogeneity layer and, in the first approximation, a scattering-induced decrease in the amplitude of the emitted wave in the medium can be disregarded. In this case, the distribution of the emitted energy has the form

$$\frac{d^2W\left(\mathbf{n},\omega\right)}{d\omega\, d\Omega} = 2\pi^6\frac{\omega^2}{c}\left|n_0\alpha\left(\omega\right)\int d^3p\left[\mathbf{k}\mathbf{E}_0(p)\right] S\left(\mathbf{p}-\mathbf{k}\right)\delta\left(\omega - p_z v\right)\right|^2, \tag{5.10}$$

where $S\left(\mathbf{p}-\mathbf{k}\right)$ is given by Eq. (5.6). Note that when $\varsigma\left(Y,Z\right) = \varsigma\left(Y\right)$, i.e., the surface does not change along the z axis, $S\left(\mathbf{p}-\mathbf{k}\right)$ is proportional to $\delta\left(p_z - k_z\right)$ and $\delta\left(\omega - p_z v\right)\delta\left(p_z - k_z\right) = 0$ if Cherenkov radiation in the medium is absent at this frequency. Thus, at $\varsigma(Y, Z) = \varsigma(Y)$, radiation is not generated because the transfer of the longitudinal momentum to the medium is impossible.

Let us now consider diffraction radiation from the medium whose surface is specified in the form $X = \varsigma(Z)$. In this case, the integral with respect to p in Eq. (5.6) contains the product of the delta functions, $\delta(p_y - k_y)$ and $\delta(\omega - p_z v)$, which ensure the integration with respect to p_y and p_z. As a result, Eq. (5.5) takes the form

$$\frac{d^2W(\mathbf{n},\omega)}{d\omega\, d\Omega} = \frac{\omega^2}{c}(2\pi)^2\left|n_0\alpha(\omega)\int_{-\infty}^{\infty} dZ \exp\left\{-i\left(k_z - \frac{\omega}{v}\right)Z\right\}\right|^2 \times$$
$$\times\left|\int_0^{\varsigma(Z)} dX \exp(ik_x X)\int_{-\infty}^{\infty} dp_x\left[\mathbf{k}\mathbf{E}_0(p_x, k_y, \omega/v)\right]\exp(-ip_x X)\right|^2 \tag{5.11}$$

The field of the uniformly moving charge is longitudinal; hence, it penetrates to the medium at the resonant frequency with much smaller losses than the transverse field. This means that the longitudinal field decreases due to interaction with the medium at the distances much larger than the absorption length of the resonant transverse field, $1/g$. However, diffraction radiation is formed in the surface layer of the thickness $1/g$. Therefore, diffraction radiation can be calculated disregarding a decrease in the self field of the fast particle that is attributed to scattering from the molecules of the medium.

The Fourier transform of the field of a charge whose motion is described by the law $x = a$, $y = 0$, and $z = vt$ can be represented in the form

$$\mathbf{E}_0(\mathbf{q},\omega) = \mathbf{E}_0(\mathbf{q})\,\delta(\omega - q_z v);$$
$$\mathbf{E}_0(q) = -\frac{ie}{2\pi^2}\frac{\omega\mathbf{v} - \mathbf{q}c^2}{q^2c^2 - \omega^2}\exp(-iq_x a). \tag{5.12}$$

The integral with respect to p_x in Eq. (5.11) has the form ($\mathbf{e}_x$ and $\mathbf{e}_y$ are the unit vectors of the x and y axes, respectively)

$$\int_{-\infty}^{\infty} dp_x \mathbf{E}_0\left(p_x, k_y, \frac{\omega}{v}\right)\exp\{-ip_x(X_a - a)\} = \frac{ie}{2\pi Q}\mathbf{L}\exp\{-Qa + QX_h\}, \tag{5.13}$$

where

$$\mathbf{L} = \mathbf{e}_y k_y - iQ\mathbf{e}_x - \mathbf{v}\frac{\omega}{v^2\gamma^2},\quad Q = \sqrt{k_y^2 + \frac{\omega^2}{c^2\beta^2\gamma^2}},\quad \beta = \frac{v}{c}. \tag{5.14}$$

Thus, after integration, Eq. (5.11) takes the form

$$\frac{d^2W(\mathbf{n},\omega)}{d\omega\, d\Omega} = e^2\frac{\omega^2}{Q^2c}\exp\left(-2aQ\right)\left|n_0\alpha\left(\omega\right)\left[\mathbf{kL}\left(\mathbf{k}\right)\right]\right|^2\times$$
$$\times\left|\frac{1}{Q-ik_x}\int_{-\infty}^{\infty}dZ\exp\left\{-i\left(k_z-\omega/v\right)Z\right\}\exp\left\{\left(Q+ik_x\right)\varsigma\left(Z\right)\right\}\right|^2 \quad (5.15)$$

If the wavelength is much larger than the thickness of the surface inhomogeneity layer, $k_x\varsigma \ll 1$, $Q\varsigma \ll 1$ and Eq. (5.15) can be simplified to the form

$$\frac{d^2W(\mathbf{n},\omega)}{d\omega\, d\Omega} = e^2\frac{\omega^2}{Q^2c}\exp\left(-2aQ\right)\left|n_0\alpha\left(\omega\right)\left[\mathbf{kL}\left(\mathbf{k}\right)\right]\right|^2\times$$
$$\times\left|\left(Q-ik_x\right)\int_{-\infty}^{\infty}dZ\varsigma\left(Z\right)\exp\left\{-i\left(k_z-\omega/v\right)Z\right\}\right|^2 \quad (5.16)$$

The distribution of radiation in this case is determined by the Fourier transform of the surface profile $\varsigma\left(k_z-\omega/v\right)$.

5.2 Diffraction Radiation at the Resonant Frequency from a Wedge

Let the uniform motion of a charged particle in the $x = a$ plane in vacuum be specified as $\mathbf{r} = \mathbf{a} + \mathbf{v}t$. Taking the z axis along $\mathbf{v}$ and the x axis along a, we can represent the field generated in this motion in the form of Eq. (5.12).

We analyze diffraction radiation generated by the particle moving near a homogeneous dielectric wedge containing n_0 molecules per unit volume whose surface is given by the relation $(\xi, \eta > 0)$

$$\begin{cases} x = \xi z, & \text{for } z < 0; \\ x = -\eta z, & \text{for } z > 0. \end{cases} \quad (5.17)$$

As mentioned above, the Fourier component of the frequency ω of the field of the uniformly moving charge in the medium decreases exponentially in the x direction perpendicular to the velocity as $\exp\left(-\kappa x\right)$, where $\kappa \simeq \left(\omega/v\right)\sqrt{1-\left(v/c\right)^2\varepsilon\left(\omega\right)}$. Therefore, diffraction radiation is primarily formed in a small part near the edge of the wedge close to the trajectory of the particle. The width of the wedge at this distance from the edge is on the order of $\kappa^{-1}\left(\eta^{-1}+\xi^{-1}\right)$.

Let us take into account that the amplitude of the resonant transverse wave formed in the process of the scattering of the fast-particle field from a molecule of the wedge decreases in the process of motion owing to the absorption in the wedge medium. Hence, diffraction radiation from the wedge at the resonant frequency is formed near the edge of the wedge in the thin surface layer of the thickness on the order of $1/g$.

Let a transverse resonant wave appear in the process of the scattering of the particle field from a molecule located at the point $\mathbf{R}_a$. The observation point direction is specified by the unit vector $\mathbf{n}$. This direction intersects the wedge surface at the point $\mathbf{R}_a + \mathbf{R}'\left(X', Y', Z'\right)$, where $X'/Z' = n_x/n_z$ and $Y'/Z' = n_y/n_z$. The radiation path length to the emission point is equal to $R' = \left|Z'/n_z\right|$. Radiation is emitted from a certain edge of the wedge, depending on the emission direction and the position of the molecule. Forward radiation from the molecule with $Z_a > 0$ is emitted through the right edge of the wedge at $n_z > 0$, whereas forward radiation from the molecule with $Z_a < 0$ is emitted through the right edge only under the condition $n_z > n_x\,(Z_a/X_a)$. Considering forward radiation from the width wedge, we can neglect the contribution to radiation from the molecules with $Z_a < 0$, except the case where η vanishes, because diffraction radiation from the molecules with $Z_a < 0$ also vanishes.

Relation (5.1) for the coordinates of the radiation emission point from the wedge has the form

$$X_a + X' = -\eta\left(Z_a + Z'\right); \tag{5.18}$$

therefore,

$$Z' = -\frac{X_a + \eta Z_a}{\eta + n_x/n_z}. \tag{5.19}$$

Similarly, considering radiation from a wide wedge, one can neglect the contribution from the molecule with $Z_a > 0$. Backward radiation at $n_z < 0$ is emitted to vacuum through the $x = \xi z$ wedge plane

$$Z' = \frac{X_a - \xi Z_a}{\xi - n_x/n_z}. \tag{5.20}$$

A decrease in the resonant wave amplitude in the path inside the medium can be taken into account by multiplying the intermolecular current density by $e^{-g\left|\frac{Z'(\mathbf{R}_a)}{2n_z}\right|}$, where g is the absorption coefficient. The absorption of the resonant transverse wave inside the wedge can be taken into account by changing the ordinary expression for the Fourier transform of the polarization current density in the single scattering approximation,

$$\mathbf{j}\left(\mathbf{q}, \omega\right) = -i\frac{\omega}{(2\pi)^3}\alpha\left(\omega\right)\sum_a \int d^3p\mathbf{E}_0\left(\mathbf{p}, \omega\right)\exp\left\{-i\left(\mathbf{p} - \mathbf{q}\right)\mathbf{R}_a\right\}, \tag{5.21}$$

to the expression

$$\mathbf{j}(\mathbf{q}, \omega) = -i\frac{\omega}{(2\pi)^3}\alpha(\omega)\sum_h \int d^3p\mathbf{E}_0(\mathbf{p}, \omega)\exp\left\{-i(\mathbf{p}-\mathbf{q})\mathbf{R}_a - g\left|Z'(\mathbf{R}_a)/2n_z\right|\right\}. \tag{5.22}$$

Here, $\alpha(\omega)$ is the polarizability of the molecule, $\mathbf{R}_a$ is the radius vector of the center of mass of the molecule, and the summation is performed over all the molecules of the medium.

Assuming that the wedge medium is homogeneous and that the wavelength is much larger than the intermolecular distances, we can neglect the effect of polarization current fluctuations and change expression (5.22) to its value averaged over the coordinates of the molecules. For forward radiation

$$\mathbf{j}(\mathbf{q}, \omega) = -i\frac{\omega\alpha(\omega)}{(2\pi)^3}\int d^3p\mathbf{E}_0(\mathbf{p}, \omega)\left\langle\sum_a \exp\left\{-i(\mathbf{p}-\mathbf{q})\mathbf{R}_a - g\left|Z'(\mathbf{R}_a)/2n_z\right|\right\}\right\rangle$$
$$= -i\frac{\omega n_0\alpha(\omega)}{(2\pi)^3}\int_{-\infty}^{0} dX\int d^3p\mathbf{E}_0(\mathbf{p}, \omega)\int_{\frac{X}{\xi}}^{\frac{-X}{\eta}} dz\exp\left\{-i(\mathbf{p}-\mathbf{q})\mathbf{R} - g\left|\frac{Z'(\mathbf{R}_a)}{2n_z}\right|\right\} \tag{5.23}$$

In the microscopic consideration of diffraction radiation as generated due to the scattering of the self field of the particle from the atoms of the medium, the field of the charge moving in vacuum with velocity $\mathbf{v}$ that is given by expression (5.12) should be taken as the field $\mathbf{E}_0(\mathbf{p}, \omega)$.

According to Eqs. (5.19) and (5.20), $Z'(\mathbf{R}_a)$ is independent of Y_h; hence, the integration with respect to Y in Eq. (5.23) provides the delta function $\delta\left(p_y - k_y\right)$. The integration with respect to p_y and p_z is performed with the use of the delta functions $\delta(\omega - p_z v)$ and $\delta\left(p_y - k_y\right)$. After that, the integration with respect to p_x reduces to integral (5.13) ($\mathbf{e}_x$ and $\mathbf{e}_y$ are the unit vectors of the x and y axes, respectively). As a result, the Fourier transform of the polarization current density is expressed as

$$\mathbf{j}(\mathbf{k}, \omega) = \frac{e}{v}\frac{\omega}{(2\pi)^3}n_0\alpha(\omega)\frac{\mathbf{L}}{Q}\int_{-\infty}^{0} dX\exp\{(ik_x + Q)X - aQ\}\times$$
$$\times\int_{X/\xi}^{-X/\eta} dZ\exp\left\{i(k_z - \omega/v)Z - g\left|Z'(\mathbf{R})/2n_z\right|\right\} \tag{5.24}$$

For the forward radiation,

$$g\left|Z'(\mathbf{R})/2n_z\right| = \frac{g}{2}\frac{|X_a + \eta Z_a|}{n_x + \eta n_z} \equiv h\,|X_a + \eta Z_a|, \tag{5.25}$$

we have

$$\mathbf{j}(\mathbf{k},\omega) = \frac{e}{v}\frac{\omega}{(2\pi)^3} n_0\alpha(\omega)\frac{\mathbf{L}}{Q}\exp(-aQ) \times$$
$$\times \frac{(\eta+\xi)\left\{h+i\left(k_z-\frac{\omega}{v}\right)\right\}}{\left[h\eta+i\left(k_z-\frac{\omega}{v}\right)\right]\left[Q\eta+i\eta k_x-i\left(k_z-\frac{\omega}{v}\right)\right]\left[Q\xi+h(\xi+\eta)+ik_x\xi+i\left(k_z-\frac{\omega}{v}\right)\right]} \tag{5.26}$$

The energy of radiation emitted from the medium during the entire time of the passage of the particle in the frequency interval $d\omega$ near the resonant frequency to the solid-angle element $d\Omega$ in the direction of the vector $\mathbf{k}$ is obtained similarly to Eq. (5.5)

$$\frac{d^2W(\mathbf{n},\omega)}{d\omega\, d\Omega} = \frac{e^2}{c}\frac{\omega^2}{Q^2v^2}\left|n_0\alpha(\omega)\left[\mathbf{kL}(\mathbf{k})\right]\right|^2\exp(-2aQ)\frac{1}{\left[h^2\eta^2+(k_z-\omega/v)^2\right]}\times$$
$$\times \frac{(\eta+\xi)^2\left\{h^2+(k_z-\omega/v)^2\right\}}{\left[Q^2\eta^2+(\eta k_x-k_z+\omega/v)^2\right]\left[\left[Q\xi+h(\xi+\eta)\right]^2+(k_x\xi+k_z-\omega/v)^2\right]} \tag{5.27}$$

The dependence of the emitted energy on the distance a between the particle trajectory and the edge of the wedge is given by the exponential

$$\exp(-2aQ) = \exp\left(-2a\sqrt{k_y^2+\frac{\omega^2}{v^2\gamma^2}}\right), \tag{5.28}$$

which decreases with an increase in the distance in the nonrelativistic case much faster that in the ultrarelativistic case and this behavior is similar to the behavior of the self field of the fast particle. This exponential is also substantial for the dependence of the radiation intensity on the azimuth angle φ. In the nonrelativistic case, $Q \simeq \omega/v \gg k_y$ and the exponential $\exp(-2aQ)$ is on the same order of magnitude for various φ values. In the ultrarelativistic case, $Q \simeq \omega/\gamma v$ for very low k_y values and $Q \simeq k_y \gg \omega/\gamma v$ for k_y values that are not low. For this reason, the exponential $\exp(-2aQ)$ in the ultrarelativistic case strongly suppresses radiation at finite k_y values. As a result, when ultrarelativistic particle moves perpendicularly to the edge of the wedge, radiation is almost entirely concentrated near the xz plane.

As known, radiation generated by the ultrarelativistic particle is concentrated in the region of small angles near the direction of the particle velocity. Therefore, $k_x^2+k_y^2 \ll k_z^2$, $Q \sim k/\gamma$, and the difference $k_z-\omega/v \simeq k/\gamma^2$ in the ultrarelativistic case is very small. Hence, the denominator in Eq. (5.27) is small when η and ξ are smaller than or on the order of γ^{-1}. However, if $\eta \ll \gamma^{-1}$ and $\xi \ll \gamma^{-1}$, the numerator in Eq. (5.27) becomes very small. Therefore, the maximum of the Fourier transform of the current density, together with the maximum radiation intensity as a function of the wedge angle, corresponds to the values $k\eta \simeq \omega/v - k_z$ and $k\xi \simeq \omega/v - k_z$.

If the wedge is narrow, i.e., $\xi \gg 1$, $\eta \gg 1$, and $\kappa/g \equiv 2kh\,(\eta n_z + n_x) > \eta^{-1} + \xi^{-1}$, the formation of diffraction radiation occurs in the region near the edge over the entire width of the wedge. Since the absorption length in this case is larger than the sizes of the radiation formation region, the inclusion of the absorption of resonant waves is not necessary. In this case, the difference of resonant waves from nonresonant ones disappears, so that the intensity of diffraction radiation for all the frequencies can be obtained by microscopically considering the generation of radiation as a result of the scattering of the self field of the fast particle. In this case, Eq. (5.27) takes the form

$$\begin{aligned}\frac{d^2W(\mathbf{n},\omega)}{d\omega\, d\Omega} &= \frac{e^2}{c}\frac{\omega^2}{Q^2v^2}\left|n_0\alpha(\omega)\left[\mathbf{kL}(\mathbf{k})\right]\right|^2 \exp(-2aQ)\times \\ &\times\frac{(\eta+\xi)^2}{\left[Q^2\eta^2+(\eta k_x-k_z+\omega/v)^2\right]\left[Q^2\xi^2+(k_x\xi+k_z-\omega/v)^2\right]}\end{aligned} \tag{5.29}$$

For a wide wedge, when the quantities ξ and η are smaller than or about one, the formation of radiation occurs in the surface layers of each edge of the wedge. In this case, it is necessary to take into account a decrease in the amplitude of the resonant wave due to the absorption in the wedge medium and the spectral–angular distribution of diffraction radiation from the wedge has the form of Eq. (5.27).

It is interesting to compare radiation for the limiting cases $\xi = 0$ (the first edge of the wedge is parallel to the particle velocity) and $\xi = \infty$ (the first edge of the wedge is perpendicular to the particle velocity).

For the case $\xi = 0$, the distribution of the emitted energy given by Eq. (5.27) is written in the form

$$\begin{aligned}\frac{d^2W(\mathbf{n},\omega)}{d\omega d\Omega} &= \frac{e^2}{c}\frac{\omega^2}{Q^2v^2}\left|n_0\alpha(\omega)\left[\mathbf{kL}(\mathbf{k})\right]\right|^2 \exp(-2aQ)\times \\ &\times\frac{(\eta+\xi)^2\left\{h^2+(k_z-\omega/v)^2\right\}}{\left[h^2\eta^2+(k_z-\omega/v)^2\right]\left[Q^2\eta^2+(\eta k_x-k_z+\omega/v)^2\right]}\end{aligned} \tag{5.30}$$

In the limiting case $\xi = \infty$, the distribution of radiation specified by Eq. (5.27) has the form

$$\begin{aligned}\frac{d^2W(\mathbf{n},\omega)}{d\omega\, d\Omega} &= \frac{e^2}{c}\frac{\omega^2}{Q^2v^2}\left|n_0\alpha(\omega)\left[\mathbf{kL}(\mathbf{k})\right]\right|^2 \exp(-2aQ)\times \\ &\times\frac{h^2+(k_z-\omega/v)^2}{\left[h^2\eta^2+(k_z-\omega/v)^2\right]\left[Q^2\eta^2+(\eta k_x-k_z+\omega/v)^2\right]\left[(Q+h)^2+k_x^2\right]}\end{aligned} \tag{5.31}$$

The comparison of Eqs. (5.30) and (5.31) shows that the particle energy dependence of the diffraction radiation intensity at $h\eta < \omega/v - k_z \sim k/\gamma^2$ is stronger when the first edge of the wedge is parallel to the particle velocity.

References

1. Fermi, E.: On the Reflection and Diffusion of Resonance Radiation. Collection of Papers. Moscow (in Russian) (1971)
2. Bolotovskiy, B.M., Voskresenskiy, G.V.: Diffraction Radiation. Phys.-Uspekhi **9**, 73 (1966)
3. Shestopalov, V.P.: Diffraction Electronics. Kharkov, Ukraine (1976)
4. Bolotovskiy, B.M., Galst'yan, E.A.: Diffraction and Diffraction Radiation. Phys.-Uspekhi **43**, 755 (2000)
5. Potylitsyn, P.: Transition Radiation and Diffraction Radiation. Similarities and Differences. Nucl. Instrum. Methods Phys. Res. B **145**, 169 (1998)
6. Fermi E.: On the Reflection and Diffusion of Resonance Radiation. Rend. Lincei. **33**(1), 90 (1924)
7. Ryazanov, M.I.: Diffraction radiation of a fast particle at resonant frequency. Sov. Phys. JETP **127**, 528 (2005)

Chapter 6
Diffraction Radiation from Media with Periodic Surfaces

6.1 Smith—Purcell Radiation

A charge moving in vacuum with a constant velocity near a perfectly conducting periodically deformed target (grating, see Fig. 6.1) induces on its surface a time-varying charge and current, which is a cause of the appearance of diffraction radiation.

Considering the charge moving along a rectilinear trajectory with velocity βc and its image moving along a periodic trajectory with period d (see Fig. 6.1) as an electric dipole with eigenfrequency $\Omega = 2\pi\beta c/d$, one can obtain an expression for the radiation frequency $\omega(\theta)$ at a fixed angle θ on the basis of the Doppler formula. Frank [1] pointed to the possibility of existence of such an effect and presented a formula for frequency $\omega(\theta)$ (see [1, expression (3.7)]). In the absence of refraction, Frank's formula reduces to the expression

$$\omega(\theta) = \frac{\Omega}{1 - \beta\cos\theta} = \frac{2\pi\beta c}{d(1 - \beta\cos\theta)} \tag{6.1}$$

or, after the passage to the wavelength,

$$\lambda = d\left(\frac{1}{\beta} - \cos\theta\right). \tag{6.2}$$

In 1953, Smith and Purcell observed for the first time monochromatic electromagnetic radiation in the visible range in transmitting an electron beam with an energy of $\sim$ 300 keV near a periodic target (standard optical grating) with period $d = 1.67\,\mu$m [2]. The electron beam with a diameter of about 150 μm passed tightly to an optical grating 48 mm in length. Radiation was detected at an angle of $\theta \simeq 20°$ with the use of the photographic method in a simple optical scheme (with a collimating lens and an analyzing grating). For 309 keV electrons, a bright line with wavelength $\lambda \simeq 0.56\,\mu$m is observed in the radiation spectrum.

Smith and Purcell derived a formula for the wavelength of the line in the radiation spectrum, using vacuum relation for wave vector and obvious phase relations (see Fig. 6.2)

A.P. Potylitsyn et al., *Diffraction Radiation from Relativistic Particles*, STMP 239, 149–195, DOI 10.1007/978-3-642-12513-3_6,

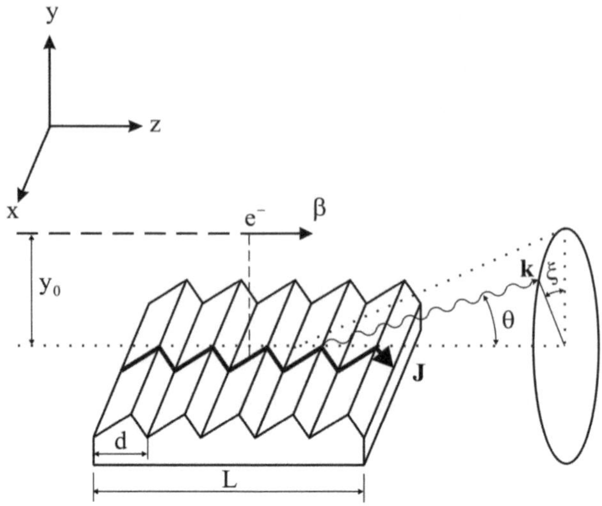

Fig. 6.1 Geometry of generating Smith—Purcell radiation

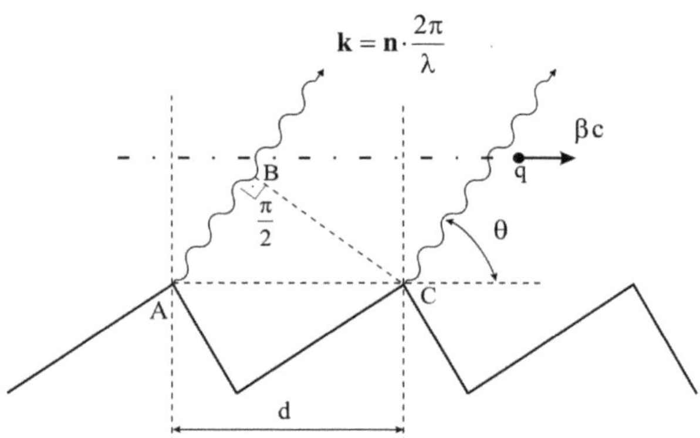

Fig. 6.2 Illustration of the derivation of the Smith—Purcell relation

$$\mathbf{k} = \mathbf{n}\frac{2\pi}{\lambda}. \tag{6.3}$$

Electromagnetic radiation is generated by the field of the passing particle in each element of the grating. For the sake of simplicity, we consider plane waves emitted at angle θ from two successive rulings of the grating, which are induced by the same charged particle passing in parallel to the grating with velocity βc. These waves have the same phase (i.e., interfere constructively) if the time in which the plane wave emitted from point A (see Fig. 6.2) reaches point B ($t_1 = \frac{d\cos\theta}{c}$) is related to the time necessary for a particle to pass distance d and to excite a plane wave at point C ($t_2 = \frac{d}{\beta c}$) as

$$c\,(t_1 - t_2) = n\lambda, \quad n = \pm 1, \pm 2, \ldots, \tag{6.4}$$

where λ is the wavelength. In other words, the phase shift of waves with frequency ω that are emitted from points A and C should satisfy the relation

$$\Delta\varphi = k\Delta z - \omega\Delta t = 2\pi n, \quad n = \pm 1, \pm 2, \ldots \tag{6.5}$$

Here, $k = 2\pi/\lambda$ is the absolute value of the wave vector and $\omega = 2\pi c/\lambda$. From this relation, we obtain the so-called "dispersion" relation

$$\lambda_n = \frac{d}{n}\left(\cos\theta - \frac{1}{\beta}\right), \tag{6.6}$$

where n is the diffraction order.

The latter relation for $n = -1$ ("fundamental" harmonic) was written by Smith and Purcell in the form of the formula

$$\lambda = d\left(\frac{1}{\beta} - \cos\theta\right), \tag{6.7}$$

which coincides with the formula obtained by Frank from other consideration.

Formula (6.7) for the wavelength of the fundamental harmonic provides the value $\lambda = 0.566\,\text{mm}$ for $\beta = 0.782$ ($E = 309\,\text{keV}$) in good agreement with the experimental data. When the electron energy is increased to $E = 340\,\text{keV}$, a shift toward lower wavelengths is observed in the experiment. The experimental data reported in [2] are well described by formula (6.7).

After this first observation of monochromatic radiation generated in uniform rectilinear motion of the charge in vacuum near the periodic target (periodic grating), this phenomenon is called the *Smith—Purcell effect*.

In one of the last experiments, Izhizuka et al. [3] investigated in detail Smith—Purcell radiation that is generated in the optical range by an electron beam with kinetic energy $E < 60\,\text{keV}$ and current $I < 10\,\text{mA}$. The beam diameter was smaller than $200\,\mu\text{m}$ and radiation was detected at angle $\theta = 80°$ by a photomultiplier. The radiation spectrum was analyzed with the use of a monochromator with a resolution (FWHM) of $\Delta\lambda = 20\,\text{nm}$.

Figure 6.3 shows the measured spectrum of Smith—Purcell radiation from the optical grating with a triangle profile and period $d = 0.556\,\text{mm}$. According to the presented results, the radiation orders up to $n = -5$ are experimentally observed. Figure 6.4 shows the dependence of the position of spectral lines of various orders on the electron energy. The solid lines are the results of calculation by formula (6.6).

For nonrelativistic Smith—Purcell effect it is easily to receive from (6.6) the relation for the wavelength shift depending on electron kinetic energy E_k:

$$\frac{\Delta\lambda_n}{\lambda_n} = -\frac{1}{(1 - \beta\cos\Theta)}\frac{\Delta E_k}{2E_k} \tag{6.8}$$

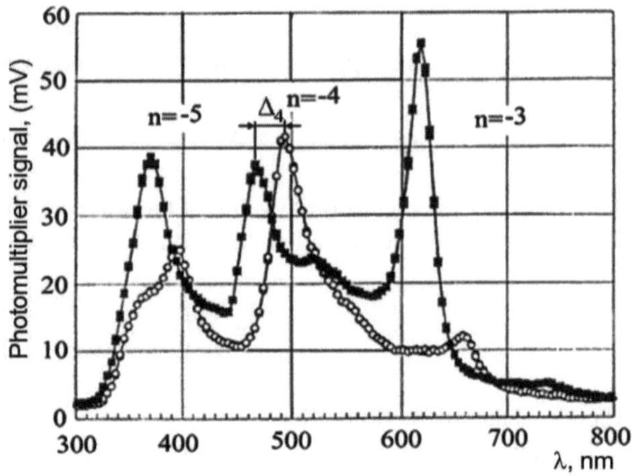

Fig. 6.3 Spectrum of Smith—Purcell radiation generated by an electron beam with an energy of (*open circles*) 20 and (*rectangles*) 22.5 keV

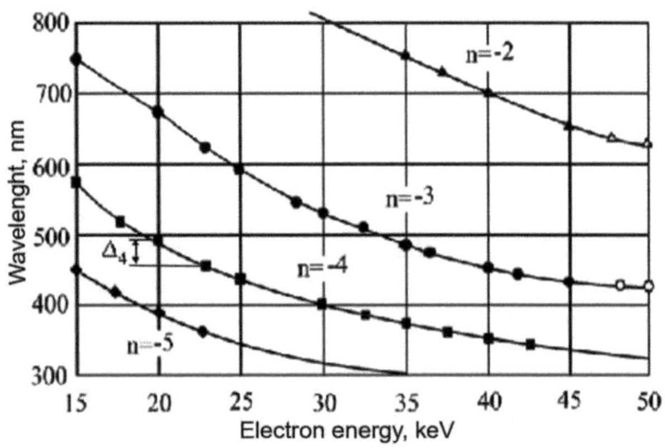

Fig. 6.4 (*Solid lines*) Calculations of the Smith—Purcell radiation wavelength as compared to the experimental points

For the geometry in the experiment [3] ($\theta = 80°$) the last relation may be written as

$$\frac{\Delta\lambda_n}{\lambda_n} = \frac{\Delta E_k}{2E_k} \tag{6.9}$$

which is in good agreement with experimental results (see, for instance, the spectral line shift for $n = -4$ at Figs. 6.3 and 6.4)

Until to the 1990s, the basic direction of both theoretical and experimental investigations of Smith—Purcell radiation was associated with radiation generated by nonrelativistic electrons.

Numerous experimental results concerning basically the centimeter wavelength range, as well as the results of numerical methods of the calculation of the characteristics of radiation in the same range, were presented in [4–6]. The creation of an orotron that is a high-power source of monochromatic radiation in the millimeter and centimeter ranges [7] and generators of diffraction radiation [8] with the use of intense low-energy electron beams (see also [9]) is among noticeable applications of the Smith—Purcell effect.

In the last decade, interest in studying Smith—Purcell radiation increased noticeably in view of the possibility of using it for diagnostics of relativistic electron beams [10], as well as in view of the possibility of creating a free-electron laser based on the Smith—Purcell effect [11, 12]. In this chapter, we describe approaches for calculating the characteristics of Smith—Purcell radiation in the single-particle approximation, i.e., for a moving point charge. In contrast to the models presented in [5, 6], such an approach makes it possible to naturally includes the characteristics of the electron beam (e.g., divergence) in the calculation of the spectral—angular distribution of Smith—Purcell radiation in application to the problems of diagnostics or the time structure of a modulated electron beam if the Smith—Purcell effect is planned to be used to create the free-electron laser.

6.2 Scalar Theory of the Diffraction of the Self Field of an Electron from a Plane Semitransparent Grating

One of the first models for calculating the spectral—angular distribution of Smith—Purcell radiation was proposed in [13] (see also review [4]). That model involved a grating with period d and number of elements $N \gg 1$ made of perfectly conducting infinitely thin strips of width a (see Fig. 6.5) located in plane. Charge q moves over the grating at distance h with velocity $v = \beta c$. The characteristics of radiation are calculated in the half-space below the grating (for this reason, the authors called such a grating plane semitransparent grating plane). The authors assumed that, according to the Helmholtz—Kirchhoff theorem, the field in the half-space below the grating is determined by its values in the gaps (in the grating plane):

$$g\left(x,y,z\right)=\int\frac{kg_0\left(x',0,z'\right)}{2\pi iR}e^{ikR}dS_n. \tag{6.10}$$

Here, $k = 2\pi/\lambda$, g_0 and g are the field components describing radiation with frequency $\omega = 2\pi c/\lambda$ from the plane grating at distance R below the grating, and dS_n is the projection of a grating area element on the direction of the wave vector. In this approximation, the effect of grating itself on the field characteristics in gaps is neglected (Kirchhoff approximation).

The lines of the real field of the charge near the perfectly conducting surface are shown in Fig. 6.6b. Note that the degree of field distortion by the conductor is small in the region near the perpendicular on the surface passing through the charge location point. In other words, for the charge at rest in the region $\rho \ll h$ (where h is the distance from the charge to the surface, see Fig. 6.6a), the electric field is almost "radial" and coincides with the vacuum distribution of the field of the initial charge. Let the charge moves with velocity $v = \beta c$ along the z axis. In this case, following [4], its field in the grating plane can be described by the scalar function

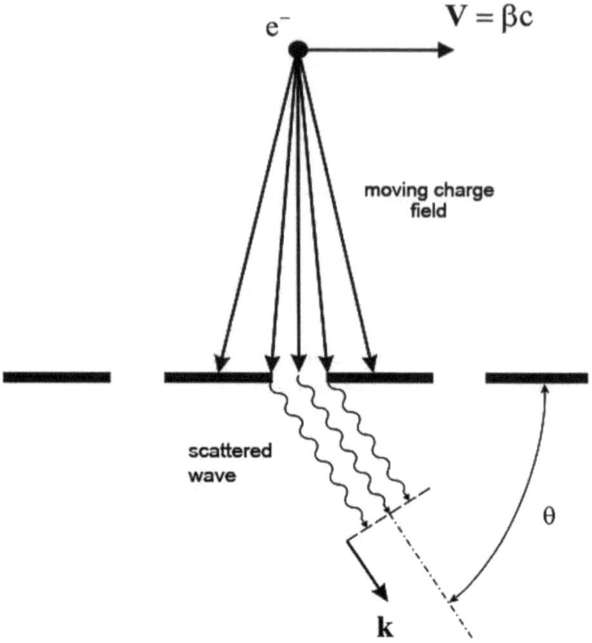

Fig. 6.5 Scheme of the generation of Smith—Purcell radiation on the plane semitransparent grating

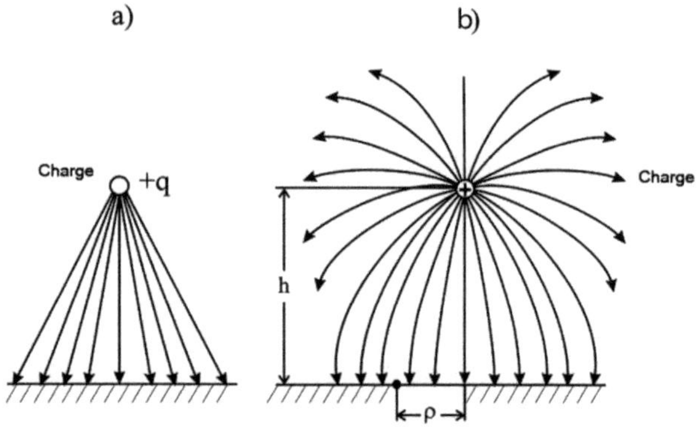

Fig. 6.6 Lines of the electric field of the charge near the conducting plane (**a**) in disregard of the effect of the conducting plane and (**b**) for the real distribution of the electric field

$$g_0(x, 0, z) = \frac{iq}{2\pi c} \int \frac{dk_x}{\sigma(k_x, \omega)} e^{i(\sigma h + k_x \kappa + z\omega/v)},$$
$$\sigma(k_x, \omega) = i\sqrt{\frac{\omega^2}{\gamma^2 v^2} + k_x^2}. \tag{6.11}$$

After the calculation of integral (6.10) with allowance for the periodicity of the grating, the absolute value squared of the function $g(x, y, z)$ provides the spectral—angular distribution of Smith—Purcell radiation in the form

$$\frac{d^2W}{d\omega d\Omega} = \frac{q^2(d-a)^2}{2\pi c}\frac{vT}{d}\frac{\sin^2\left[\frac{d-a}{2}\frac{\omega}{v}(1-\beta\cos\theta)\right]}{\left[\frac{d-a}{2}\frac{\omega}{v}(1-\beta\cos\theta)\right]^2}\times$$
$$\times\sin^2\theta\exp\left\{-2h\frac{\omega}{v}\sqrt{1-\beta^2\left(1-\sin^2\theta\sin^2\xi\right)}\right\} \tag{6.12}$$
$$\sum_n \delta\left[\frac{\omega}{v}(1-\beta\cos\theta)d - 2\pi n\right].$$

Here, $T = N\frac{d}{\beta c}$ is the time of flight of the charge over the grating and n is the diffraction order. In order to obtain the angular distribution of the radiation energy for the nth diffraction order, it is necessary to integrate the last expression with respect to the frequency. This integration is easily performed with the use of the δ function:

$$\frac{dW_n(\theta,\xi)}{d\Omega} = \frac{2\pi q^2(d-a)^2}{c^3}N\left(\frac{v}{d}\right)\frac{\sin^2\left(n\pi\frac{d-a}{d}\right)}{\left(\pi\frac{d-a}{d}\right)^2}\times$$
$$\times\frac{\sin^2\theta}{(1-\beta\cos\theta)^3}\exp\left\{-\frac{4\pi nh}{d(1-\beta\cos\theta)}\sqrt{\gamma^{-2}+\beta^2\sin^2\theta\sin^2\xi}\right\} \tag{6.13}$$

According to the resulting formula for the first diffraction order (fundamental harmonic), the radiation yield is maximal for a strip width of $a = d/2$. In this case ($n = 1$), formula (6.13) for radiation generated by the electron ($q^2 = e^2 = \alpha\hbar c$) is written in the form

$$\frac{dW_1(\theta,\xi)}{d\Omega} = \frac{2\alpha}{\pi}\beta^3\frac{\hbar c}{d}N\frac{\sin^2\theta}{(1-\beta\cos\theta)^3}\times$$
$$\times\exp\left\{-\frac{4\pi h}{\gamma d(1-\beta\cos\theta)}\sqrt{1+\gamma^2\beta^2\sin^2\theta\sin^2\xi}\right\}. \tag{6.14}$$

Thus, the simple obvious model made it possible to obtain an analytical formula providing important physical consequences. The exponential can be written in the form

$$\exp\left\{-\frac{4\pi h}{\gamma\beta\lambda}\sqrt{1+\gamma^2\beta^2\sin^2\theta\sin^2\xi}\right\} \tag{6.15}$$

providing the following stringent constraint on the effective impact parameter:

$$h \le h_{\text{eff}} \simeq \frac{\gamma\beta\lambda}{4\pi}. \tag{6.16}$$

It is obviously almost impossible to satisfy condition (6.16) in the nonrelativistic case ($\gamma\sim 1$, $\beta\ll 1$) in the optical range ($\lambda\le 1\,\mu$m). As a rule, in such experiments, $h\gg h_{\text{eff}}$, which leads to the exponential suppression of the yield of Smith—Purcell radiation. As the energy of the electrons increases (with an increase in the Lorentz factor) or as the wavelength increases from the optical range to the infrared (or submillimeter) range, condition (6.16) becomes less stringent.

If the impact parameter satisfies the condition $h \leq h_{\text{eff}}$, Smith—Purcell radiation in the relativistic case is concentrated in a plane perpendicular to the grating, i.e., in the azimuth angle range $\xi \leq \gamma^{-1}$, because the exponential suppression of the yield is again observed for large azimuth angles (owing to an increase in the radicand in the argument of the exponential).

In the last years, more real models were proposed for describing the Smith—Purcell effect in the relativistic case [14, 15]; however, most results obtained (including those obtained by numerical methods) are in qualitative agreement with the above conclusions based on the formula obtained by B.M. Bolotovskii and A.K. Burtsev more than four decades ago.

6.3 Smith—Purcell Effect As Radiation Generated by Induced Surface Currents

One of the simple models describing the Smith—Purcell effect is the surface current model [15]. In this model, the charge uniformly moving near the periodically deformed surface (grating) induces a current varying in space and time on this surface, which generates Smith—Purcell radiation (see Fig. 6.1).

According to the formulas presented in Sect. 1.1, the spectral—angular distribution of radiation is expressed in terms of the induced current density $\mathbf{J}(\mathbf{r}, t)$ as

$$\frac{d^2W}{d\omega d\Omega} = \frac{\omega^2}{4\pi^2 c^3}\left|\int dt \int d^3r \left[\mathbf{n}\left[\mathbf{nJ}(\mathbf{r}, t)\right] e^{i(\omega t - \mathbf{kr})}\right]\right|^2, \tag{6.17}$$

where $\mathbf{n} = \{\sin\theta\sin\xi,\ \sin\theta\cos\xi,\ \cos\theta\}$ is the unit vector in the photon emission direction and $\mathbf{k} = \mathbf{n}\omega/c$ is the photon momentum. Since the current $\mathbf{J}(\mathbf{r}, t)$ induced on periodically located elements of the grating is a periodic function of the variables z and t,

$$\mathbf{J}(\mathbf{r}, t) = \sum_{m=1}^{N} \mathbf{J}_0\left(x, y, z - md,\ t - \frac{md}{v}\right), \tag{6.18}$$

integral (6.17) reduces to the expression

$$\frac{d^2W}{d\omega d\Omega} = \frac{\omega^2}{4\pi^2 c^3}\left|\sum_{m=1}^{N} e^{imd\omega\left(\frac{1}{v} - \frac{n_z}{c}\right)}\right|^2 \left|\int dt \int dxdydz \mathbf{J}_0(\mathbf{r}, t)\, e^{i(\omega t - \mathbf{kr})}\right|^2. \tag{6.19}$$

In Eqs. (6.18) and (6.19), subscript 0 of vector $\mathbf{J}$ means that the integration is performed over the volume of one element and over the time interval corresponding to the passage of the particle through one period of the grating. The sum squared in Eq. (6.19) reduces to the standard expression

$$F_3 = \left| \sum_{m=1}^{N} e^{imd\omega\left(\frac{1}{v} - \frac{n}{c}\right)} \right|^2 = \frac{\sin^2\left[\left(\frac{1}{\beta} - \cos\theta\right)\frac{N\omega d}{2c}\right]}{\sin^2\left[\left(\frac{1}{\beta} - \cos\theta\right)\frac{\omega d}{2c}\right]} \approx \sum_{n\neq 0} \frac{N\omega}{|n|}\delta\left(\omega - \omega_n\right). \tag{6.20}$$

The Smith—Purcell dispersion relation is determined from the argument of the δ function:

$$\omega_n = \frac{2\pi\,|n|\,c}{d\left(1/\beta - \cos\theta\right)}, \qquad n - \text{integer}. \tag{6.21}$$

In such a formulation, the problem is solved most simply when the grating is a set of $N \gg 1$ parallel, perfectly conducting strips separated by vacuum gaps and the distortion of the field of the moving charge that is induced by the presence of the conducting surface near the charge is neglected. The same approximation was used by Bolotovskii and Burtsev; however, the model proposed in [15] allows the calculation of the spectral—angular distribution of radiation in the half-space above the grating (see Fig. 6.7), because current **J** is induced on the surface of the strips directed to the passing charge.

The "flattening" of the electric field of the moving charge in the direction of motion increases with the particle energy. The longitudinal, E_z, and transverse, E_y, components of the field of charge q moving with velocity v along the z axis at distance ρ from its trajectory are given by the expressions

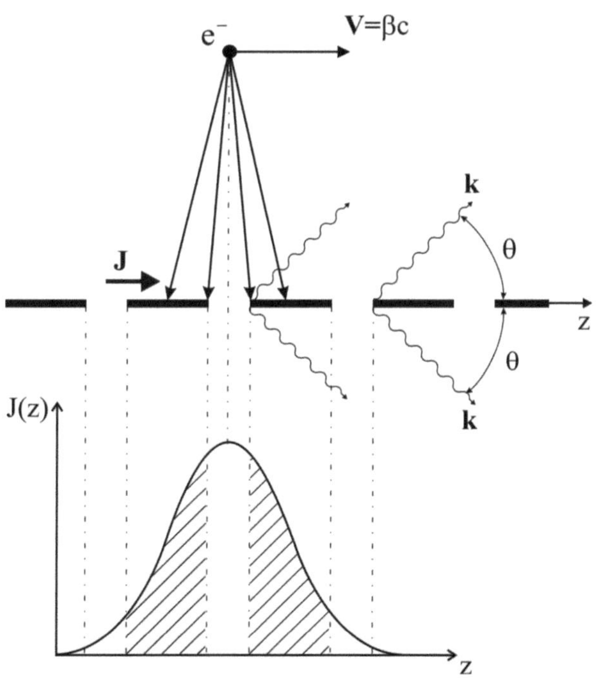

Fig. 6.7 Geometry of Smith—Purcell radiation generated by an induced current on the plane grating

$$E_z(\mathbf{r},t) = \frac{q\gamma vt}{\left(\rho^2+\gamma^2v^2t^2\right)^{3/2}}$$
$$E_y(\mathbf{r},t) = \frac{q\gamma\rho}{\left(\rho^2+\gamma^2v^2t^2\right)^{3/2}}. \tag{6.22}$$

Expressions (6.22) are valid for the field (i) in the laboratory reference frame at the point $\mathbf{r} = \{\boldsymbol{\rho}, 0\}$, where $\boldsymbol{\rho} \equiv \{x, y\}$, and (ii) at the point $\mathbf{r} = \{\boldsymbol{\rho}, vt\}$ with respect to the instantaneous charge position (see, e.g., [16] and [17]).

Atoms and molecules of the medium that are direct sources of polarization radiation are at rest in the laboratory coordinate frame. Therefore, the characteristic values of the time t_{eff} of the action of the field at a given point $\mathbf{r}$ in the laboratory coordinate frame are determined by the denominators of expressions (6.22) and are equal to

$$t_{\text{eff}} \sim \frac{\rho}{\gamma v}. \tag{6.23}$$

Hence, the relation between the field components is determined by the Lorentz factor of the particle:

$$\frac{E_z}{E_y} \sim \gamma^{-1}. \tag{6.24}$$

In other words, the field configuration for ultrarelativistic particles differs only slightly from the transverse field of electromagnetic radiation (for which $E_z = 0$).

Figure 6.8 schematically shows the effective region on the conductor surface on which the surface charge is induced (this figure corresponds to a fixed time instant). It is clear that the effective size of the region in the direction of motion is much smaller than the impact parameter (if $\gamma \gg 1$); i.e., the distortion of the initial charge field can be neglected in this case.

As known, the surface charge density induced in the perfectly conducting plane is proportional to the perpendicular component of the electric field strength:

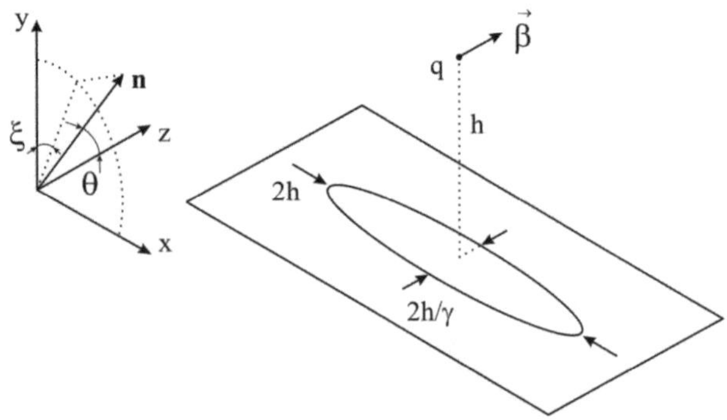

Fig. 6.8 (*Left panel*) Angular variables describing the direction of the wave vector and (*right panel*) the effective region of the target in which the surface charge is induced

$$\sigma(x,y,z) = \text{const}\, E_y(x,y,z) = \frac{\text{const}\, q\gamma\,(y-y_0)\,\delta\,(y-y_1)}{\left[(x-x_0)^2+(y-y_0)^2+\gamma^2\,(z-z_0-vt)^2\right]^{3/2}}. \tag{6.25}$$

The numerical value of the constant depends on the system of units. We take const $=(2\pi)^{-1}$. Here, $\mathbf{r}_0 = \{x_0, y_0, z_0\}$ are the coordinates of the charge at time $t = 0$. The delta function reflects the fact that the conducting plane is parallel to the xz coordinate plane at the distance y_1 from the origin of the coordinates. If the charged particle moves with constant velocity $\mathbf{v} = \{0, 0, c\beta\}$ along the z axis, the induced current is determined by the expression

$$\mathbf{J}_0(\mathbf{r}, t) = \sigma c\beta\hat{\mathbf{z}}\,. \tag{6.26}$$

Here, $\hat{\mathbf{z}}$ is the unit vector of the z axis. The Fourier component of the induced current is determined by the standard expression

$$\begin{aligned}\mathbf{J}_0(\omega,\mathbf{n}) = \frac{q\gamma}{2\pi}\int_{-\infty}^{\infty} dt \int_{z_1}^{z_2} dz \int_{-\infty}^{\infty} dy \int_{-\infty}^{\infty} dx\, e^{i(\omega t-\mathbf{kr})}\times \\ \times\frac{(y-y_o)\,\delta\,(y-y_1)}{\left[(x-x_0)^2+(y-y_0)^2+\gamma^2\,(z-z_0-vt)^2\right]^{3/2}}\,c\beta\hat{\mathbf{z}}\end{aligned} \tag{6.27}$$

Owing to the presence of the delta function, one integral is easily calculated. In order to calculate the remaining integrals, we introduce the new variables

$$\begin{aligned} u &= \gamma\,(vt - z + z_0)\,, \\ \bar{x} &= x_0 - x, \\ h &= (y_0 - y_1)\,, \\ \kappa &= \frac{\omega}{v} - k_z. \end{aligned} \tag{6.28}$$

Thus, it is necessary to calculate the triple integral

$$\begin{aligned}\mathbf{J}_0(\omega,\mathbf{n}) = \hat{\mathbf{z}}\frac{q}{2\pi}\int_{z_1}^{z_2} dz \int_{-\infty}^{\infty} d\bar{x} \int_{-\infty}^{\infty} du \frac{h}{\left[h^2+\bar{x}^2+u^2\right]^{3/2}}\times \\ \times \exp\left\{i\left[\tfrac{\omega u}{\gamma v} - \tfrac{\omega z_0}{v} + k_x\,(\bar{x}-x_0) + \kappa z - k_y y_1\right]\right\}.\end{aligned} \tag{6.29}$$

The integration with respect to z is performed over the strip width ($z_2 - z_1 = a$). The inner double integral (with respect to $d\bar{x}\,du$) reduces to a table integral, which is written in the polar coordinates as

$$I = \int_{-\infty}^{\infty} d\bar{x} \int_{-\infty}^{\infty} du \frac{\exp\left\{i\left[\frac{\omega u}{\gamma v} + k_x\bar{x}\right]\right\}}{\left(h^2+\bar{x}^2+u^2\right)^{3/2}} = \int_0^{2\pi} d\varphi \int_0^{\infty} \rho\, d\rho\, \frac{\exp\{i\mu\rho\cos\varphi\}}{\left(h^2+\rho^2\right)^{3/2}}, \tag{6.30}$$

where $\mu = \sqrt{\left(\frac{\omega}{\gamma v}\right)^2 + k_x^2}$. The azimuthal integral in Eq. (6.30) is expressed in terms of the zeroth order Bessel function and, therefore,

$$I = 2\pi \int_0^\infty \rho\, d\rho \frac{J_0(\mu\rho)}{[h^2+\rho^2]^{3/2}} = \frac{2\pi}{h} \exp\left[-h\sqrt{\left(\frac{\omega}{\gamma v}\right)^2 + k_x^2}\right]. \tag{6.31}$$

For radiation in vacuum, $k_x = \frac{\omega}{c} \sin\theta \sin\xi$ (see Fig. 6.7); thus, the Fourier components of the induced current are expressed as

$$\mathbf{J}_0(\mathbf{n},\omega) = \hat{\mathbf{z}}\, q \exp\left(-\frac{\omega h}{\gamma\beta c}\sqrt{1+\gamma^2\beta^2\sin^2\theta\sin^2\xi}\right) \exp\left(-i\, k_y y_1\right) \frac{e^{i\kappa z_2} - e^{i\kappa z_1}}{i\kappa}. \tag{6.32}$$

After squaring, instead of expression (6.19), we obtain the following formula for $n = 1$:

$$\begin{aligned} \frac{d^2W_1}{d\omega d\Omega} &= q^2 \frac{\omega^3}{4\pi^2 c^3} N\delta(\omega-\omega_1)\sin^2\theta \frac{\left|e^{i\kappa z_2} - e^{i\kappa z_1}\right|^2}{\kappa^2} \times \\ &\times \exp\left(-\frac{2\omega_1 h}{\gamma\beta c}\sqrt{1+\gamma^2\beta^2\sin^2\theta\sin^2\xi}\right) \end{aligned} \tag{6.33}$$

For strip width a, the angular distribution of Smith—Purcell radiation generated by a single electron ($q = -e$) for $n = 1$ can be obtained from Eq. (6.33) in the form

$$\begin{aligned} \frac{dW_1}{d\Omega} &= \frac{\alpha}{2\pi} N \frac{\hbar c}{d} \frac{\sin^2\theta}{\left(\beta^{-1}-\cos\theta\right)^3} 4\sin^2\left(\frac{\pi a}{d}\right) \times \\ &\times \exp\left(-\frac{4\pi h}{\gamma\beta d\left(\beta^{-1}-\cos\theta\right)}\sqrt{1+\gamma^2\beta^2\sin^2\theta\sin^2\xi}\right) \end{aligned} \tag{6.34}$$

where the relation $\alpha = e^2/\hbar c$ is used.

The resulting expression is symmetric with respect to the change $\xi \rightleftarrows \pi \pm \xi$, i.e., describes the spectral—angular distribution of radiation not only above the grating, but also below it. This circumstance makes it possible to perform comparison with Bolotovskii's model described in the preceding section. Note that the derivation of the resulting formula is based on an approach in which the notion of the induced current on the surface of the plane perfect grating is consistently used.

Both in the model based on the scalar theory of diffraction and in the model described above, the distortion of the field of the conducting grating is disregarded. For this reason, the results obtained completely coincide in the ultrarelativistic case (formula (6.14) with expression (6.34) in the limit $\beta \to 1$). The model of induced currents can be applied not only to the plane grating, but also to the volume grating each element in which can be represented in the form of a set of variously oriented strips. In particular, for the grating consisting of the strips perpendicular to the particle trajectory, Brownell et al. [21] obtained the following expression for the spectral—angular distribution of Smith—Purcell radiation for the kth diffraction order:

$$\frac{d^2 W_n}{d\omega\, d\Omega} = \frac{q^2}{4\pi^2} \frac{\omega^3}{c^3} \frac{N d^2}{|n|} \exp\left(-\frac{2h}{\lambda_e}\right) \delta\left(\omega - \omega_n\right) \left|\left[\mathbf{n}\left[\mathbf{n}\mathbf{G}_n^{\perp}\right]\right]\right|^2, \tag{6.35}$$

where

$$\begin{aligned} \omega_n &= \frac{2\pi c}{\lambda_n}, \\ \mathbf{n} &= \{\sin\theta\, \sin\xi,\ \sin\theta\cos\xi,\ \cos\theta\}, \\ \lambda_e &= \frac{\gamma\beta\lambda_n}{2\pi\sqrt{1+\gamma^2\beta^2\sin^2\theta\sin^2\xi}}. \end{aligned} \tag{6.36}$$

In Eq. (6.35), vector $\mathbf{G}_n^{\perp}$ is defined as follows:

$$\begin{aligned} \mathbf{G}_n^{\perp} &= A\left\{1,\ i\lambda_e \frac{\omega_n}{c} \sin\theta\sin\xi,\ 0\right\}, \\ A &= \frac{\exp\left[\left(-\frac{1}{\lambda_e} - i\frac{\omega_n}{c}\sin\theta\cos\xi\right)a\right] - 1}{\left(\frac{1}{\lambda_e} - i\frac{\omega_n}{c}\sin\theta\cos\xi\right)d}. \end{aligned} \tag{6.37}$$

Calculating the absolute value squared in Eq. (6.35), we arrive at the expression

$$\begin{aligned} \left|\left[\mathbf{n}\left[\mathbf{n}\mathbf{G}_n^{\perp}\right]\right]\right|^2 &= \left[1+\cos^2\theta - \frac{\sin^2\theta\cos^2\xi + \cos^2\theta}{1+\gamma^2\beta^2\sin^2\theta\sin^2\xi}\right] \times \\ \times \frac{\sinh^2\left(\frac{a}{2\lambda_e}\right) + \sin^2\left(\frac{a\omega}{2c}\sin\theta\cos\xi\right)}{\left(\sin^2\theta + \frac{1}{\gamma^2\beta^2}\right)} &\left(\frac{c}{\omega d}\right)^2 4\exp\left(-\frac{a}{\lambda_e}\right). \end{aligned} \tag{6.38}$$

After the substitution of this expression into Eq. (6.35), it is easy to perform the integration with respect to the frequency with allowance for the δ function. Thus, for the fundamental harmonic ($|n| = 1$) in the ultrarelativistic approximation, the formula for the angular distribution of Smith—Purcell radiation is obtained in the form

$$\frac{dW_1}{d\Omega} = \frac{2\alpha}{\pi} \frac{\hbar c}{d\left(\beta^{-1} - \cos\theta\right)} N \exp\left\{ -\frac{4\pi\,(h + a/2)\sqrt{1 + \gamma^2\beta^2 \sin^2\theta \sin^2\xi}}{\gamma\beta d\left(\beta^{-1} - \cos\theta\right)} \right\} \times$$
$$\times \frac{1 + \cos^2\theta - \frac{\sin^2\theta\cos^2\xi + \cos^2\theta}{1 + \gamma^2\beta^2\sin^2\theta\sin^2\xi}}{\sin^2\theta + \gamma^{-2}\beta^{-2}} \times$$
$$\times \left\{ \sinh^2\left(\frac{\pi a \sqrt{1 + \gamma^2\beta^2\sin^2\theta\sin^2\xi}}{\gamma\beta d\left(\beta^{-1} - \cos\theta\right)} \right) + \sin^2\left(\frac{\pi a \sin\theta\cos\xi}{d\left(\beta^{-1} - \cos\theta\right)} \right) \right\} \tag{6.39}$$

The dependence on the azimuth angle ξ is more complex than that for the angular distribution of radiation from the plane grating (see Eq. (6.34)), where the dependence on the azimuth angle ξ appears only in the argument of the exponential and thereby the azimuthal distribution is single-modal with a maximum at $\xi = 0$. However, expression (6.39) remains unchanged, as in the case of the plane grating, under the change $\xi \rightleftarrows \pi \pm \xi$; i.e., the characteristics of radiation in the half-space "above the grating" is the same as that "below the grating" for $\theta = \text{const}$.

6.4 Smith—Purcell Effect As Resonant Diffraction Radiation

It is easy to reveal a relation between Smith—Purcell radiation and resonant diffraction radiation on an example of the perfect plane grating considered in Sect. 6.2. With the use of the results reported in Chap. 3, one can obtain the field of diffraction radiation generated when a particle moves in parallel to a conducting strip of width a. In this case, the total field of radiation for N periodically located strips can be obtained in standard way. If strip width a is much larger than the wavelength, field $\mathbf{E}$ generated in the half-plane (see expression (3.49) for the parallel motion, i.e., $\varphi_0 = 0$) can be used as a reasonable approximation for the radiation field generated at the input boundary of the strip:

$$\mathbf{E}_{\text{in}} = \mathbf{E}_{\infty}. \tag{6.40}$$

In this approximation, the radiation field generated at the "output" boundary of the strip is related to field (6.40) as [14]

$$\mathbf{E}_{\text{out}} = -\mathbf{E}_{\text{in}} \exp\left(i\varphi_S\right). \tag{6.41}$$

Phase shift φ_S on the strip is calculated in complete analogy with the phase shift on a gap (formula (3.107) for $\theta_0 = 0$):

$$\varphi_S = \frac{2\pi a}{\lambda}\left(\cos\theta - \frac{1}{\beta}\right), \tag{6.42}$$

where θ is the polar angle of the photon emission with respect to the electron momentum. In this case, the field of diffraction radiation on a single strip is written in the form

$$\mathbf{E}_{\text{strip}} = \mathbf{E}_{\infty}\left(1 - e^{i\varphi_S}\right). \tag{6.43}$$

The resulting field from N periodically located strips (grating) can be represented in the form

$$\begin{aligned} \mathbf{E}_N &= \mathbf{E}_{\text{strip}} + \mathbf{E}_{\text{strip}}e^{i\varphi_0} + \mathbf{E}_{\text{strip}}e^{i2\varphi_0} + \cdots + \mathbf{E}_{\text{strip}}e^{i(N-1)\varphi_0} = \\ &= \mathbf{E}_{\text{strip}}\left\{1 + \exp\left(i\varphi_0\right) + \exp\left(i2\varphi_0\right) + \cdots + \exp\left(i\left(N-1\right)\varphi_0\right)\right\}. \end{aligned} \tag{6.44}$$

Here, φ_0 is the phase associated with the displacement of a strip by period d:

$$\varphi_0 = \frac{2\pi d}{\lambda}\left(\cos\theta - \frac{1}{\beta}\right). \tag{6.45}$$

The sum of the first N terms of a geometric progression in the braces in formula (6.44) can be calculated analytically as

$$\mathbf{E}_N = \mathbf{E}_{\text{strip}}\frac{1 - \exp\left(iN\varphi_0\right)}{1 - \exp\left(i\varphi_0\right)}. \tag{6.46}$$

In this case, in view of the above result, the expression for the spectral–angular distribution of resonant diffraction radiation on the grating takes the form

$$\begin{aligned} \frac{d^2W_{\text{RDR}}}{d\omega d\Omega} &= cr^2\left|\mathbf{E}_N\right|^2 = \left|E_{\text{strip}}\right|^2\left|\frac{1 - \exp\left(iN\varphi_0\right)}{1 - \exp\left(i\varphi_0\right)}\right|^2 = \\ &= \left|\mathbf{E}_{\infty}\right|^2 F_2F_3 = \frac{d^2W_{\text{DR}}}{d\omega d\Omega}F_2F_3. \end{aligned} \tag{6.47}$$

Here, $\frac{d^2W_{\text{DR}}}{d\omega d\Omega}$ is the spectral–angular distribution of diffraction radiation in the parallel half-plane, F_2 is the factor describing interference fields from the "input" and "output" boundaries of the strip, and F_3 is the factor describing interference from N identical elements of the grating. The spectral–angular distribution of radiation is given by formula (3.56) at $\beta_y = 0$ and $\beta_x = \beta$:

$$\begin{aligned} \frac{d^2W}{d\omega d\Omega} &= \frac{\alpha}{4\pi^2}\frac{\exp\left(-\frac{4\pi h}{\lambda\beta\gamma}\sqrt{1+\beta^2\gamma^2\cos^2\psi}\right)}{\beta\sin\psi}\times \\ &\frac{\cos^2\psi\cos^2\frac{\varphi}{2}\left(1-\beta\sin\psi\right) + \left(\gamma^{-2}+\beta^2\cos^2\psi\right)\sin^2\frac{\varphi}{2}\left(1+\beta\sin\psi\right)}{\left(\gamma^{-2}+\beta^2\cos^2\psi\right)\left(\sin\psi\cos\varphi - 1/\beta\right)^2} \end{aligned} \tag{6.48}$$

Angular variables ψ and φ are shown in Fig. 6.9a.

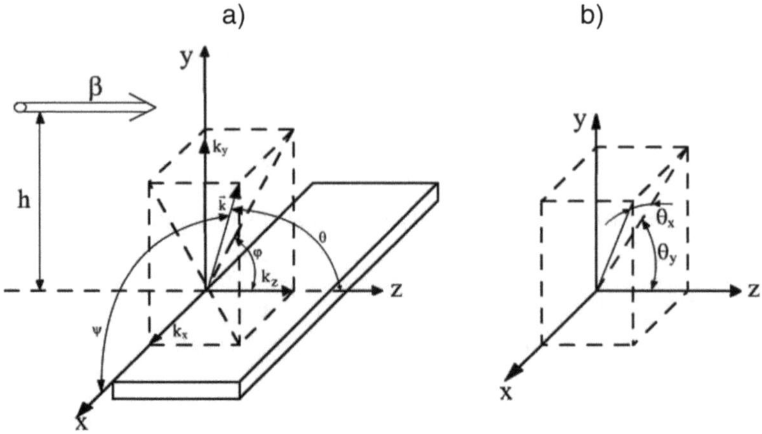

Fig. 6.9 Angular variables for the description of diffraction radiation in the parallel motion of the charge over the conducting half-plane

In view of relation (6.42), the factor F_2 is written as follows:

$$F_2 = 4\sin^2\left(\frac{\varphi s}{2}\right) = 4\sin^2\left[\frac{\pi a}{\lambda}\left(\cos\theta - \frac{1}{\beta}\right)\right]. \tag{6.49}$$

Similarly, taking into account Eqs. (6.45) and (6.47), we can obtain the following expression for the factor F_3:

$$F_3 = \frac{\sin^2(N\varphi_0/2)}{\sin^2(\varphi_0/2)} \approx 2\pi N\delta(\varphi_0 - 2\pi k). \tag{6.50}$$

The last relation is valid for a sufficiently "long" grating ($N \gg 1$).

Diffraction radiation generated by ultrarelativistic particles is concentrated near a plane perpendicular to the target plane in an angular range of $\sim \gamma^{-1}$. When passing to angular variables θ_x and θ_y, which present the specificity of the problem (see Fig. 6.9b),

$$\psi = \frac{\pi}{2} - \theta_x, \quad \varphi = \theta_y \tag{6.51}$$

expression (6.48) is certainly simplified.

When the terms of the order of γ^{-2}, θ_x^2 are neglected, Eq. (6.48) is replaced by the expression

$$\frac{d^2W_{DR}}{\hbar\, d\omega\, d\theta_x d\theta_y} = \frac{\alpha}{2\pi^2}\frac{1}{1-\cos\theta_y}\exp\left(-\frac{4\pi h}{\gamma\lambda}\sqrt{1+\gamma^2\theta_x^2}\right). \tag{6.52}$$

The last expression is valid for radiation angles $\theta_y \gg \gamma^{-1}$, whereas it follows from Eq. (6.48) for angles $\theta_y \sim \gamma^{-1}$ that

$$\frac{d^2 W_{\mathrm{DR}}}{\hbar\, d\omega d\theta_x d\theta_y} = \frac{\alpha}{\pi^2} \exp\left(-\frac{4\pi h}{\gamma\lambda}\sqrt{1+\gamma^2\theta_x^2}\right) \frac{\theta_x^2+\theta_y^2}{\gamma^{-2}+\theta_x^2+\theta_y^2}. \tag{6.53}$$

Note that the structure of the above formula is similar to the formula describing the spectral–angular distribution of transition radiation when intersecting the vacuum—perfectly conducting medium boundary if $h \ll \gamma\lambda/4\pi$.

Let us write the relations between frequently used photon emission angles with respect to the electron motion direction (with respect to the z axis, see Fig. 6.9):

$$\begin{aligned} \cos\psi &= \sin\theta_x = \sin\theta\sin\xi, \\ \tan\varphi &= \tan\theta_y = \tan\theta\cos\xi, \end{aligned} \tag{6.54}$$

as well as the inverse relations

$$\begin{aligned} \cos\theta &= \cos\theta_x\cos\theta_y = \sin\psi\cos\varphi, \\ \tan\xi &= \frac{\sin\theta_x}{\cos\theta_x\sin\theta_y} = \frac{\cos\psi}{\sin\psi\sin\varphi}. \end{aligned} \tag{6.55}$$

In Eq. (6.54), θ and ξ are the polar and azimuth angles, respectively, in the system associated with the electron momentum. The phases φ_S, φ_0 are expressed in terms of the variables θ_x, θ_y as

$$\begin{aligned} \varphi_S &= \frac{2\pi}{\lambda} a\left(\cos\theta_x\cos\theta_y - \frac{1}{\beta}\right), \\ \varphi_0 &= \frac{2\pi}{\lambda} d\left(\cos\theta_x\cos\theta_y - \frac{1}{\beta}\right). \end{aligned} \tag{6.56}$$

The argument of the δ function in Eq. (6.50) provides the standard relation

$$\lambda_n = \frac{d\,(\cos\theta - 1/\beta)}{n}. \tag{6.57}$$

In order to avoid the use of the negative diffraction orders ($n < 0$), the last formula is often written in the following form proposed by Smith and Purcell:

$$\lambda_n = \frac{d\,(1/\beta - \cos\theta)}{n}. \tag{6.58}$$

The integration of Eq. (6.47) with the use of the δ function yields the angular density of the energy emitted at the wavelength corresponding to the nth diffraction order:

$$\frac{dW_n}{d\theta_x d\theta_y} = \frac{4\alpha\hbar c}{\pi} N \frac{1}{\lambda_n\left(1-\cos\theta_y\right)} \exp\left\{-\frac{4\pi h}{\gamma\lambda_n}\sqrt{1+\gamma^2\theta_x^2}\right\}. \tag{6.59}$$

This formula is written for the relation $a = d/2$ at which the factor F_2 is maximal. For $|n| = 1$, Eq. (6.59) provides the angular density of energy in the ultrarelativistic case:

$$\frac{dW_1}{d\Omega} = \frac{4\alpha\hbar c}{\pi d} N \frac{1}{(1-\cos\theta)^2} \exp\left\{-\frac{4\pi h\sqrt{1+\gamma^2\sin^2\theta\sin^2\xi}}{\gamma d(1-\cos\theta)}\right\}. \quad (6.60)$$

Comparing this formula with similar expression (6.34) derived with the use of the induced current model, one can verify that the results coincide after multiplying expression (6.60) by a factor of $(1+\cos\theta)/2$.

Therefore, different approaches provide close results for photon emission angles $\theta \leq \pi/4$, whereas the scalar theory of diffraction, as well as the induced current model [15], for larger values of angle θ gives a smaller value for the spectral–angular density as compared to the model of resonant diffraction radiation.

In recent work [18] authors showed that well known analytical solution of diffraction radiation problem while electron travels in the vicinity of inclined ideally conducting target (Kazantsev—Surdutovich solution [19]) has limited region of validity. For half-plane incination angles $\gg \gamma^{-1}$ the solution obtained in [18] practically coincides to Kazantsev—Surdutovich one; however for parallel flight the difference increases while Lorentz-factor increases. Thus, for instance, for the parallel flight the following solution written in variables of (6.54) and (6.55) was found

$$\begin{aligned}
\frac{dW_{\mathrm{DR}}}{\hbar d\omega\, d\Omega} &= \frac{\alpha}{4\pi^2 c} \frac{\exp\left\{-\frac{4\pi h}{\gamma\beta\lambda}\sqrt{1+(\gamma\beta\sin\theta_x)^2}\right\}}{\left[1+(\gamma\beta\sin\theta_x)^2\right]\left[1/\beta-\cos\theta_x\cos\theta_y\right]^2} \times \\
&\times \left\{\frac{1}{\gamma^2\beta^2} + \gamma^2\sin^2\theta_x - \sin^2\theta_x\left(\frac{1}{\gamma\beta} - \gamma\cos\theta_x\cos\theta_y\right)\right\} = \\
&= \frac{\alpha}{4\pi^2} \frac{\exp\left\{-\frac{4\pi h}{\gamma\beta\lambda}\sqrt{1+(\gamma\beta\sin\theta\sin\xi)^2}\right\}}{\left[1+(\gamma\beta\sin\theta\sin\xi)^2\right](1/\beta-\cos\theta)^2} \times \\
&\times \left\{\frac{1}{\gamma^2\beta^2} + (\gamma\sin\theta\sin\xi)^2 - (\sin\theta\sin\xi)^2\left(\frac{1}{\gamma\beta} - \gamma\cos\theta\right)^2\right\}.
\end{aligned} \quad (6.61)$$

Substituting Eq. (6.61) into (6.47) and integrating over photon energy one may obtain the formula describing the angular distribution of intensity for $n = 1$:

$$\begin{aligned}
\frac{dW_1}{d\Omega} &= \frac{2\alpha}{\pi} N \frac{\hbar c}{d(1/\beta-\cos\theta)^3} \frac{\exp\left\{-\frac{4\pi h\sqrt{1+(\gamma\beta\sin\theta\sin\xi)^2}}{\gamma\beta d(1/\beta-\cos\theta)}\right\}}{1+(\gamma\beta\sin\theta\sin\xi)^2} \times \\
&\times \left[\frac{1}{\gamma^2\beta^2} + (\gamma\sin\theta\sin\xi)^2 - (\sin\theta\sin\xi)^2\left(\frac{1}{\gamma\beta} - \gamma\cos\theta\right)^2\right].
\end{aligned} \quad (6.62)$$

For polar angles $\theta \sim \gamma^{-1} \ll 1$ the Eq. (6.62) drastically differs from the formula (6.60) based on the Kazantsev—Surdutovich model.

In contrast to abovementioned models of Smith—Purcell radiation based on induced surface currents model and on resonant diffraction radiation model that follows from the solution by Kazantsev—Surdutovich [19], the Eq. (6.62) obviously depends on the Lorentz—factor. As follows from Eq. (6.62), the Smith—Purcell radiation intensity decreases as γ^{-2} while the particle energy increases while radiating in the plane perpendicular to the grating $\xi = 0$.

The maximum of angular distribution displaces to smaller polar angles while the particle energy increases (Fig. 6.10). One may see in the Fig. 6.10 in the region of angles that considerably exceed the angle of maximum the Smith—Purcell radiation yield decreases with increase of γ. Such a dependence is very similar to Van-den-Berg's model one [20].

One should note that the dependence of formula 6.62 vs. azimuthal angle ξ is two-modal.

Figures 6.11a, b show the 3-dimensional angular distributions of Smith—Purcell radiation $dW_1/d\Omega$ for the grating with the same parameters as in Fig. 6.10. One may see in Fig. 6.11c that Smith—Purcell radiation intensity may be rather high for azimuthal angles $\xi \neq 0$.

Comparing Fig. 6.11a, b one may expect that the radiation intensity increases in global maxima while Lorentz—factor increases that may compensate the decrease of intensity in the plane perpendicular to the grating while γ increases.

Unfortunately it is impossible to obtain the analytical formula for the total energy radiated by a particle (radiation losses) even for the simplest plane grating.

This formula for radiation losses of Smith—Purcell radiation while $n = 1$ may be written as:

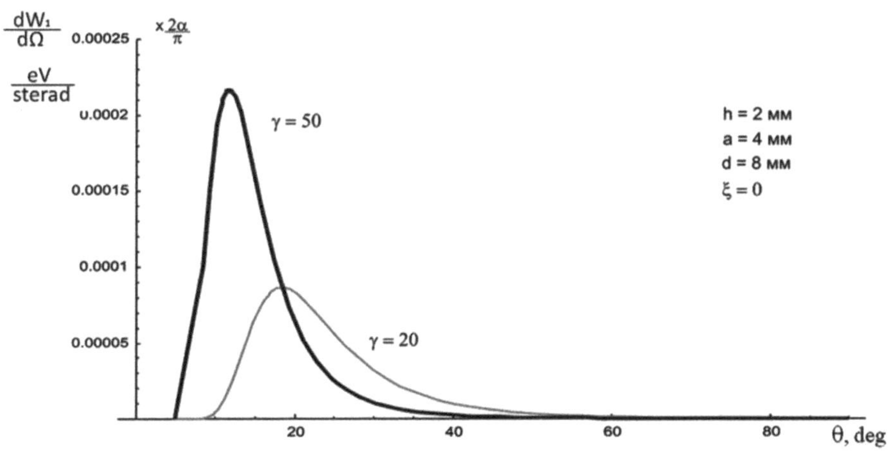

Fig. 6.10 Angular distributions of Smith—Purcell radiation intensity for the plane grating calculated using the model [18]

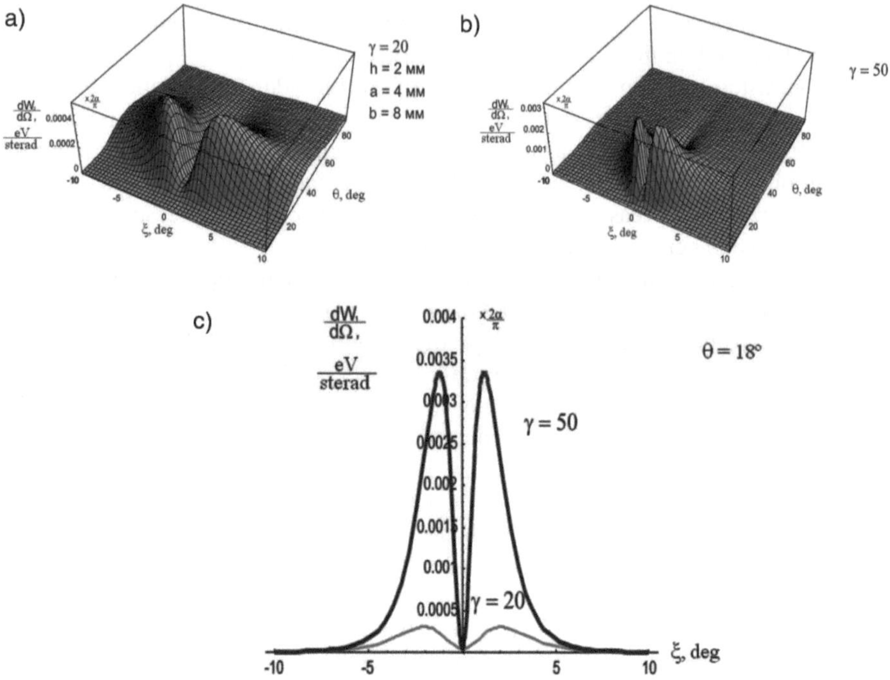

Fig. 6.11 Smith—Purcell radiation angular distributions calculated for different γ using the model from [18]: (**a**) $\gamma = 20$, (**b**) $\gamma = 50$ and azimuthal distribution for both cases while $\theta = 18°$ (**c**)

$$W_1 = \int \frac{dW_1}{d\Omega} d\Omega = N \frac{\hbar c}{d} Q\left(h/d, \gamma\right), \tag{6.63}$$

where $Q\left(h/d, \gamma\right)$ is the function that depends only on h/d ratio and the Lorentz–factor.

Figure 6.12 shows the results of calculation of $W_1(\gamma)$ for two values of $h/d = 0.25, 0.125$. One should point that the dependence with accuracy of several percent may be described with the function

$$W_1(\gamma) \sim \gamma^{1/2} \tag{6.64}$$

One should point out that the surface current model gives stronger dependence $W_1(\gamma) \sim \gamma^{3/2}$ [22].

In the complete analogy with the developed approach, we can consider a so-called "volume" grating whose strips are inclined at angle $\theta_0 \gg \gamma^{-1}$ with respect to the central grating plane. The characteristics of resonant diffraction radiation for such a grating can be calculated by the same formula (6.47), where the first factor $dW/\hbar d\omega d\Omega$ with a good accuracy can be calculated from Eq. (3.56) for the particle velocity components $\beta_x = \beta \cos\theta_0$ and $\beta_y = \beta \sin\theta_0$. The second factor F_2 is determined from a simple scheme (see Fig. 6.13).

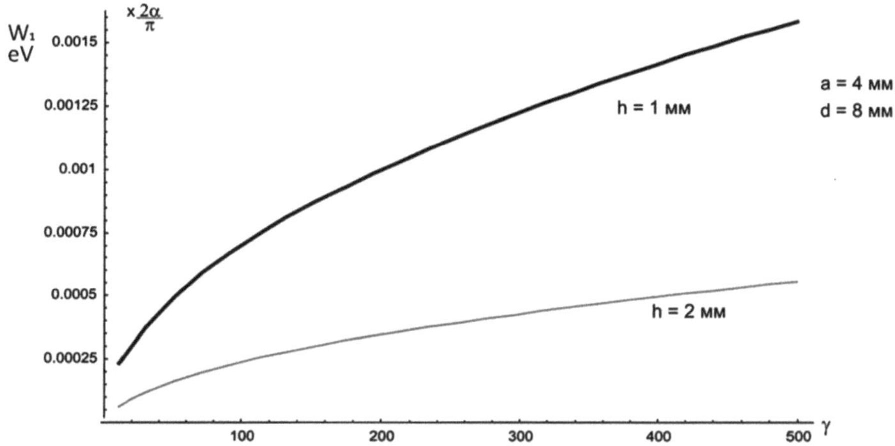

Fig. 6.12 The dependence of radiation losses through Smith—Purcell radiation mechanism while $n = 1$ vs. Lorentz–factor for different impact-parameters

Fig. 6.13 Scheme of the generation of diffraction radiation on the inclined strip

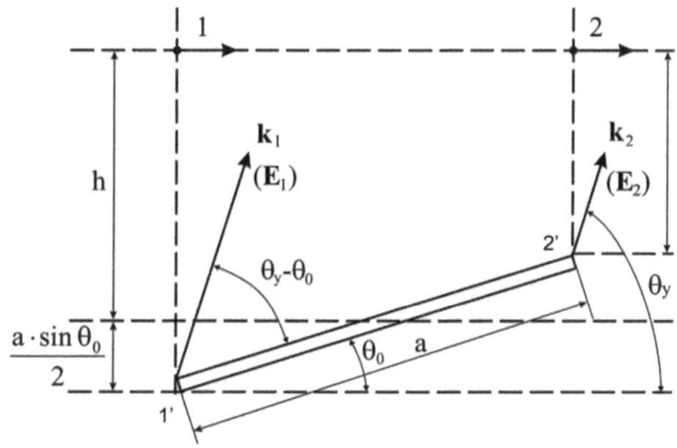

According to the superposition principle, the field strength of diffraction radiation from an inclined strip is the sum of the fields of diffraction radiation from the input (lower) and output (upper) edges of the strip with the corresponding phase factor:

$$\begin{aligned} \mathbf{E}_{\text{strip}} &= \mathbf{E}_{\text{lower}} + \mathbf{E}_{\text{upper}}\, e^{i\Delta\varphi}, \\ \mathbf{E}_{\text{lower}} &= \mathbf{E}_{\text{DR}}\left(h + \frac{a\sin\theta_0}{2}\right), \\ \mathbf{E}_{\text{upper}} &= -\mathbf{E}_{\text{DR}}\left(h - \frac{a\sin\theta_0}{2}\right). \end{aligned} \tag{6.65}$$

Here, phase shift $\Delta\varphi$ depends on phase difference $\Delta\varphi_1$ between the waves emitted in the direction specified by vector $\mathbf{n} = \{\sin\theta\sin\xi,\ \sin\theta\cos\xi,\ \cos\theta\}$ from points

$1'$ and $2'$, which are the projections of points 1 and 2 lying on the trajectory of the charge on the strip surface (see Fig. 6.13):

$$\Delta\varphi_1 = -\frac{\omega}{c}\mathbf{nr} + \frac{\omega}{c}\mathbf{n}(\mathbf{r} + \mathbf{S}_0) = \frac{\omega}{c}\mathbf{nS}_0, \tag{6.66}$$

where

$$\mathbf{S}_0 = \{0,\ a\sin\theta_0,\ a\cos\theta_0\}, \tag{6.67}$$

is the vector connecting points $1'$ and $2'$. In this case,

$$\Delta\varphi_1 = \frac{a\omega}{c}(\sin\theta\cos\xi\sin\theta_0 + \cos\theta\cos\theta_0). \tag{6.68}$$

The additional phase shift $\Delta\varphi_2 = 2\pi c\Delta t/\lambda$ depends on the time Δt necessary for an electron to pass a distance from point 1 to point 2 with finite velocity βc:

$$\Delta t = \frac{a\cos\theta_0}{\beta c}. \tag{6.69}$$

Therefore, the resulting phase shift has the form

$$\Delta\varphi = \Delta\varphi_2 - \Delta\varphi_1 = \frac{2\pi a}{\lambda}\left(\frac{\cos\theta_0}{\beta} - \cos\theta\cos\theta_0 - \sin\theta\sin\theta_0\cos\xi\right). \tag{6.70}$$

For the most interesting case, where the electron momentum is perpendicular to the edges of the strip (perpendicular geometry),

$$\mathbf{E}_{\mathrm{DR}}\left(h + \frac{a\sin\theta_0}{2}\right) = \mathbf{E}_{\mathrm{DR}}(h)\exp\left(-\frac{\pi a\sin\theta_0}{\lambda\beta_\perp}\sqrt{1-\beta_\perp^2\sin^2\psi}\right). \tag{6.71}$$

In order to simplify the further expressions, the subscript $\perp$ of the quantity β will be omitted. With the use of the last formula, expression (6.65) can be written in the more symmetric form

$$\begin{aligned}\mathbf{E}_{\mathrm{strip}} = \mathbf{E}_{\mathrm{DR}}(h)\, e^{i\frac{\Delta\varphi}{2}}\left\{\exp\left[-\frac{\pi a\sin\theta_0}{\lambda\beta}\sqrt{1-\beta^2\sin^2\psi}\right]e^{-i\frac{\Delta\varphi}{2}} - \right.\\ \left. -\exp\left[\frac{\pi a\sin\theta_0}{\lambda\beta}\sqrt{1-\beta^2\sin^2\psi}\right]e^{i\frac{\Delta\varphi}{2}}\right\}.\end{aligned} \tag{6.72}$$

The square of the absolute value of this expression gives the spectral–angular distribution of diffraction radiation from an inclined strip:

$$\frac{d^2W_{\mathrm{strip}}}{d\Omega\, d\omega} = \frac{d^2W_{\mathrm{DR}}}{d\Omega\, d\omega}F_2. \tag{6.73}$$

Here, $\frac{d^2 W_{DR}}{d\Omega\, d\omega}$ is the spectral–angular distribution of diffraction radiation from the perfect inclined half-plane (see expression (3.56)) and the factor F_2 is written in the form

$$F_2 = \left| \exp\left(-\chi - i\frac{\Delta\varphi}{2}\right) - \exp\left(\chi + i\frac{\Delta\varphi}{2}\right) \right|^2 = 4\left(\sinh^2\chi + \sin^2\frac{\Delta\varphi}{2}\right). \tag{6.74}$$

Here,

$$\chi = \frac{\pi a \sin\theta_0}{\lambda\beta}\sqrt{1-\beta^2\sin^2\psi} = \frac{\pi a \sin\theta_0}{\lambda\beta\gamma}\sqrt{1+\gamma^2\beta^2\sin^2\theta\sin^2\xi}. \tag{6.75}$$

Let us write spectral–angular distribution (3.56) in terms of the standard angular variables θ, ξ (see Fig. 6.8). For the strip turned by angle θ_0 about the x axis (see Figs. 6.9 and 6.13), relations (6.54) change to

$$\begin{aligned} \cos\psi &= \sin\theta\sin\xi, \\ \tan(\varphi+\theta_0) &= \tan\theta\cos\xi. \end{aligned} \tag{6.76}$$

From the latter equation, it is easy to obtain the following expression for $\cos\varphi$:

$$\cos\varphi = \frac{\cos\theta\cos\theta_0 + \sin\theta\sin\theta_0\cos\xi}{\sqrt{1-\sin^2\theta\sin^2\xi}}. \tag{6.77}$$

Expressing $\sin\psi$ and $\sin\varphi$ from Eqs. (6.76) and (6.77) and substituting the corresponding expressions into formula (3.56), we obtain the desired dependence $d^2W(\theta,\xi,\theta_0)/d\omega d\Omega$. For the sake of convenience, the following dimensionless spectral–angular distribution $d^2W/\hbar d\omega d\Omega$ will be used below:

$$\frac{d^2W}{\hbar d\omega\, d\Omega} = \frac{\alpha}{4\pi^2}\beta\exp\left(-\frac{4\pi h}{\gamma\beta\lambda}\sqrt{1+\gamma^2\beta^2\sin^2\theta\sin^2\xi}\right)\Lambda(\theta,\xi,\theta_0), \tag{6.78}$$

where

$$\begin{aligned} \Lambda(\theta,\xi,\theta_0) = {} & \frac{1}{\sqrt{1-\sin^2\theta\sin^2\xi}}\frac{1}{(\gamma^{-2}+\beta^2\sin^2\theta\sin^2\xi)\ (1-\beta\cos\theta)}\times \\ & \times\frac{1}{1-\beta(\cos 2\theta_0\cos\theta+\sin 2\theta_0\sin\theta\cos\xi)}\times \\ & \times\left\{(\gamma^{-2}+\beta^2\sin^2\theta\sin^2\xi)\left(1+\beta\cos\theta_0\sqrt{1-\sin^2\theta\sin^2\xi}\right)\left(1-\frac{\cos\theta\cos\theta_0+\sin\theta\sin\theta_0\cos\xi}{\sqrt{1-\sin^2\theta\sin^2\xi}}\right)+\right. \\ & \left.+\sin^2\theta\sin^2\xi\left(1-\beta\cos\theta_0\sqrt{1-\sin^2\theta\sin^2\xi}\right)\left(1+\frac{\cos\theta\cos\theta_0+\sin\theta\sin\theta_0\cos\xi}{\sqrt{1-\sin^2\theta\sin^2\xi}}\right)\right\} \end{aligned} \tag{6.79}$$

Here, as usual, $d\Omega = \sin\theta\, d\theta d\xi$. The dependence on the azimuth angle ξ appears in the expression for the spectral–angular density of diffraction radiation (see the above formula), as well as in the factor F_2 (see formulas (6.70) and (6.75)).

For the "long" grating consisting of inclined strips ($N \gg 1$), the angular distribution of resonant diffraction radiation of the nth order can be obtained from the spectral—angular distribution of resonant diffraction radiation (see formula (6.78)) with the use of the δ function, which appears in F_3. The integration with respect to the frequency with the use of the formula

$$d\omega = \frac{c}{d\left(\frac{1}{\beta} - \cos\theta\right)} d\varphi_0 \tag{6.80}$$

yields

$$\frac{dW}{d\Omega} = \frac{\alpha}{2\pi}\frac{\hbar c}{d} N \frac{\beta \exp\left(-\frac{4\pi h}{\gamma\beta\lambda_n}\sqrt{1+\gamma^2\beta^2\sin^2\theta\sin^2\xi}\right)}{\left(\frac{1}{\beta} - \cos\theta\right)} \Lambda\left(\theta, \xi, \theta_0\right) F_2\left(\lambda_n\right). \tag{6.81}$$

Note that formula (6.81) contains the dependence on the diffraction order only in the wavelength $\lambda_n = \frac{d}{n}\left(\frac{1}{\beta} - \cos\theta\right)$.

For zero azimuth angle $\xi = 0$, the following formula can be obtained from Eq. (6.81):

$$\frac{dW}{d\Omega} = \frac{2\alpha}{\pi}\frac{\hbar c}{d} N\beta^2 \exp\left(-\frac{4\pi h}{\gamma\beta\lambda_n}\right) \frac{(1+\beta)(1-\cos\theta)}{(1-\beta\cos\theta)^3} \sin^2\left(\frac{\pi a}{d} n\right). \tag{6.82}$$

Let us consider the grating consisting of strips perpendicular to the particle trajectory ($\theta_0 = \pi/2$). For the partial case $\xi = 0$, formula (6.82) provides the following formula for the angular distribution of Smith—Purcell radiation of the first order:

$$\begin{aligned} \frac{dW_\perp}{d\Omega} &= \frac{2\alpha}{\pi}\frac{\hbar c}{d} N\beta^2 \frac{1-\sin\theta}{(1-\beta\cos\theta)^2(1+\beta\cos\theta)} \times \\ &\times \exp\left(-\frac{4\pi h}{\gamma\beta\lambda_1}\right)\left\{\sinh^2\left[\frac{\pi a}{\gamma\beta\lambda_1}\right] + \sin^2\left[\frac{\pi a \sin\theta}{\lambda_1}\right]\right\}. \end{aligned} \tag{6.83}$$

The ratio of the expressions describing the angular distribution of Smith—Purcell radiation in the $\xi = 0$ plane that are calculated in the model of induced surface currents (see formula (6.39)) and in the model of resonant diffraction radiation (formula (6.83)) in the ultrarelativistic approximation has the form:

$$\eta_\perp = 1 + \sin\theta. \tag{6.84}$$

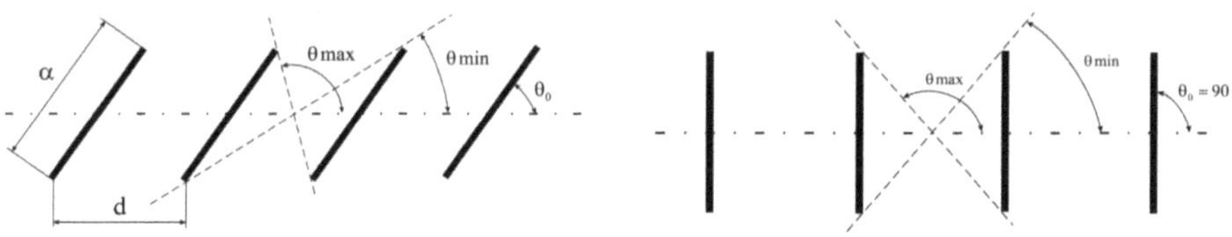

Fig. 6.14 Schemes of the gratings consisting of inclined conducting strips separated by vacuum gaps

Thus, as well as for the plane grating, both models give close results for angles $\theta \leq \pi/4$.

Note that h in formula (6.39) denotes (following [22]) the distance between the particle trajectory and strip edge, whereas the impact parameter h in this section means the distance between the particle trajectory and plane passing through the central lines of the strips (see Fig. 6.13).

For the grating consisting of inclined strips, whose scheme is shown in Fig. 6.14, the approach developed above is applicable for polar angles $\theta_{\min} \leq \theta \leq \theta_{\max}$, which are determined from the elementary geometric construction:

$$\tan\theta_{\min} = \frac{a \sin \theta}{d + a \cos \theta_0}; \quad \tan\theta_{\max} = \frac{a \sin \theta_0}{d - a \cos \theta_0}. \tag{6.85}$$

In particular, for the ratio of the strip width to the period, $a/d = 1/2$, and a strip inclination angle of $\theta_0 = 45°$(see Fig. 6.14 left),

$$\theta_{\min} = 14.64°; \quad \theta_{\max} = 151.33°. \tag{6.86}$$

For the perpendicular strips (see Fig. 6.14 right),

$$\theta_{\min} = 26.56°; \quad \theta_{\max} = 153.44°. \tag{6.87}$$

Figure 6.15 shows the results of the calculation of the angular distribution of Smith—Purcell radiation for the first diffraction order $dW_1/d\Omega$ as a function of the polar angle θ in the plane perpendicular to the mean grating plane ($\xi = 0$). The calculation was performed for two types of gratings: with strip inclination angles $\theta_0 = 45°$ and $90°$ (see Fig. 6.14). The remaining parameters were as follows $\gamma = 100$; $d = 6\,\text{mm}$, $a = 3\,\text{mm}$, $h = 15\,\text{mm}$, $\xi = 0$.

For the perpendicular grating, the additional minima are attributed to vanishing the second term in expression (6.74)):

$$\frac{\Delta\varphi}{2} = k\pi, \quad k - \text{integer}. \tag{6.88}$$

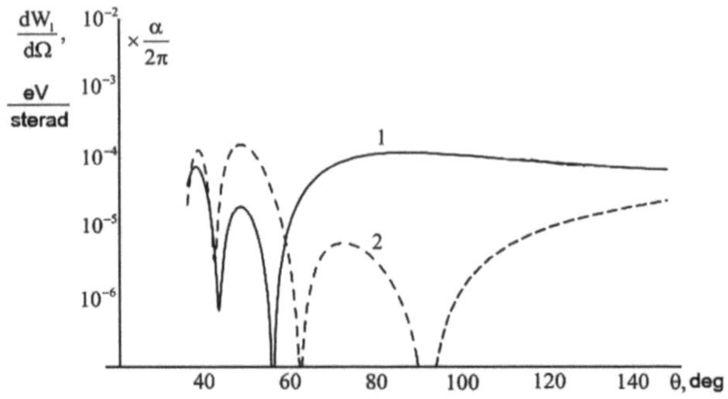

Fig. 6.15 Angular distribution of the first harmonic of Smith—Purcell radiation in the plane perpendicular to the grating ($\xi = 0$) with strip inclination angles (*line 1*) $\theta_0 = \pi/4$ and (*line 2*) $\theta_0 = \pi/2$. The calculation was performed with the following parameters: $\gamma = 100$, $d = 6\,\text{mm}$, $a = 3\,\text{mm}$, and $h = 15\,\text{mm}$

For the angles $\theta_0 = \pi/2$, $\xi = 0$ and the first diffraction order one may obtain

$$\frac{a}{\lambda_1}\sin\theta_k = \frac{a\sin\theta_k}{d\left(\frac{1}{\beta} - \cos\theta_k\right)} = k. \tag{6.89}$$

In the ultrarelativistic case, polar angles θ_k at which the angular density of radiation, $dW_1/d\Omega$, reaches minima can be obtained from this equation in the form

$$\theta_{\min,k} = 2\arctan\left(\frac{a}{dk}\right). \tag{6.90}$$

For the case under consideration, $a/d = 1/2$ and the last relation gives the values (see Fig. 6.15) obtained from this equation in the form

$$\theta_{\min,\,1} = 53.13°; \quad \theta_{\min,\,2} = 28.07°, \tag{6.91}$$

Figure 6.16 shows the two-dimensional distributions $dW_1/d\Omega$ for the gratings under consideration. At $\theta_0 \neq 0$, the angular distribution of radiation in the azimuth angle is usually two-modal in the presence of local maxima with respect to the polar angle. The angles $\theta_{\max}$ corresponding to these maxima can be approximately determined from the following consideration.

The sine squared in expression (6.83) determines not only the minima, but also the positions of the maxima (with good accuracy), because the other factors vary more smoothly. The approximate maximum condition for the angular density $dW_1/d\Omega$ can be written as:

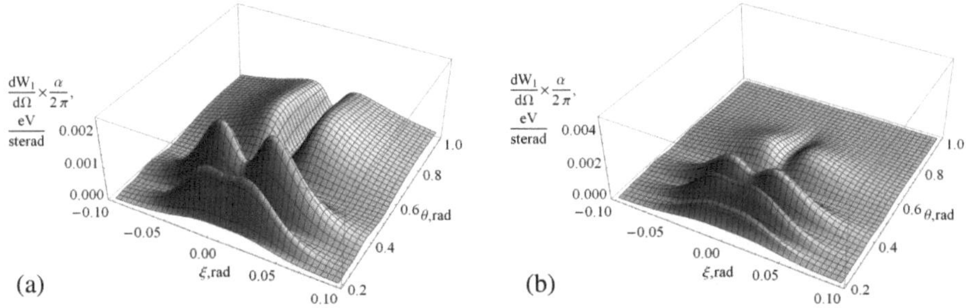

Fig. 6.16 Two-dimensional distribution of the intensity of Smith—Purcell radiation in the polar and azimuth angles for the parameters $\gamma = 100$, $d = 6\,\text{mm}$, $a = 3\,\text{mm}$, $h = 5\,\text{mm}$, and (**a**) $\theta_0 = 0$; and (**b**) $\theta_0 = \pi/4$

$$\frac{\Delta\varphi}{2} = (2k+1)\,\frac{\pi}{2}, \tag{6.92}$$

where k is an integer, or

$$\frac{a}{\lambda_1}\sin\theta_k = \frac{2k+1}{2}. \tag{6.93}$$

From this equation, we obtain

$$\theta_{\max,\,k} \approx 2\arctan\left(\frac{2}{2k+1}\frac{a}{d}\right). \tag{6.94}$$

For the ratio $a/d = 1/2$ and $k = 0,\ 1$, this expression yields

$$\theta_{\max,\,0} = 90^\circ;\, \theta_{\max,\ 1} = 36.84^\circ. \tag{6.95}$$

Figure 6.17 shows the results of the calculations of $dW_1/d\Omega$ for azimuth angles $\xi = 0$ and $\xi = 0.013$ and various values of the impact parameter.

According to this figure, the positions of the maxima in the angular scale are almost independent of the impact parameter h. The exact values of the angle $\theta_{\max,1}$ for both cases are

$$\begin{aligned} &\theta_{\max,\,1}\,(h = 15,\ \xi = 0) = 37.7^\circ,\\ &\theta_{\max,\,1}\,(h = 3,\ \xi = 0) = 36.2^\circ, \end{aligned} \tag{6.96}$$

in good agreement with the estimate obtained by approximate formula (6.94). It is easy to show that approximate values of the angles at which the yield of Smith—Purcell radiation is maximal is generally given by the formula

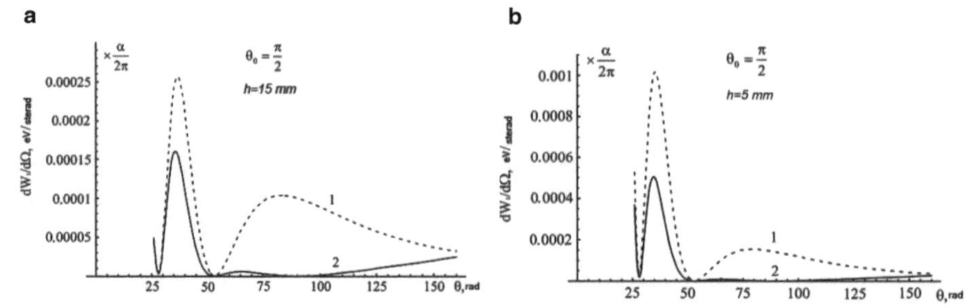

Fig. 6.17 Intensity of Smith—Purcell radiation $dW_1/d\Omega$ versus the polar angle for azimuth angles (*line 1*) $\xi = 0$ and (*line 2*) $\xi = 0.013$ for the parameters $\gamma = 100$, $d = 6$ mm, $a = 3$ mm, and (**a**) $h = 15$ mm and (**b**) $h = 5$ mm

$$\theta_{\max,\,m} = 2\arctan\left[\frac{\sin\theta_0}{\cos\theta_0 + \frac{2m+1}{2}\frac{d}{a}}\right], \quad m - \text{integer}. \tag{6.97}$$

For the main maximum ($m = 0$) in the case $a/d = 1/2$, the relation between $\theta_{\max,0}$ and θ_0 follows from Eq. (6.97) in the form

$$\theta_{\max,\,0} = \theta_0, \tag{6.98}$$

whereas $\theta_{\max,\,m}$ decreases with an increase in the order m. Thus, the simple dependence $\theta_{\max} \sim \theta_0/2$ indicated by Brownell and Doucas [22] is generally incorrect, because several maxima appear in the distribution $dW_1/d\Omega$.

The maxima in the azimuthal dependence of radiation in the planes $\pm\xi_{\max} \neq 0$ appear, because the exponential factor decreases with an increase in ξ^2, whereas the function $\Lambda\,(\theta, \xi)$ (see formula (6.79)) increases. For the typical case (see Fig. 6.16b) where $\theta_{\max} \approx \theta_0$, the function $\Lambda\,(\theta, \xi)$ up to the γ^{-2} terms inclusively can be simplified to the form

$$\Lambda\,(\theta = \theta_0, \xi) \approx \frac{1}{1-\cos\theta_0}\,\frac{\gamma^2\sin^2\theta\sin\xi^2}{1+\gamma^2\sin^2\theta\sin\xi^2}. \tag{6.99}$$

In terms of the projection angle $\theta_x \approx \gamma\sin\theta\sin\xi \approx \xi\gamma\sin\theta$, the azimuthal dependence of the yield of Smith—Purcell radiation has form

$$\frac{dW_n}{d\Omega} \Rightarrow \exp\left(-\frac{4\pi h}{\gamma\lambda_n}\sqrt{1+\gamma^2\theta_x^2}\right)\frac{\gamma^2\theta_x^2}{1+\gamma^2\,\theta_x^2} \tag{6.100}$$

According to this dependence, the position of the maximum, ξ_m, in the azimuthal dependence of $dW_1/d\Omega$ is determined only by the parameter $z = \frac{4\pi h}{\gamma\lambda}$.

Figure 6.18 shows the dependence of the quantity $\xi_m\gamma\sin\theta$ on parameter z, which is well approximated by the dependence

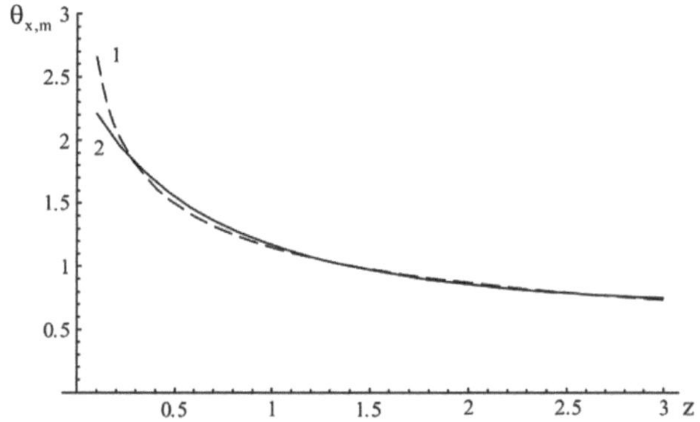

Fig. 6.18 Quantity $\xi_m \gamma \sin\theta$ versus parameter $z = 4\pi h/\gamma\lambda$. Line 1 is the exact calculation and line 2 is the approximation by Eq. (6.100)

$$\gamma \xi_m \sin\theta = \frac{1}{0.4 + 0.528z - 0.072z^2}. \tag{6.101}$$

In conclusion, we note that the two-modal azimuthal distribution of Smith—Purcell radiation is not an attribute of this type of radiation as stated in [23]. The distribution for the volume grating in the ultrarelativistic case becomes single-modal for small polar angles, where approximate formula (6.100) cannot be used (see Fig. 6.16).

6.5 Resonant Diffraction Radiation Generated by Electrons Moving Near a Tilted Planar Grating

It was shown in the preceding section that resonant diffraction radiation from a planar grating in the framework of the models considered above has a number of advantages over radiation from a grating with inclined strips (high intensity and single-modal azimuthal distribution). For a given observation angle θ_y with respect to an electron beam moving in parallel to the grating, the position of the spectral line is determined by the period and particle velocity in agreement with the Smith—Purcell formula. To change the wavelength of Smith—Purcell radiation, the observation angle should be changed, but such a change is not necessarily convenient. Another method for "the line shifting" was proposed in [24] by varying the grating inclination angle θ_0 at the fixed observation angles θ, ξ (see Fig. 6.19):

$$\mathbf{n} = \{\sin\theta \sin\xi,\ \sin\theta\cos\xi,\ \cos\theta\}. \tag{6.102}$$

We consider resonant diffraction radiation for the geometry shown in Fig. 6.19, i.e., in the oblique motion of electrons under the planar grating. Let us introduce the field strength in the strip nearest to the electron trajectory (this strip will be marked by index 1). As previously, the impact parameter h is measured from the central line of the strip. Writing formula (6.72) in terms of the variables θ and ξ (the angles

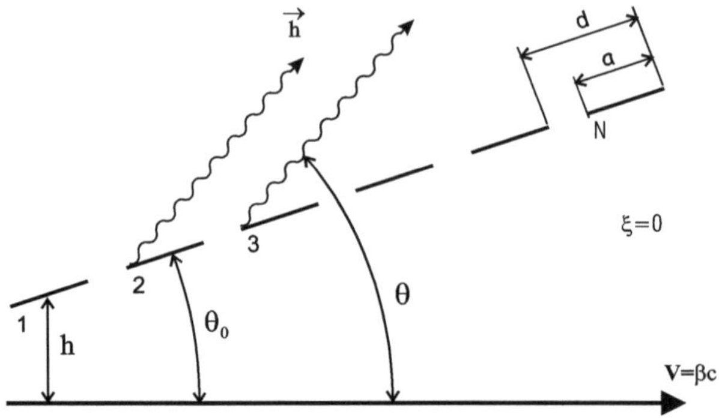

Fig. 6.19 Geometry of resonant diffraction radiation generated by electrons passing near the grating tilted

characterizing the wave vector $\mathbf{k} = \frac{\omega}{c}\mathbf{n}$ in the reference frame associated with the electron momentum), we obtain

$$\mathbf{E}_2 = \mathbf{E}_{\text{strip}} = \mathbf{E}_{\text{DR}}\,(h)\exp\,(i\varphi_0)\times$$

$$\times\left\{\exp\left[-\frac{\pi a\sin\theta_0}{\gamma\beta\lambda}\sqrt{1+\gamma^2\beta^2\sin^2\theta\,\sin^2\xi}-i\varphi_a\right]-\right. \tag{6.103}$$

$$\left.-\exp\left[\frac{\pi a\sin\theta_0}{\gamma\beta\lambda}\sqrt{1+\gamma^2\beta^2\sin^2\theta\,\sin^2\xi}+i\varphi_a\right]\right\}.$$

Here, the phase shift φ_a is given by expression (6.70):

$$\varphi_a = \frac{\pi a}{\lambda}\left(\frac{\cos\theta_0}{\beta}-\cos\theta\,\cos\theta_0-\sin\theta\sin\theta_0\cos\xi\right). \tag{6.104}$$

The field strength in the next strip (no. 2) is calculated for the impact parameter $h + d\sin\theta_0$ and for different phase shifts φ_0:

$$\mathbf{E}_2 = \mathbf{E}_1\,(h+d\sin\theta_0)\exp\,(-i\varphi_0) = \mathbf{E}_{\text{strip}}\,(h)\exp\,(-\chi_0-i\varphi_0)\,, \tag{6.105}$$

where

$$\chi_0 = \frac{2\pi d\sin\theta_0}{\gamma\beta\lambda}\sqrt{1+\gamma^2\beta^2\sin^2\theta\,\sin^2\xi}, \tag{6.106}$$

$$\varphi_0 = \frac{2\pi d}{\lambda}\left(\frac{\cos\theta_0}{\beta}-\cos\theta\cos\theta_0-\sin\theta\sin\theta_0\cos\xi\right). \tag{6.107}$$

The parameter χ_0 characterizes the “decay” in the strength of the self field of the electron with an increase in the distance between the next strip and electron trajectory. The field for the kth strip is calculated similarly:

$$\mathbf{E}_k = \mathbf{E}_{\text{strip}}(h) \exp\left[-(k-1)(\chi_0 + i\varphi_0)\right]. \tag{6.108}$$

The resulting field from the grating consisting of N strips is determined as the sum

$$\mathbf{E}_{\text{RDR}} = \mathbf{E}_1 + \mathbf{E}_2 + \cdots + \mathbf{E}_N = \mathbf{E}_{\text{strip}} \sum_{k=1}^{N} \exp\left[-(k-1)(\chi_0 + i\varphi_0)\right]. \tag{6.109}$$

As usual, the spectral–angular density of resonant diffraction radiation is determined by the absolute value squared of the field specified by Eq. (6.46):

$$\frac{d^2 W_{\text{RDR}}}{\hbar d\Omega d\omega} = \frac{d^2 W_{\text{DR}}}{\hbar d\omega d\Omega} F_2 F_N. \tag{6.110}$$

In expression (6.110), the first and second factors describe diffraction radiation from one inclined strip (see formulas (6.73) and (6.74)), whereas the factor F_N is written in the form of the geometric progression

$$F_N = \left| \sum_{k=1}^{N} \exp\left[-(k-1)(\chi_0 + i\varphi_0)\right] \right|^2 = \left| \frac{1-c^N}{1-c} \right|^2, \tag{6.111}$$

where $c = \exp(-\chi_0 - i\varphi_0)$. Simple algebra reduces Eq. (6.111) to the form

$$F_N = \exp\left[-(N-1)\chi_0\right] \frac{\sinh^2\left(\frac{N\chi_0}{2}\right) + \sin^2\left(\frac{N\varphi_0}{2}\right)}{\sinh^2\left(\frac{\chi_0}{2}\right) + \sin^2\left(\frac{\varphi_0}{2}\right)}. \tag{6.112}$$

Note that the structure of expression (6.112) coincides with a similar formula describing resonant transition radiation with taking account absorption processes [25]. Substituting into Eq. (6.112) exponential form instead both sines, one can obtain the following more convenient form of expression (6.112):

$$F_N = \frac{1 + e^{-2N\chi_0} - 2e^{-N\chi_0}\cos(N\varphi_0)}{1 + e^{-2\chi_0} - 2e^{-\chi_0}\cos\varphi_0}. \tag{6.113}$$

It is easy to see that Eq. (6.112) for parallel motion ($\theta_0 = 0$, $\chi_0 = 0$) reduces to the following standard resonant factor characterizing particular Smith—Purcell radiation:

$$F_N(\theta_0 = 0) = \frac{\sin^2(N\varphi_0/2)}{\sin^2(\varphi_0/2)}. \tag{6.114}$$

In other words, monochromatic radiation in this case is determined by the number of grating periods N. However, in the general case for $\theta_0 \neq 0$, the monochromaticity of the line of resonant diffraction radiation depends not only on N, but also on other parameters of the problem (on angle θ_0, Lorentz factor, wavelength, etc.).

Let us consider the characteristics of resonant diffraction radiation from the so-called "long" grating:

$$N\varphi_0 \gg 1. \tag{6.115}$$

In this limiting case, Eq. (6.113) changes to

$$F_N = \frac{1}{1 - 2\,e^{-\chi_0}\cos\varphi_0 + e^{-2\chi_0}}. \tag{6.116}$$

It is clear that expression (6.116) reaches a maximum under the condition

$$\cos\varphi_0 = 1, \quad \varphi_0 = 2n\pi, \quad n - \text{integer}. \tag{6.117}$$

Under this condition, the resonance factor F_N depends only on the parameter χ_0:

$$F_N^{\max} = \frac{1}{\left(1 - e^{-\chi_0}\right)^2}. \tag{6.118}$$

For small values of the parameter $\chi_0 \ll 1$, Eq. (6.118) reduces to the simpler formula

$$F_N^{\max} \approx \frac{1}{\chi_0^2}. \tag{6.119}$$

Resonance condition (6.117) also determines the wavelength of resonant diffraction radiation that corresponds to the radiation maximum, which has the simplest form for $\xi = 0$:

$$\lambda_n = \frac{d\left[(\cos\theta_0/\beta) - \cos\left(\theta - \theta_0\right)\right]}{n}, \quad n - \text{integer} \tag{6.120}$$

This expression is a generalization of Smith—Purcell formula (6.6), to which it is transformed at $\theta_0 = 0$.

Figure 6.20 shows the spectral distribution of resonant diffraction radiation for parallel motion over the grating at the observation angle $\theta = 0.075\,\text{rad} = 4.3°$ (calculation was performed with the parameters $\gamma = 1000$, $d = 400\,\mu\text{m}$, $a = 200\,\mu\text{m}$, $h = 100\,\mu\text{m}$, $\xi = 0$, and the number of periods is equal to (a) 50 and (b) 10.

According to Fig. 6.21, the intensity of the spectral–angular distribution in the maximum is proportional to N^2, whereas the width of the spectral line, to $1/N$. Indeed, the full width at half maximum (FWHM) for $N = 50$ is $\Delta\hbar\omega = 0.02\,\text{eV}$. As the number of periods decreases to 10, the FWHM increases to $\Delta\hbar\omega = 0.1\,\text{eV}$. Thus, the monochromaticity of the resonant diffraction radiation line for the first order is

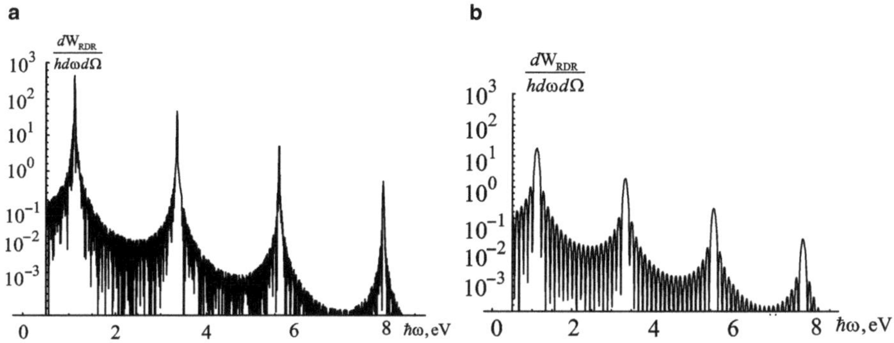

Fig. 6.20 Spectral–angular distribution of resonant diffraction radiation for parallel motion ($\theta_0 = 0$) at given observation angles and number of periods (**a**) $N = 50$ and (**b**) $N = 10$; the kinematic variables are $\gamma = 1000$, $d = 400\,\mu$m, $a = 200\,\mu$m, $h = 100\,\mu$m, $\theta = 0.075$ rad, and $\xi = 0$

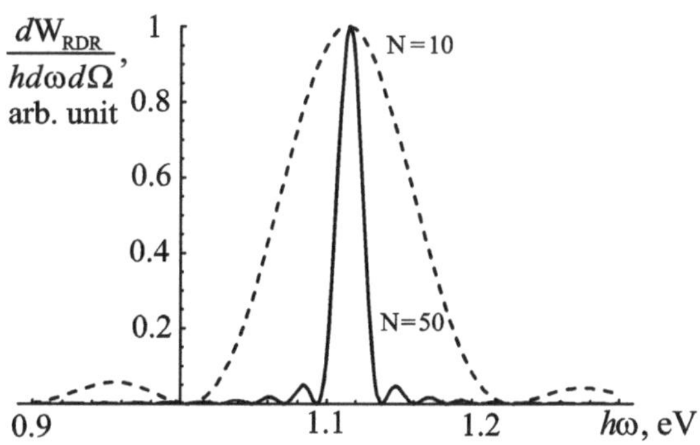

Fig. 6.21 Shape of the spectral line for the first diffraction order ($n = 1$) for the same parameters as in Fig. 6.20. For convenience of comparison, the intensity of the spectral line with $N = 50$ is reduced by a factor of 25

$$\frac{\Delta\hbar\omega}{\hbar\omega_1}(N = 10) = \frac{0.1}{1.115} \approx 9\,\%;\ \frac{\Delta\hbar\omega}{\hbar\omega_1}(N = 50) = 1.8\,\%. \tag{6.121}$$

As the diffraction order n increases, monochromaticity is improved as $1/n$. For example, for the seventh order one may estimate,

$$\frac{\Delta\hbar\omega}{\hbar\omega_7}(N = 10) = 1.4\,\%. \tag{6.122}$$

The spectral distributions of resonant diffraction radiation from the grating under consideration with $N = 50$ at the same impact parameter $h = 100\,\mu$m and various inclination angles are shown in Fig. 6.22.

As seen in this figure, the continuous component in the radiation spectrum increases with the inclination angle θ_0. Further, in agreement with formula (6.120), the positions of the spectral maxima are given by the expression

$$\hbar\omega_n = \frac{2\pi\hbar c}{\lambda_n} = \frac{2\pi\hbar c n}{d\left[\cos(\theta - \theta_0) - \frac{\cos\theta_0}{\beta}\right]}. \tag{6.123}$$

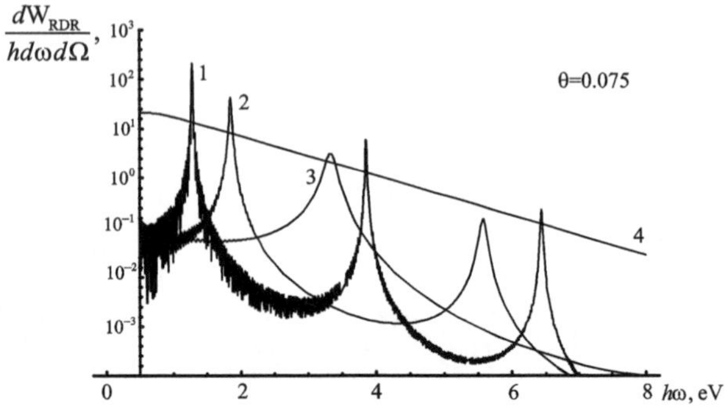

Fig. 6.22 Spectral distribution of resonant diffraction radiation for $h = 100\,\mu\text{m}$, $N = 50$, and the same parameters as in Fig. 6.20. The inclination angle is *(1)* $\theta_0 = 0.005$, *(2)* $\theta_0 = 0.015$, *(3)* $\theta_0 = 0.025$, and *(4)* $\theta_0 = 0.0375$

For the ultrarelativistic case, this formula can be represented in the form

$$\hbar\omega_n = \frac{\pi\hbar cn}{d\sin\frac{\theta}{2}\sin\left(\theta_0 - \frac{\theta}{2}\right)}. \tag{6.124}$$

It is easy to see that, as the inclination angle θ_0 approaches half the polar observation angle, $\theta/2$, the position of the maximum is shifted toward the hard part of the spectrum and the line intensity at the maximum decreases strongly (see Fig. 6.22). Under the condition $\theta_0 = \theta/2$, the spectral intensity of resonant diffraction radiation increases noticeably, but any maxima are absent in the spectrum (see line 4 in Fig. 6.22). With a further increase in the inclination angle θ_0, maxima again appear in the spectrum, but their intensities are much lower (see Fig. 6.23).

Note that the intensity of resonant diffraction radiation (along with the factor $dW/d\Omega\hbar d\omega$) vanishes under the condition $\theta = \theta_0$ (i.e., when radiation is directed along the grating surface). With a further increase in the inclination angle ($\theta_0 > \theta$), maxima against the continuous "background" again appear in the spectrum, but the intensity of resonant diffraction radiation is generally much lower.

Figure 6.24 shows the spectral distributions of resonant diffraction radiation for $\theta_0 = 0.03$ and $\theta_0 = 0.045$ at the same parameters as above. For chosen values of the

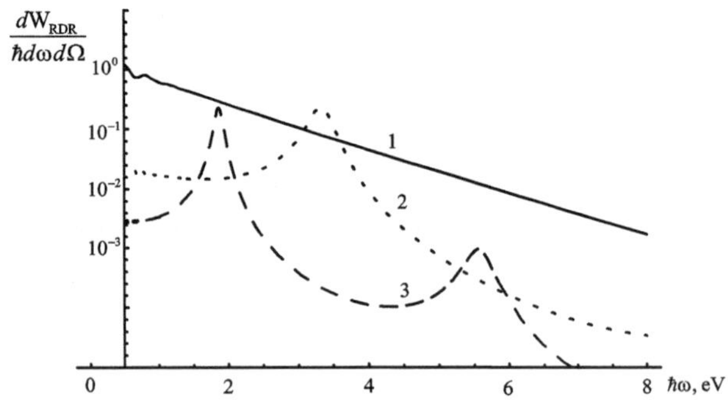

Fig. 6.23 Same as in Fig. 6.22, but for (*line 1*) $\theta_0 = 0.04$, (*line 2*) $\theta_0 = 0.05$, and (*line 3*) $\theta_0 = 0.06$

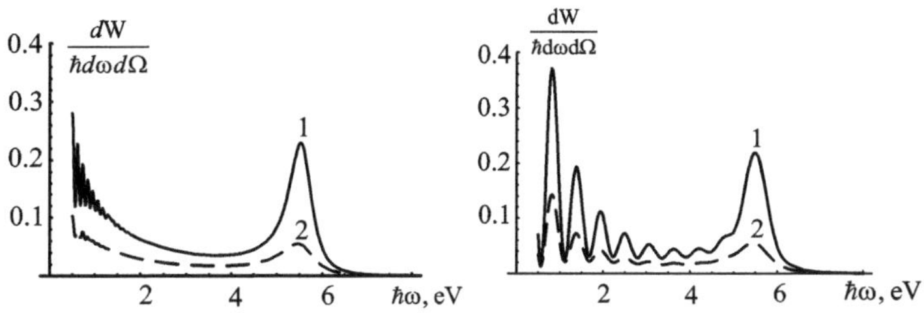

Fig. 6.24 Same as in Fig. 6.20, but for (left) $\theta_0 = 0.03$, (right) $\theta_0 = 0.045$, $N = 50$, and $N = 10$

inclination angle θ_0, the positions of the peaks in the spectrum of resonant diffraction radiation coincide, because differences $\left|\theta_0 - \frac{\theta_y}{2}\right|$ coincide, but the intensity of resonant diffraction radiation decreases with increasing θ_0, as mentioned above. Further, the monochromaticity of the spectral peak in this case ($\Delta\hbar\omega/\hbar\omega \sim 10\,\%$), as well as the radiation intensity in the maximum, is almost independent of the number of the grating periods (see Fig. 6.24).

The approach developed above makes it possible to take into account the effect of a finite emittance of the electron beam on the characteristics of Smith–Purcell radiation. Let us first consider the effect of the angular divergence of the electron beam. For this, it is convenient to use the angular variables θ_x, θ_y defined in the reference frame associated with the grating (see Fig. 6.9b):

$$\begin{aligned} \sin\theta_x &= \sin\theta \sin\xi; \\ \tan\theta_y &= \tan\theta \cos\xi. \end{aligned} \tag{6.125}$$

Using these variables, three terms entering into initial formula (6.110) are expressed as

$$\frac{d^2 W_{\mathrm{DR}}}{\hbar d\omega d\Omega} = \frac{\alpha}{4\pi^2}\frac{\beta}{\cos\theta_0}\exp\left\{-\frac{4\pi h}{\gamma\beta\lambda}\sqrt{1+\gamma^2\beta^2\sin^2\theta_x}\right\}\times$$
$$\times\frac{\left(\gamma^{-2}+\beta^2\sin^2\theta_x\right)(1+\beta\cos\theta_0\cos\theta_x)+\sin^2\theta_x\,(1-\beta\cos\theta_0\cos\theta_x)\left(1+\cos\theta_y\right)}{\left(\gamma^{-2}+\beta^2\sin^2\theta_x\right)\left[1-\beta\cos\theta_x\cos\left(\theta_y-\theta_0\right)\right]\left[1-\beta\cos\theta_x\cos\left(\theta_y+\theta_0\right)\right]}; \tag{6.126}$$

$$\begin{aligned} F_2 &= 4(\sinh^2\chi_0 + \sin^2\varphi_a); \\ F_N &= \frac{[1+\exp(-2N\chi_0) - 2\exp(-N\chi_0)\cos(N\varphi_0)]}{[1+\exp(-2\chi_0) - 2\exp(-\chi_0)\cos\varphi_0]} \end{aligned} \tag{6.127}$$

where

$$\chi = \frac{\pi a \sin\theta_0}{\gamma\beta\lambda}\sqrt{1+\gamma^2\beta^2\sin^2\theta_x}; \qquad \chi_0 = \frac{\pi d \sin\theta_0}{\gamma\beta\lambda}\sqrt{1+\gamma^2\beta^2\sin^2\theta_x};$$

$$\varphi_a = \frac{\pi a}{\lambda}\left(\frac{\cos\theta_0}{\beta} - \cos\theta_x\cos(\theta_y-\theta_0)\right); \ \varphi_0 = \frac{2\pi d}{\lambda}\left(\frac{\cos\theta_0}{\beta} - \cos\theta_x\cos(\theta_y-\theta_0)\right). \tag{6.128}$$

In the first formula, as earlier, h is the distance between the central line of the nearest strip and electron trajectory. However, in particular calculations for a given beam emittance, it is more convenient to use the parameter h_0, i.e., the distance between the grating and "waist" of the electron beam (see Fig. 6.25).

In order to increase the efficiency of interaction between the beam and grating, the "beam waist" is placed above the middle of the grating. For an even number of the grating periods N, the relation between the impact parameters h and h_0 is given by the formula

$$h_0 = h + \frac{d}{2}(N-1)\tan\theta_0. \tag{6.129}$$

In the geometry under consideration, i.e., when the observation angle θ_y is fixed with respect to the grating, the dependence of the wavelength of resonant diffraction radiation on the kinematic variables under the resonance condition given by Eq. (6.117) has the form

$$\lambda_n(\theta_0) = \frac{d}{n}\left(\frac{\cos\theta_0}{\beta} - \cos\theta_y\cos\theta_x\right). \tag{6.130}$$

One should have in mind the relation between angles θ'_y(Fig. 6.25) and θ_y (Eq. (6.125)), $\theta'_y = \theta_y - \theta_0$.

The shift of the line frequency in the spectrum of resonant diffraction radiation generated by the electrons moving over the grating at angle $\theta_{0\,max}$ with respect to the case of parallel motion is easily determined from the above formula:

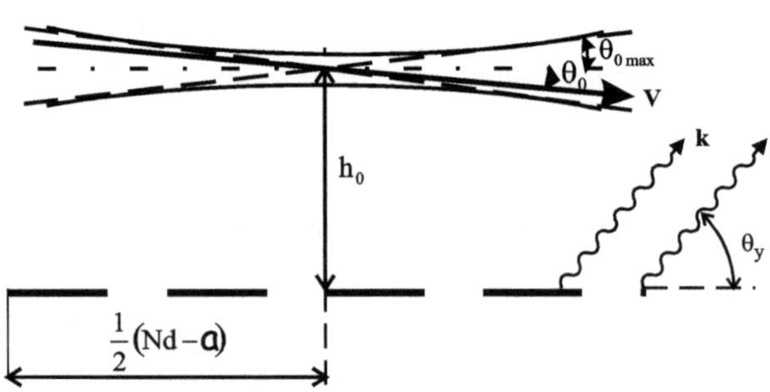

Fig. 6.25 Generation of resonant diffraction radiation by a divergent beam

$$\Delta\omega_n = \omega_n(\theta_0) - \omega_n(0) = 2\pi c\left(\frac{1}{\lambda(\theta_{0\,\max})} - \frac{1}{\lambda(0)}\right). \tag{6.131}$$

In the relativistic case, where the beam divergence is not too large ($\theta_{0\,\max} \ll 1$) and for detection angle $\theta_y \gg \gamma^{-1}$, $\theta_{0\,\max}$, formula (6.131) provides the simple relation

$$\frac{\Delta\omega_n}{\omega_n(\theta_{0\,\max})} \approx \frac{\theta_{0\,\max}^2}{2(1-\cos\theta_y)}. \tag{6.132}$$

In other words, the shift and, therefore, broadening of the line are quadratic functions of the beam divergence in the plane perpendicular to the grating and increase with a decrease in the observation angle. Further, note that the vertical divergence shifts the line toward the hard part of the spectrum. As shown in [26], when the electron moves in the plane parallel to the grating, but at angle $\frac{\pi}{2} - \theta_x$ slightly differed from right one ($\theta_x \ll 1$) with respect to the strips (rulings) of the grating, the line is shifted toward the soft part of the spectrum, because Smith—Purcell radiation in this case corresponds to radiation from the grating with period $d/\cos\theta_y'$. In other words, the "horizontal" divergence of the electron beam gives rise to line broadening:

$$\frac{\Delta\omega_n}{\omega_n(\theta)} \approx -\frac{\theta_x^2}{2}, \tag{6.133}$$

which is independent of the observation angle.

The spectra of Smith—Purcell radiation detected at a given angle θ_y for motion in parallel, $\theta_0 = 0$, and at angle $\theta_{0\,\max} = 0.02$ rad with respect to gratings with various numbers of periods are shown in Figs. 6.26 and 6.27. The calculations are performed with the parameters $\gamma = 20$, $d = 2$ mm, $a = 1$ mm, $h_0 = 1$ mm, $\theta_x = 0$, and $\theta_y = 20°$.

According to Fig. 6.27, in the process of oblique motion near a long grating, electrons more efficiently interact with the nearest part of the grating. This circumstance finally leads to an increase in the radiation intensity and, simultaneously, to worsening of monochromaticity.

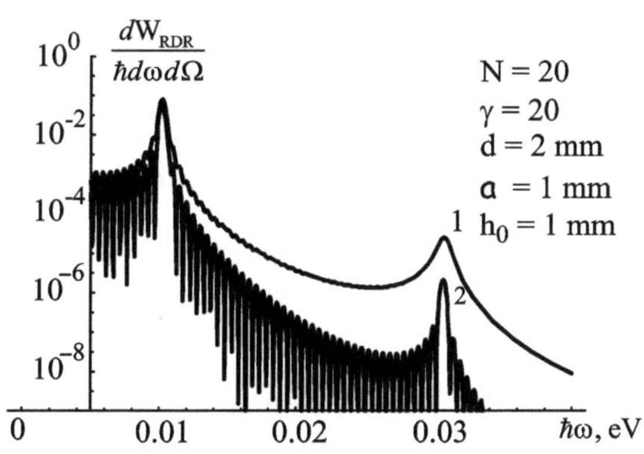

Fig. 6.26 Spectral distribution of Smith—Purcell radiation for observation angle $\theta_y = 20°$ generated by an electron moving (*line 2*) in parallel and (*line 1*) at angle $\theta_{0\,\max} = 0.02$ rad with respect to a grating with the number of periods $N = 20$

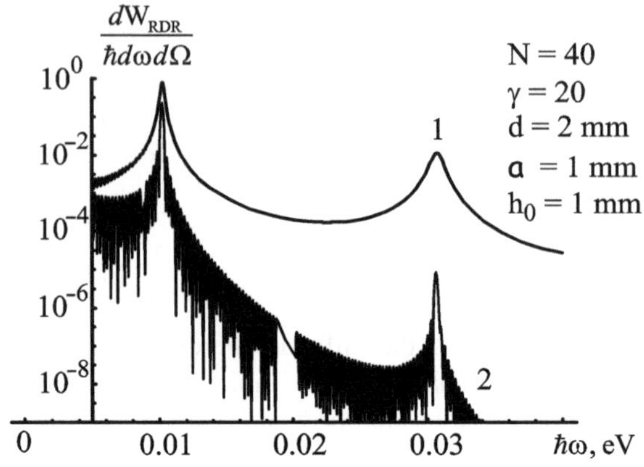

Fig. 6.27 Same as in Fig. 6.26, but for the number of periods $N = 40$

Figure 6.28 shows the normalized shape of the line of Smith—Purcell radiation for the case of the "long" grating ($N = 40$) for parallel ($\theta_0 = 0$) and oblique ($\theta_0 = 0.02$) motions. As seen in this figure, the shift of the maximum is no more than 0.35% in agreement with formula (6.132). The monochromaticity of the line at $\theta_0 = 0$ is equal to 2.2%, whereas at $\theta_0 = 0.02$, the line is broadened by almost a factor of 1.5–3.2%. It is seen that the divergence of the electron beam in the plane perpendicular to the grating gives rise to a noticeable broadening of the line of Smith—Purcell radiation and to insignificant energy shift.

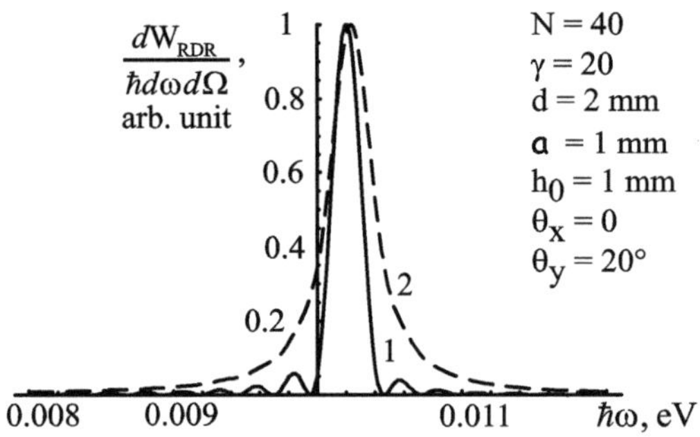

Fig. 6.28 Line shape for the spectra of Smith—Purcell radiation shown in Fig. 6.27

6.6 Smith—Purcell Radiation from a Thin Dielectric Layer on a Conducting Substrate

Let us consider radiation appearing when a charged particle moves above a thin dielectric layer located on a metal substrate following [27]. In addition, we point to work [28], where the same problem was considered. Approaches used in [27] and [28] differ in details.

As known, the charge uniformly moving in vacuum in parallel to the planar surface of a homogeneous medium emits only under the Cherenkov condition [29].

If the Cherenkov condition is not satisfied, radiation does not appear, because the transfer of the longitudinal (along the velocity direction) momentum is impossible. In the case of the inhomogeneous surface, the transfer of the longitudinal momentum to the inhomogeneities of the medium becomes possible and diffraction radiation appears. In what follows, we assume that the Cherenkov condition is not satisfied.

Let us consider a thin dielectric layer on a perfectly conducting substrate. Since radiation does not appear in the absence of the layer, radiation appears due exclusively to the inhomogeneous layer. The boundary condition on the metal surface is satisfied by introducing an image charge [30]. The field outside the conductor coincides with the sum of the fields of two charges, real charge e and fictitious image charge $-e$, uniformly moving in vacuum. This field polarizes the inhomogeneous layer and polarization currents appearing in the layer become sources of diffraction radiation. The magnetic component of the field can be determined from the equation

$$\Delta \mathbf{H}(\mathbf{r}, \omega) + k^2 \mathbf{H}(\mathbf{r}, \omega) = -\frac{4\pi}{c} \operatorname{rot} \mathbf{j}(\mathbf{r}, \omega). \tag{6.134}$$

After that, the electric component is determined as

$$\mathbf{E}(\mathbf{r}, \omega) = i \frac{c}{\omega} \operatorname{rot} \mathbf{H}(\mathbf{r}, \omega) - \frac{4\pi i}{\omega} \mathbf{j}(\mathbf{r}, \omega). \tag{6.135}$$

Here $k^2 = \omega^2/c^2$ and $\mathbf{j}(\mathbf{r}, \omega)$ is the Fourier image of the total current density including the current of the charged particle and currents induced in the medium by the field of this particle:

$$\mathbf{j} = \mathbf{j}_0 + \mathbf{j}_0^{\mathrm{im}} + \mathbf{j}_{\mathrm{layer}} + \mathbf{j}_{\mathrm{layer}}^{\mathrm{im}}, \tag{6.136}$$

where $\mathbf{j}_0$ is the current density corresponding to the motion of the charged particle, $\mathbf{j}_0^{\mathrm{im}}$ is the current density of the image charge, $\mathbf{j}_{\mathrm{layer}}$ is the current density induced in the dielectric layer, and $\mathbf{j}_{\mathrm{layer}}^{\mathrm{im}}$ is the current density of the image layer. The second and fourth terms correspond to the inclusion of the boundary conditions on the surface of the perfect conductor.

In the absence of the layer, $\mathbf{j}_{\mathrm{layer}} = \mathbf{j}_{\mathrm{layer}}^{\mathrm{im}} = 0$. Let $\mathbf{H}_0$ be a solution of equation (6.134) for this case:

$$\Delta \mathbf{H}_0(\mathbf{r}, \omega) + k^2 \mathbf{H}_0(\mathbf{r}, \omega) = -\frac{4\pi}{c} \operatorname{rot} \left(\mathbf{j}_0(\mathbf{r}, \omega) + \mathbf{j}_0^{\mathrm{im}}(\mathbf{r}, \omega) \right). \tag{6.137}$$

Field $\mathbf{H}_0$ cannot be the field of radiation outside the substrate, because radiation is absent in the case of the uniform motion of the charge over the plane surface of the substrate half-space. Hence, field $\mathbf{H}_0$ can be subtracted from field $\mathbf{H}$:

$$\mathbf{H}' = \mathbf{H} - \mathbf{H}_0, \tag{6.138}$$

without the loss of the radiation effects. Subtracting Eq. (6.137) from Eq. (6.134), we obtain

$$\Delta \mathbf{H}'(\mathbf{r},\omega) + k^2 \mathbf{H}'(\mathbf{r},\omega) = -\frac{4\pi}{c} \mathrm{rot} \left(\mathbf{j}_{\mathrm{layer}}(\mathbf{r},\omega) + \mathbf{j}^{\mathrm{im}}_{\mathrm{layer}}(\mathbf{r},\omega) \right). \quad (6.139)$$

The Fourier image of the polarization current in the dipole approximation has the form

$$\mathbf{j}_{\mathrm{layer}}(\mathbf{r},\omega) = -i\omega n_{\mathrm{mic}}(\mathbf{r})\,\alpha(\omega)\,\mathbf{E}_{\mathrm{act}}(\mathbf{r},\omega), \quad (6.140)$$

where $\alpha(\omega)$ is the polarizability of a single molecule and $n_{\mathrm{mic}}(\mathbf{r})$ is the microscopic density of the number of molecules in the layer. According to the superposition principle, the field $\mathbf{E}_{\mathrm{act}}$ acting on the molecules of the layer can be represented as the sum of the fields induced by independent sources:

$$\mathbf{E}_{\mathrm{act}} = \mathbf{E}_0 + \mathbf{E}_0^{\mathrm{im}} + \mathbf{E}_{\mathrm{layer}} + \mathbf{E}^{\mathrm{im}}_{\mathrm{layer}}, \quad (6.141)$$

where $\mathbf{E}_{\mathrm{layer}}$ determines the contribution from all the other molecules of the layer; i.e., the interaction between different parts of the layer that are polarized by the field external sources and $\mathbf{E}^{\mathrm{im}}_{\mathrm{layer}}$ represents the effect of the substrate. For a sufficiently thin layer with not too large relative permittivity, the terms $\mathbf{E}_{\mathrm{layer}} + \mathbf{E}^{\mathrm{im}}_{\mathrm{layer}}$ in the first approximation can be neglected as compared to $\mathbf{E}_0 + \mathbf{E}_0^{\mathrm{im}}$ (the explicit form of the inequality will be given below). Thus, the polarization current density responsible for radiation can be written in the form

$$\mathbf{j}_{\mathrm{layer}}(\mathbf{r},\omega) = -i\omega n_{\mathrm{mic}}(\mathbf{r})\,\alpha(\omega) \left[\mathbf{E}_0(\mathbf{r},\omega) + \mathbf{E}_0^{\mathrm{im}}(\mathbf{r},\omega) \right]. \quad (6.142)$$

The field $\mathbf{E}_0(\mathbf{r},t)$ appears in the $y > 0$ half-space in the process of the uniform motion of the charge e according to the law $\mathbf{r} = \mathbf{a} + \mathbf{v}t$ in parallel to the $y = 0$ surface of the perfect conductor (vector $\mathbf{a}$ is perpendicular to the surface of the medium). The field $\mathbf{E}_0^{\mathrm{im}}$ is determined with the use of the image method as the field of the fictitious image charge $-e$ whose motion is described by the law $\mathbf{r} = -\mathbf{a} + \mathbf{v}t$. Choosing the z axis along $\mathbf{v}$ and the y axis along $\mathbf{a}$, we can represent the Fourier image of the total field of the real and image charges in the form

$$\mathbf{E}_0(\mathbf{r},\omega) + \mathbf{E}_0^{\mathrm{im}}(\mathbf{r},\omega) = i \int d^3q \exp(i\mathbf{q}\mathbf{r})\,\mathbf{L}\left(q_x, q_y\right) \delta(\omega - q_z v) \sin\left(a q_y\right), \quad (6.143)$$

where

$$\mathbf{L}\left(q_x, q_y\right) = \frac{ie}{\pi^2} \frac{\mathbf{Q} + \mathbf{v}\frac{\omega}{v^2\gamma^2}}{Q^2 + \left(\frac{\omega}{v\gamma}\right)^2}, \quad \gamma = \left(1 - v^2/c^2\right)^{-1/2}, \quad Q = \left(q_x, q_y\right). \quad (6.144)$$

The substitution of Eq. (6.143) into Eq. (6.142) yields

$$\mathbf{j}_{\mathrm{layer}}(\mathbf{k},\omega)=\frac{\omega\alpha(\omega)}{(2\pi)^3}\int d^3q\mathbf{L}\left(q_x,q_y\right)\delta\left(\omega-q_zv\right)\sin\left(aq_y\right)\sum_a\exp\left\{i\left(\mathbf{q}-\mathbf{k}\right)\mathbf{R}_a\right\},\tag{6.145}$$

where vector $\mathbf{R}_a=\{X_a,Y_a,Z_a\}$ specifies the position of the ath molecule and the summation is performed over all the molecules of the layer.

Let us write the expression for the polarization current of image molecules $\mathbf{j}^{\mathrm{im}}_{\mathrm{layer}}$. Bound electrons accelerated by the external field emit in the real molecule and, similarly, the images of the bound electrons emit in the image molecule. The Fourier image of the current of the image molecules can be found from Eq. (6.145) by changing $Y_a\to -Y_a$, $j_y\to j_y$, $j_x\to -j_x$, and $j_z\to -j_z$. The first of these changes provides the mirror-symmetric position of the image molecule with respect to the conductor surface. The remaining changes are due to the application of the image method to the electrons bound in the molecule. They correspond to change in the sign of the velocity component perpendicular to the surface and the sign of the charges bound in the image molecule. Therefore,

$$\begin{aligned}\mathbf{j}^{\mathrm{im}}_{\mathrm{layer}}(\mathbf{k},\omega)=&\frac{\omega\alpha(\omega)}{(2\pi)^3}\int d^3q\mathbf{L}\left(q_x,q_y\right)\delta\left(\omega-q_zv\right)\sin\left(aq_y\right)\times\\&\times\sum_a\exp\left\{i\left(\mathbf{q}-\mathbf{k}\right)\mathbf{R}_a+2i\left(\mathbf{ek}\right)\left(\mathbf{eR}_a\right)\right\}\end{aligned}\tag{6.146}$$

where $\mathbf{e}$ is the unit vector perpendicular to the surface, i.e., $\mathbf{e}\equiv\mathbf{e}_y$. The total energy of radiation in frequency interval $d\omega$ in solid angle element $d\Omega$ is determined as

$$\frac{d^2W(\mathbf{n},\omega)}{d\omega d\Omega}=(2\pi)^6\left(1/c\right)\left|\left[\mathbf{kj}_{\mathrm{layer}}(k,\omega)\right]+\left[\mathbf{kj}^{\mathrm{im}}_{\mathrm{layer}}(\mathbf{k},\omega)\right]\right|^2.\tag{6.147}$$

Expression (6.147) is microscopic, because it depends on the coordinates of individual molecules of the layer. However, the inclusion of the microscopic structure of the layer for long-wavelength radiation in the wave zone is unnecessary; hence, it is appropriate to average over all the molecules of the layer. Let the coordinate dependence of the average number density of the conduction electrons in layer be known as $n(r)$. The Fourier images of the currents in macroscopic electrodynamics $\mathbf{j}_{\mathrm{layer}}(\mathbf{k},\omega)$ and $\mathbf{j}^{\mathrm{im}}_{\mathrm{layer}}(\mathbf{k},\omega)$ in Eq. (6.147) can be changed to their average values $\left\langle\mathbf{j}_{\mathrm{layer}}(\mathbf{k},\omega)\right\rangle$ and $\left\langle\mathbf{j}^{\mathrm{im}}_{\mathrm{layer}}(k,\omega)\right\rangle$, respectively:

$$\begin{aligned}\left\langle\mathbf{j}_{\mathrm{layer}}(\mathbf{k},\omega)\right\rangle&=\alpha(\omega)\int d^3q\mathbf{L}\left(q_x,q_y\right)\delta\left(\omega-q_zv\right)\sin\left(aq_y\right)n\left(\mathbf{k}-\mathbf{q}\right)\\\left\langle\mathbf{j}^{\mathrm{im}}_{\mathrm{layer}}(\mathbf{k},\omega)\right\rangle&=\alpha(\omega)\int d^3q\mathbf{L}\left(q_x,q_y\right)\delta\left(\omega-q_zv\right)\sin\left(aq_y\right)n\left(\mathbf{k}-\mathbf{q}-2\mathbf{e}\left(\mathbf{ek}\right)\right)\end{aligned}\tag{6.148}$$

where

$$n(\mathbf{k}-\mathbf{q}) = \int \frac{d^3r}{(2\pi)^3} n(\mathbf{r}) \exp\{-i\mathbf{r}(\mathbf{k}-\mathbf{q})\} \tag{6.149}$$

is the Fourier image of the molecule number density. Thus, the angular–spectral distribution of the emitted energy can be written in the form

$$\frac{d^2W(\mathbf{n},\omega)}{d\omega d\Omega} = \frac{\omega^2}{c} |\alpha(\omega)|^2 |[\mathbf{kJ}(\mathbf{k},\omega)]|^2, \tag{6.150}$$

where

$$\mathbf{J}(\mathbf{k},\omega) = (2\pi)^3 \int d^3q \mathbf{L}(q_x, q_y) \sin(aq_y) \delta(\omega - q_z v) \{n(\mathbf{k}-\mathbf{q}) + n(\mathbf{k}-\mathbf{q}-2\mathbf{e}(\mathbf{ek}))\}. \tag{6.151}$$

Resulting radiation distribution (6.150) is valid at an arbitrary coordinate dependence of the molecule number density in the layer.

Using the expression obtained, we calculate radiation from the surface layer in the form of a diffraction grating on the surface of the perfect conductor filling the $y < 0$ half-space (see Fig. 6.29). Let the grating be homogeneous along the x axis and its profile along the z axis consists of N periodically located strips. The period is d, the width of an individual strip is w, and its height and shape are specified by the function $f(z)$. The average molecule number density in all the strips is the same and is n_0.

The average molecule number density is determined by averaging the microscopic density over the coordinates of all the molecules of the layer:

$$\begin{aligned} n(\mathbf{r}) &= \left\langle n^{\text{mic}}(\mathbf{r}) \right\rangle = \left\langle \sum_{s=0}^{N-1} \sum_b \delta(\mathbf{r} - \mathbf{R}_s - \mathbf{R}_{bs}) \right\rangle = \\ &= n_0 \sum_{s=0}^{N-1} \int_{-\infty}^{\infty} dX_{bs} \int_0^w dZ_{bs} \int_0^{f(Z_{bs})} dY_{bs} \delta(\mathbf{r} - \mathbf{R}_s - \mathbf{R}_{bs}) \end{aligned} \tag{6.152}$$

where the angular brackets mean averaging.

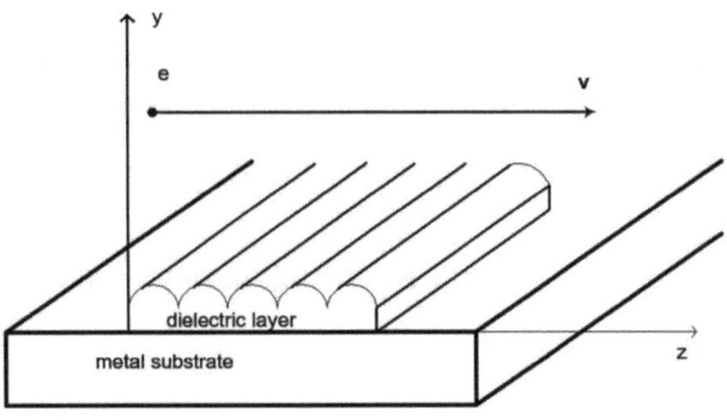

Fig. 6.29 Geometry of the problem. The metal substrate fills the lower half-space. The dielectric layer consists of N individual strips with the profile described by an arbitrary function $f(z)$. The strips are located along the z axis with period d and forms a diffraction grating uniform along the x axis. The charge moves in parallel to the z axis

Vector $\mathbf{R}_s = \{0, 0, sd\}$ in the argument of the delta function represents the periodicity of the positions of the strips, s is the strip number, and vector $\mathbf{R}_{bs}$ specifies the bth molecules of the sth strip (see Fig. 6.30). Therefore,

$$n(\mathbf{r}) = n_0 \sum_{s=0}^{N-1} \xi(y, z - sd)$$
$$\xi(y, z - sd) = \int_0^w dz' \delta(z - sd - z') \int_0^{f(z')} dy' \delta(y - y') \tag{6.153}$$

so that the Fourier image of the molecule density in the layer has the form

$$n(\mathbf{q}) = n_0 \delta(q_x) \xi(q_y, q_z) \sum_{s=0}^{N-1} \exp(-isdq_z). \tag{6.154}$$

The substitution of Eq. (6.154) into Eq. (6.151) and integration with respect to q_z and q_x yield

$$\mathbf{J}(\mathbf{k}, \omega) = \frac{n_0}{v} \sum_{s=0}^{N-1} \exp\{-isd\varphi\} \int_{-\infty}^{\infty} dq_y L(q_y, k_x) k \sin(aq_y) [\xi(k_y - q_y, -\varphi)$$
$$+ \xi(-k_y - q_y, -\varphi)]. \tag{6.155}$$

In terms of the following factors governed by the profile of an individual tooth:

$$F_1 = \int_0^w dz e^{iz\varphi} \{\eta \cos[k_y f(z)] \sinh[\eta f(z)] + k_y \sin[k_y f(z)] \cosh[\eta f(z)]\}$$
$$F_2 = \int_0^w dz e^{iz\varphi} \{\eta \cos[k_y f(z)] \cosh[\eta f(z)] - \eta + k_y \sin[k_y f(z)] \sinh[\eta f(z)]\}, \tag{6.156}$$

where φ and η are given by the respective formulas

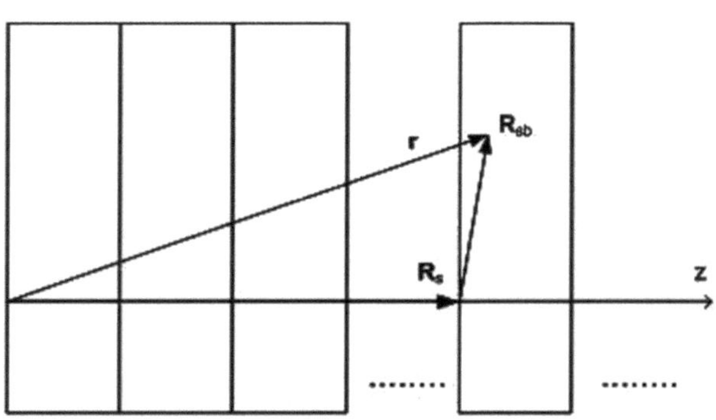

Fig. 6.30 Strips (individual inhomogeneities) of the dielectric layer, plan view

$$\varphi \equiv \frac{\omega}{v} - k_z = k\left(\beta^{-1} - n_z\right), \tag{6.157}$$

$$\eta = \frac{\omega}{c\beta\gamma}\sqrt{1+\gamma^2\beta^2 n_x^2}, \tag{6.158}$$

the expression for **J** can be reduced to the form

$$\mathbf{J}(\mathbf{k},\omega) = \frac{4en_0}{v}\frac{\exp\{-a\eta\}}{\eta^2+k_y^2}\left(iF_1\mathbf{e}_y - \frac{F_2}{\eta}\left(k_x\mathbf{e}_x + \frac{\omega}{v\gamma^2}\mathbf{e}_z\right)\right)\sum_{s=0}^{N-1}\exp\{isd\varphi\}. \tag{6.159}$$

The factors F_1 and F_2 are easily calculated for a diffraction grating consisting of rectangular strips. In this case, the profile of the teeth is a rectangle with sides b (strip thickness) and w (strip width). In this case, $f(z) = b$ and expression (6.156) reduces to the form

$$\begin{aligned} F_1 &= \frac{e^{iw\varphi}-1}{i\varphi}\left\{\eta\cos\left(bk_y\right)\sinh(b\eta) + k_y\sin\left(bk_y\right)\cosh(b\eta)\right\} \\ F_2 &= \frac{e^{iw\varphi}-1}{i\varphi}\left\{\eta\cos\left(bk_y\right)\cosh(b\eta) - \eta + k_y\sin\left(bk_y\right)\sinh(b\eta)\right\} \end{aligned} \tag{6.160}$$

The substitution of Eq. (6.147) in Eq. (6.150 yields

$$\begin{aligned} \frac{d^2W(\mathbf{n},\omega)}{d\omega d\Omega} &= \frac{e^2}{c}\left(\frac{3}{\pi}\right)^2\left|\frac{\varepsilon(\omega)-1}{\varepsilon(\omega)+2}\right|^2\frac{\beta^2}{\left(1-\beta^2 n_z^2\right)^2}\frac{\sin^2(\varphi dN/2)}{\sin^2(\varphi d/2)}\times \\ &\times\left|\left[\mathbf{n}\left(F_1\mathbf{e}_y + iF_2\frac{\beta\gamma n_x\mathbf{e}_x+\gamma^{-1}\mathbf{e}_z}{\sqrt{1+\gamma^2\beta^2 n_x^2}}\right)\right]\right|^2\exp\left\{-\frac{2a\omega}{c\beta\gamma}\sqrt{1+\gamma^2\beta^2 n_x^2}\right\} \end{aligned} \tag{6.161}$$

where the polarizability of an individual molecule, $\alpha(\omega)$, is expressed in terms of the relative permittivity $\varepsilon(\omega)$ of the layer material by the Clausius—Mossotti formula [31]

$$\alpha(\omega) = \frac{3}{4\pi n_0}\frac{\varepsilon(\omega)-1}{\varepsilon(\omega)+2}. \tag{6.162}$$

Formula (6.161) describes the spectral–angular density of emitted energy in the case where a periodically inhomogeneous dielectric layer is located on the surface of a perfect conductor. The particle energy in this case can be arbitrary. The properties of the layer are characterized by:

1. relative permittivity $\varepsilon(\omega)$;

2. parameters specifying the geometry of an individual inhomogeneity, tooth: width w and tooth profile $f(z)$;
3. parameters defining the structure of the entire grating: grating period d and the number of teeth N.

The results obtained in the first order of perturbation theory by changing the acting field to the sum of the fields of the moving charge and image charge. The estimate of the second order corrections by perturbation theory implies the following restrictions on the domain of applicability of the results:

$$\begin{cases} |\varepsilon(\omega) - 1| \ll 1, b \gg \gamma\beta\lambda \\ |\varepsilon(\omega) - 1| \dfrac{b}{\gamma\beta\lambda} \ll 1, b \ll \gamma\beta\lambda \end{cases} \tag{6.163}$$

Thus, in the transparency region of the dielectric, where $|\varepsilon - 1| \approx 4n_0\alpha \ll 1$, the approximation accepted above describes radiation with good accuracy for any thickness of the layer. The domain of applicability for a sufficiently thin layer ($b < \gamma\beta\lambda$) extends when the layer thickness b decreases. In particular, the second of inequalities (6.163) with the parameters $b = 100\,\mu\text{m}$, $\lambda = 100\,\mu\text{m}$, and $\gamma = 25$ (which corresponds to an electron energy of 12.5 MeV) is satisfied sufficiently well at a relative permittivity of the order of several units.

Macroscopic averaging performed when deriving expression (6.150) implies that the number of molecules in a single inhomogeneity of the layer is much larger than one. This condition imposes a lower limit on the sizes of a single inhomogeneity. The domain of applicability of the results obtained above is also limited by the assumption of the perfect conductivity of the metal substrate. For this reason, the frequency range under consideration is limited from above by the Langmuir frequency of the material of the substrate (metal), which lies in the optical or near ultraviolet frequency range. These results are also inapplicable for resonant frequencies and disregard the possibility of Cherenkov radiation.

The spectral—angular characteristics of radiation (6.161) are determined by two factors. The first factor is $\left(1 - \beta^2 n_z^2\right)^{-2}$. In the ultrarelativistic case, this factor provides the forward radiation peak at small angles $\sim \gamma^{-1}$. The second factor is the function $\frac{\sin^2(\varphi d N/2)}{\sin^2(\varphi d/2)}$. This function has maxima at points $\varphi d = 2\pi m, m = 1, 2, 3 \ldots$. This condition provides the Smith—Purcell dispersion relation

$$m\frac{\lambda}{d} = \beta^{-1} - \cos\theta, \quad m = 1, 2, 3 \ldots, \tag{6.164}$$

where the polar angle θ is defined through the relation $n_z = \cos\theta$. As known in optics [32], the angular width of individual peaks is given by the expression $\Delta\theta = \lambda/(Nd)$. Physically, relation (6.164) appears because the momentum transferred by a moving charge to a periodically inhomogeneous medium can take only discrete values.

At $N \gg 1$, peaks are so sharp that the ratio of the squares of the sines can be changed to the sum of the delta functions:

$$\frac{\sin^2(\varphi dN/2)}{\sin^2(\varphi d/2)} \xrightarrow{N \gg 1} 2\pi N \sum_m \delta(\varphi d - 2\pi m). \tag{6.165}$$

At a given period d and a given wavelength λ, the diffraction order m is limited by the inequalities

$$\frac{d}{\lambda}\left(\beta^{-1} - 1\right) \leq m \leq \frac{d}{\lambda}\left(1 + \beta^{-1}\right), \tag{6.166}$$

which are obtained from the condition $-1 \leq \cos\theta \leq 1$ and Eq. (6.164).

In the nonrelativistic case at $\beta \ll 1$, Smith—Purcell radiation is generated at wavelengths much higher than the grating period:

$$\lambda \sim \frac{d}{\beta m} \gg d. \tag{6.167}$$

In the ultrarelativistic case at $\gamma \gg 1$,

$$\begin{aligned} \beta^{-1} - 1 &\approx \gamma^{-2}/2 \\ 1 + \beta^{-1} &\approx 2 \end{aligned} \tag{6.168}$$

Therefore, condition (6.166) can be approximately represented in the form

$$d\gamma^{-2}/2 \leq \lambda m \leq 2d. \tag{6.169}$$

It is seen that Smith—Purcell radiation from a grating with period d can be generated in the ultrarelativistic case only at wavelengths shorter than two grating periods:

$$\lambda \leq \frac{2d}{m}, \tag{6.170}$$

and the "threshold" wavelength decreases as the diffraction order m increases.

References

1. Frank, I.M.: Izv. Akad. Doppler's effect in refractive media. Nauk USSR. Fizika. **6**, 3 (1942)
2. Smith, S.J., Purcell, E.M.: Visible light from localized surface charges moving across a grating. Phys. Rev. **92**, 1069 (1953)
3. Izhizuka, H., Kawamura, Y., Yokoo, K. et al.: Smith-Purcell radiation experiment using a field-emission array cathode measurements of radiation. Nucl. Instrum. Methods A **475**, 593 (2001)
4. Bolotovskiy, B.M., Voskresenskiy, G.V.: Diffraction radiation. Phys.-Uspekhi **94**, 377 (1968)

5. Shestopalov, V.P.: Diffraction Electronics. Kharkov, Ukraine (1976)
6. Shestopalov, V.P.: The Smith-Purcell effect. Nova Science Publishers, Commack, NY (1998)
7. Rusin, F.S., Bogomolov, G.V.: Generation of oscillations in open resonator. Sov. JETP Lett. **4**, 236 (1966)
8. Shestopalov, V.P.: Physical Base of Millimeter Technique. Kiev, Naukova Dumka, 1985. in Ukraine
9. Tsimring, S.E.: Electron Beams and Microwave Vacuum Electronics. Wiley, Hoboken, NJ (2003)
10. Kube, G., Backe, H., Lauth, W. et al., Smith-Purcell radiation in view of particle beam diagnostics. Proceedings DIPAC 2003, Mainz, Germany, (2003)
11. Wachtel, J.M.: Free-electron lasers using the Smith-Purcell effect. J. Appl. Phys. **50**, 49 (1979)
12. Urata, J., Goldstein, M., Kimmitt, M.F., Naumov, A., Platt, C., Walsh, J.E.: Superradiant Smith-Purcell emission. Phys. Rev. Lett. **80**, 516 (1998)
13. Bolotovskiy, B.M., Burtsev, A.K.: Radiation of charge moving above diffraction grating. Opt. Spectrosc. **19**, 470 (1965)
14. Potylitsyn, A.P.: Resonant diffraction radiation and Smith-Purcell effect. Phys. Lett. A **238**, 112 (1998)
15. Brownell, J.H., Walsh, J., Doucas, G.: Spontaneous Smith-Purcell radiation described through induced surface currents. Phys. Rev. E **57**, 1075 (1998)
16. Landau, L.D., Lifshitz, E.M.: The Classical Theory of Fields. Pergamon Press, (1987)
17. Ter-Mikaelyan, M.L.: High-Energy Electromagnetic Processes in Condensed Media. Wiley-Interscience, New York, NY (1972)
18. Karlovets, D.V., Potylitsyn, A.P.: On the theory of diffraction radiation. JETP **107**, 755 (2008)
19. Kazantsev, A.P., Surdutovich, G.I.: Radiation of a charged particle passing close to a metal screen. Sov. Phys. Dokl. **7**, 990 (1963)
20. Van den Berg, P.M.: Smith-Purcell radiation from a point charge moving parallel to a reflection grating. J. Opt. Soc. Am. **63**, 1588 (1973)
21. Brownell, J.H., Doucas, G.: Longitudinal electron bunch profile diagnostics at 45 MeV using coherent Smith-Purcell radiation. Phys. Rev. Spec. Top. – Accel. Beams **9**, 092801 (2006)
22. Brownell, J.H., Doucas, G.: Role of the grating profile in Smith-Purcell radiation at high energies. Phys. Rev. ST-AB **8**, 091301 (2005)
23. Gover, A., Dvorkis, P., Eliska, U.J.: Angular radiation pattern of Smith-Purcell radiation. Opt. Soc. Am. B. **1**(5), 723 (1984)
24. Potylitsyn, A.P., Karataev, P.V., Naumenko, G.A.: Resonant diffraction radiation from an ultra–relativistic particle moving close to a tilted grating. Phys. Rev. E **61**, 7039 (2000)
25. Rullhusen, P., Artru, X., Dhez, P.: Novel Radiation Sources Using Relativistic Electrons. World Scientific, Singapore (1998)
26. Haeberle, O., Rullhusen, P., Salome J-M. et al.: Smith-Purcell radiation from electrons moving parallel to a grating at oblique incidence to the rulings. Phys. Rev. E. **55**, 4675 (1997)
27. Ryazanov, M.I., Strikhanov, M.N., Tishchenko, A.A.: Diffraction radiation from an inhomogeneous dielectric film on the surface of a perfect conductor. JETP **126**, 349 (2004)
28. Zhevago, N.K., Glebov, V.I.: Modified theory of Smith-Purcell radiation. Nucl. Instrum. Methods Phys. Res. A **341**, ABS 101 (1943)
29. Jelley, J.V.: Cerenkov radiation. Prog. Nucl. Phys. **3**, 84 (1956)
30. Pafomov, V.E.: Radiation of charged particle at boundary. In: D.V. Skobel'tsyn (ed.) Proceeedings of the P.N. Lebedev Physical Institute, Vol. 44, pp. 25–157. Consultants Bureau, New York, NY (1971)
31. Ryazanov, M.I.: Introduction to Electrodynamics of Condensed Medium. Fizmatlit, Moskow (in Russian) (2002)
32. Ahmanov, S.A., Nikitin, S.Yu.: Physical Optics. Moscow State University, Moscow (in Russian) (1998)

Chapter 7
Coherent Radiation Generated by Bunches of Charged Particles

7.1 Coherent Radiation Generated by Short Electron Bunches

In one of the first works [1], where synchrotron radiation generated by an electron bunch containing N_e electrons was considered, it was mentioned that, in the range of wavelengths comparable with the length of the electron bunch ℓ_B, radiation becomes coherent, i.e., the intensity of radiation generated by the bunch depends quadratically on the number of electrons in the bunch (on the "population" of the bunch). If the bunch contains N_e electrons rotating with constant angular velocity ω on an orbit of radius R, then the power of synchrotron radiation at the nth harmonic of the bunch is generally determined by the expression [2]

$$P_B^{(n)} = P_0^{(n)} \left| \sum_{k=1}^{N} \mathrm{e}^{-\mathrm{i}n\varphi_k} \right|^2, \tag{7.1}$$

where $P_0^{(n)}$ is the power of synchrotron radiation at the nth harmonic of a single electron and φ_k is the angular coordinate of the kth electron in the bunch. If the typical wavelength of synchrotron radiation at the nth harmonic is longer than the length of the bunch $l_B \sim R\varphi_{k\,\max}$, the power of synchrotron radiation generated by the bunch is calculated by the formula

$$P_B^{(n)} = P_0^{(n)} \left\{ N_e + N_e (N_e - 1) \sum_{k \neq q} \cos\left[n\left(\varphi_k - \varphi_q\right)\right] \right\} \approx P_0^{(n)} \left\{ N_e + N_e^2 f_n \right\}. \tag{7.2}$$

The first and second terms in the braces describe incoherent and coherent radiations, respectively.

The quantitative characteristic of the fraction of particles emitting coherently (i.e., in phase) is determined by the so-called "form factor" of the bunch f_n [2]. If the number of electrons in the bunch satisfies the condition $N \gg 1$, this form factor can be changed to the continuous distribution $S(\varphi)$; in this case,

A.P. Potylitsyn et al., *Diffraction Radiation from Relativistic Particles*, STMP 239, 197–220, DOI 10.1007/978-3-642-12513-3_7,

$$f_n \simeq \left[\int \cos(n\varphi) S(\varphi) d\varphi \right]^2, \tag{7.3}$$

where $S(\varphi)$ is the normalized distribution of the electrons in the bunch in the azimuth angle φ. For the electrons uniformly distributed in a given angular range

$$S(\varphi) = \begin{cases} \dfrac{1}{\alpha}, & \varphi \leq \left|\frac{\alpha}{2}\right| \\ 0, & \varphi > \left|\frac{\alpha}{2}\right| \end{cases} \tag{7.4}$$

the calculation of form factor (7.3) yields the result

$$f_n = \frac{\sin^2(n\alpha/2)}{(n\alpha/2)^2}. \tag{7.5}$$

According to the last expression, the form factor f_n varies from 1 ($\alpha \to 0$, complete coherence) to 0 ($\alpha = 2\pi$, coherence is absent). Note that form factor (7.5) does not depend explicitly on the wavelength. It is clear that this approximation is too rough. A more correct approximation was used in [3–5], where the form factor describing coherent synchrotron radiation depends on the frequency of an emitted photon:

$$f_{\mathrm{CSR}}(\mathbf{k}) = \left| \int_V \exp(-i\mathbf{k}\mathbf{r}) S(\mathbf{r}) d\mathbf{r} \right|^2. \tag{7.6}$$

Here, $S(\mathbf{r})$ is the normalized distribution of electrons in the bunch (with respect to the center of the bunch) and $\mathbf{k}$ is the wave vector. Integration is performed over the bunch volume V.

In complete analogy with expression (7.2), the spectral–angular distribution of coherent synchrotron radiation generated by the electron bunch can be represented in the form [3, 4]

$$\frac{dW_{\mathrm{CSR}}(\mathbf{n},\omega)}{\hbar d\omega d\Omega} = N_{\mathrm{e}} \left[1 + N_{\mathrm{e}} f_{\mathrm{CSR}}(\mathbf{k})\right] \frac{dW_0(\omega)}{\hbar d\omega d\Omega}. \tag{7.7}$$

Here, $\dfrac{dW_0(\mathbf{n},\omega)}{\hbar d\omega d\Omega}$ is the spectral–angular distribution of radiation generated by one electron.

The form factor written in the form of expression (7.6) generally depends on the direction of the wave vector $\mathbf{k} = 2\pi/\lambda \{\sin\theta\cos\varphi, \sin\theta\sin\varphi, \cos\theta\}$. In the ultrarelativistic case, the phase $\varphi = \mathbf{kr}$ in the argument of the exponential in expression (7.6) can be written in the form

$$\begin{aligned}\varphi = \mathbf{kr} &= \frac{2\pi}{\lambda}\left(x \sin\theta \cos\varphi + y\ \sin\theta \sin\varphi + z \cos\theta\right) = \\ &= \frac{2\pi}{\lambda}\left(\rho \sin\theta \cos\left(\varphi - \chi\right) + z \cos\theta\right).\end{aligned} \tag{7.8}$$

Here, x, y, z and ρ, χ, z are the Cartesian and cylindrical coordinates of an electron in the bunch, respectively. For ultrarelativistic beams, the length of the bunch is usually much larger than its transverse sizes. In this case, the distribution $S(\mathbf{r})$ can be represented in the form of the superposition of the longitudinal and transverse distributions

$$S(\mathbf{r}) = S_L(z) S_T(\boldsymbol{\rho}). \tag{7.9}$$

In addition, emission occurs at angles $\theta \sim \gamma^{-1}$; hence, $\cos\theta$ in expression (7.8) can be changed to unit. In this case, form factor (7.6) is factorized, i.e., reduces to the product of the longitudinal and transverse form factors:

$$f_{\mathrm{CSR}} = f_{\mathrm{CSR}}^{L} f_{\mathrm{CSR}}^{T}. \tag{7.10}$$

Integration with respect to the azimuth angle χ for the cylindrically symmetric bunch yields

$$\begin{aligned}f_{\mathrm{CSR}}^{L} &= \left|\int dz S_L(z) \exp\left(-i\frac{\omega}{c}z\right)\right|^2, \\ f_{\mathrm{CSR}}^{T} &= \left|2\pi \int \rho d\rho S_T(\rho)\, J_0\left(\frac{2\pi\rho}{\lambda}\sin\theta\right)\right|^2.\end{aligned} \tag{7.11}$$

Hereinafter, $J_n(x)$ is the nth order Bessel function.

As an example, we calculate form factor (7.11) for the simplest case where the particles in the electron bunch are uniformly distributed over a cylinder of radius r_0 and length ℓ_B:

$$\begin{aligned}S_L(z) &= \begin{cases} 1/\ell_B, & 0 \le z \le \ell_B \\ 0, & z > \ell_B \end{cases} \\ S_T(\boldsymbol{\rho}) &= \begin{cases} 1/\pi r_0^2, & 0 \le \rho \le r_0 \\ 0, & \rho > r_0. \end{cases}\end{aligned} \tag{7.12}$$

For this distribution, integrals appearing in expression (7.10) are calculated analytically:

$$\begin{aligned}f_{\mathrm{CSR}}^{L} &= \left(\frac{\lambda}{\pi\ell_B}\right)^2 \sin^2\frac{\pi\ell_B}{\lambda}, \\ f_{\mathrm{CSR}}^{T} &= 4\left(\frac{\lambda}{2\pi r_0 \sin\theta}\right)^2 J_1^2\left(\frac{2\pi r_0 \sin\theta}{\lambda}\right).\end{aligned} \tag{7.13}$$

According to expressions (7.13), the conditions of complete coherence (proportionality of the intensity of synchrotron radiation to the bunch charge squared) reduce to the expressions

$$\begin{aligned} \lambda &\gg \pi \ell_B, \\ \lambda &\gg 2\pi r_0 \sin\theta \approx 2\pi r_0/\gamma. \end{aligned} \tag{7.14}$$

For ultrarelativistic electrons, the typical polar angle of the emission of a synchrotron radiation photon is determined by the inverse Lorentz factor. Therefore, the second of conditions (7.14) is usually much weaker than the first condition, which determines the wavelength range where coherent synchrotron radiation is generated.

In order to calculate the form factor of the bunch for describing coherent effects in the case of an emission mechanism different than synchrotron radiation (e.g., transition radiation), we consider a simple scheme illustrating transition radiation generated by two electrons that move in parallel to each other with the same velocity (see Fig. 7.1) in the $x = 0$ plane and intersect an inclined target. The angle between the normal to the target and the velocity of the electrons is ψ (as above, we consider the ultrarelativistic case).

Coherent transition radiation in this geometry has no azimuthal symmetry for both radiation cones (forward coherent transition radiation and backward coherent transition radiation). First, we consider the following geometry (see Fig. 7.1). Let us determine the phase difference between the waves emitted by two electrons in the process of the generation of transition radiation if both electrons move in the figure plane ($x = 0$) and emit photons in the same plane.

At time $t_1 = 0$, the first electron e_1 moving with velocity βc intersects the interface between the media at point $0(y_1 = 0; z_1 = 0)$. The field of transition radiation generated by the first electron at long distances from point 0 can be written in the form

$$\mathbf{E}_1 = \mathbf{E}_0 e^{i\mathbf{k}\mathbf{R}_1 - i\omega t}, \tag{7.15}$$

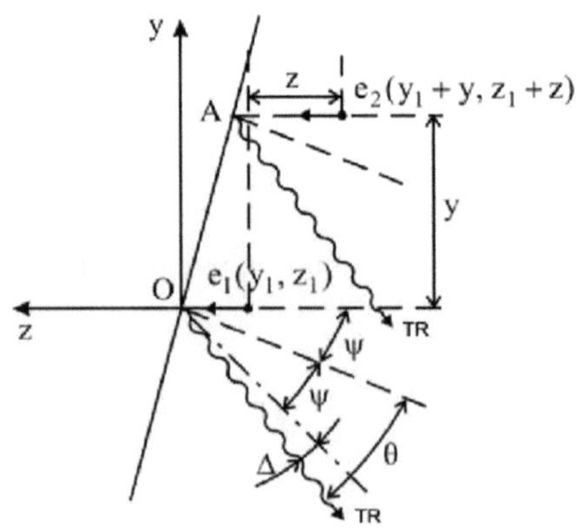

Fig. 7.1 Transition radiation generated by two electrons intersecting the inclined interface between the media

where $\mathbf{E}_0$ is the radiation field amplitude from one electron, $\mathbf{k}$ is the wave vector, $\mathbf{R}_1$ is the vector from the point of intersecting the interface between media by the charge to the observation point, and ω is the frequency of a transition radiation photon.

At time $t_2 = t_1 + \tau$, the target is intersected at point A ($y_2 = y$; $z_2 = y \tan \psi$) by the second electron e_2 whose radiation field is represented in the form

$$\mathbf{E}_2 = \mathbf{E}_0 e^{i\mathbf{k}\mathbf{R}_2 - i\omega(t-\tau)}. \tag{7.16}$$

Transition radiation in the far-field zone can be considered as radiation from a point source. For this reason, photons emitted from different points of the target arrive at the far-field zone at the same angle θ, which is measured from the normal to the target. Then, according to the elementary geometrical consideration (see Fig. 7.1), it is easy to obtain

$$|\mathbf{R}_2| = |\mathbf{R}_1| + \frac{y \sin\theta}{\cos\psi}, \quad \tau = \frac{z}{\beta c} - \frac{y \tan\psi}{\beta c}. \tag{7.17}$$

Therefore, in the case under consideration, the phase difference for radiation fields generated by electrons e_1 and e_2 is determined by the expression

$$\varphi(y, z) = \frac{\omega}{c} \left(\frac{y \sin\theta}{\cos\psi} - \frac{y \tan\psi}{\beta} + \frac{z}{\beta} \right). \tag{7.18}$$

Expression (7.18) is independent of the x coordinate. The passage to the case $x \neq 0$ may be done when phase (7.18) is calculated in the vector representation

$$\varphi = -(\mathbf{k}\Delta\mathbf{r} - \omega\Delta t), \tag{7.19}$$

where $\mathbf{k}$ is the wave vector, $\Delta\mathbf{r} = \mathbf{R}_1 - \mathbf{R}_2$, $\Delta t = \frac{z - y\tan\psi}{\beta}$. The time shift Δt is independent of the x coordinate (along the axis of target rotation by angle ψ, see Fig. 7.1).

For convenience of calculations, we use a prime reference frame, where the z' axis coincides with the mirror reflection direction, the y' axis is located in the figure plane. In this frame, the vectors under consideration are represented in the form

$$\begin{aligned} \mathbf{k} &= \frac{2\pi}{\lambda} \{\sin\theta \sin\xi,\ \sin\theta\cos\xi,\ \cos\theta\}, \\ \Delta\mathbf{r} &= \left\{-x,\ -\frac{y}{\cos\psi},\ 0\right\}. \end{aligned} \tag{7.20}$$

Then, the phase shift is generally determined by the expression

$$\varphi = \frac{2\pi}{\lambda}\left\{x\sin\theta\sin\xi + \frac{y\sin\theta\cos\xi}{\cos\psi} + \frac{z - y\tan\psi}{\beta}\right\} = \\ = \frac{2\pi}{\lambda}\left\{\rho\sin\theta\left[\sin\chi\sin\xi + \frac{\cos\chi\cos\xi}{\cos\psi}\right] + \frac{z - \rho\cos\chi\tan\psi}{\beta}\right\}. \tag{7.21}$$

In the last expression, the cylindrical coordinates ρ, χ, z are used to describe the distribution of the electrons in the bunch with respect to its center of mass.

Thus, with the use of the phase shift found for various particles of the bunch, it is possible to calculate the form factor describing coherent transition radiation for any distribution of the particles in the bunch $S(\mathbf{r})$:

$$f_{\mathrm{CTR}} = \left|\int_V d\mathbf{r} S(\mathbf{r}) \exp\{i\phi(\mathbf{r})\}\right|^2. \tag{7.22}$$

Let us consider coherent transition radiation generated by the electron bunch with the uniform distribution of the particles inside the cylindrical bunch of volume $V = \pi r_0^2 \ell_B$, which intersects the perpendicular target ($\psi = 0$). Substituting $S(\mathbf{r}) = \frac{1}{\pi r_0^2 \ell_B}$ into expression (7.22), we obtain the form factor in the form

$$f_{\mathrm{CTR}} = f_{\mathrm{CTR}}^T \times f_{\mathrm{CTR}}^L, \tag{7.23}$$

$$f_{\mathrm{CTR}}^T = \left|\frac{2}{r_0^2}\int_0^{r_0} J_0\left(\frac{\omega}{c}\rho\sin\theta\right)\rho d\rho\right|^2 = \left[\frac{\lambda}{\pi r_0 \sin\theta} J_1\left(\frac{2\pi r_0\sin\theta}{\lambda}\right)\right]^2,$$

$$f_{\mathrm{CTR}}^L = \left|\frac{1}{\ell_B}\int_0^{\ell_B} e^{i\frac{\omega z}{\beta c}} dz\right|^2 = \left|\frac{2\beta c}{\omega \ell_B}\sin\left(\frac{\ell_B\omega}{2\beta c}\right)\right|^2 = \left[\frac{\beta\lambda}{\pi\ell_B}\sin\left(\frac{\pi\ell_B}{\beta\lambda}\right)\right]^2. \tag{7.24}$$

Thus, the resulting expression is written as

$$f_{\mathrm{CTR}} = \left(\frac{\beta\lambda^2}{\pi^2 r_0 \ell_B \sin\theta_0} J_1\left(\frac{2\pi r_0 \sin\theta}{\lambda}\right)\sin\left(\frac{\pi\ell_B}{\beta\lambda}\right)\right)^2. \tag{7.25}$$

Comparison with form factor (7.13) for coherent synchrotron radiation shows that the results formally coincide. For an arbitrary (but azimuthally symmetric) shape of the bunch, the form factor is not separated into the product of the longitudinal and transverse form factors.

For coherent transition radiation generated by ultrarelativistic particles intersecting an inclined target, i.e., when radiation is concentrated in the angular cone near the mirror reflection direction, it is more natural to use the following angular variables corresponding to this geometry (see Fig. 7.1):

$$\theta = \psi + \Delta, \quad \Delta \sim \gamma^{-1}. \tag{7.26}$$

Then, disregarding the terms γ^{-2} and $\varphi \sim \gamma^{-1}$ in expression (7.22), we obtain the argument of the function f_{CTR}^{T}:

$$\varphi_T \approx \frac{2\pi}{\lambda} \left(\rho \Delta \cos \chi \right) \tag{7.27}$$

and, therefore, instead of expression (7.25), we arrive at

$$f_{\mathrm{CTR}} = \left(\frac{\beta \lambda^2}{\pi^2 r_0 \ell_B \Delta} J_1 \left(\frac{2\pi\, r_0\, \Delta}{\lambda} \right) \sin \left(\frac{\pi \ell_B}{\lambda} \right) \right)^2 . \tag{7.28}$$

It is easy to see that form factor (7.28) formally coincides with expression (7.13) for $f_{\mathrm{CSR}} = f_{\mathrm{CSR}}^{L} \cdot f_{\mathrm{CSR}}^{T}$ in the approximation $\sin\theta = \Delta$, but the polar angle θ in expression (7.13) is measured from the tangent to the electron trajectory, whereas the polar angle θ in expressions (7.22) and (7.28) can be arbitrary up to $\theta \sim \pi/2$.

In [6], the following expression was obtained for the form factor of a uniform cylindrical bunch of length ℓ_B and radius r_0 that intersect a perpendicular target:

$$f_{\mathrm{CTR}} = \left(\frac{4\beta J_1 \left[r_0 \omega \cos \vartheta \right] \sin \left(\frac{\omega \ell_B}{2\beta} \right)}{r_0 \ell_B \omega^2 \cos \vartheta} \right)^2 . \tag{7.29}$$

In that book, the authors used the angle ϑ related to the variable Δ in formula (7.28) as

$$\vartheta = \frac{\pi}{2} - \Delta . \tag{7.30}$$

It is easy to see that form factor (7.29) obtained for the case of the transverse passage of the bunch through the target coincides with formula (7.24), but more general formula (7.22) with phase (7.21) is valid for any target inclination angle.

If the distribution of the particles is described by an arbitrary function $S(\mathbf{r})$ or the shape of the bunch is such that the form factor cannot be represented in the form of the product of the transverse and longitudinal form factors even in the case of azimuthal symmetry, it is necessary to use general formula (7.22).

For illustration, we calculate the form factor describing the generation of coherent transition radiation by a spherical uniform bunch of radius R ($V = \frac{4\pi}{3} R^3$) moving through a perpendicular target:

$$S(\mathbf{r}) = \begin{cases} \dfrac{1}{V}, & r \le R \\ 0, & r > R \end{cases} \tag{7.31}$$

For the sake of generality with the above calculations, the calculation is performed in the cylindrical coordinate system (although the spherical coordinate system is more convenient for this problem). In view of the azimuthal symmetry of the problem, the form factor is represented in the form

$$S(\rho, z) = \begin{cases} \dfrac{1}{V}, & \rho^2 + z^2 \le R \\ 0, & \rho^2 + z^2 > R \end{cases} \tag{7.32}$$

The standard integration with respect to the azimuth angle in expression (7.22) yields

$$f_{\mathrm{CTR}} = \left| \frac{2\pi}{V} \int_{-R}^{R} dz \int_0^{a(z)} \rho d\rho J_0 \left(\frac{2\pi\rho \sin\theta}{\lambda} \right) \exp\left(i \frac{2\pi}{\lambda} z \right) \right|^2 . \tag{7.33}$$

After the integration with respect the transverse radius of the bunch from its center to its boundary $a(z)$, we obtain

$$f_{\mathrm{CTR}} = \left| \frac{\lambda}{V \sin\theta} \int_{-R}^{R} dz\, a(z) J_1 \left(\frac{2\pi \sin\theta}{\lambda} a(z) \right) \exp\left(i \frac{2\pi}{\lambda} z \right) \right|^2 . \tag{7.34}$$

For a sphere, $a(z) = \sqrt{R^2 - z^2}$. For a relativistic case ($\sin\theta \sim \gamma^{-1} << 1$) it is possible to expand the Bessel function in a series and retain only the first term of the expansion. Then, expression (7.34) reduces to the simple integral

$$f_{\mathrm{CTR}} = \left| \frac{\pi}{V} \int_{-R}^{R} dz \left(R^2 - z^2 \right) \exp\left(i \frac{2\pi}{\lambda} z \right) \right|^2 = \left(3 \frac{\sin x - x \cos x}{x^3} \right)^2 , \quad x = \frac{2\pi R}{\lambda}. \tag{7.35}$$

This expression coincides with the results obtained in [7] for a spherical bunch (see formula (13) with $\alpha = 1$ and $\theta \ll 1$).

Figure 7.2 shows the form factor for the (line 1) spherical and (line 2) cylindrical bunches. In the latter case, the calculation is performed by formula (7.25) for a cylinder of length $2R$ and radius R.

Owing to a sharp (stepwise) decrease in the charge density on the input and output surfaces of the cylindrical bunch, the form factor exhibits local maxima in the region of small wavelengths ($\lambda < 2R$). These maxima are less pronounced for the spherical bunch owing to a more smooth decrease in the charge distribution

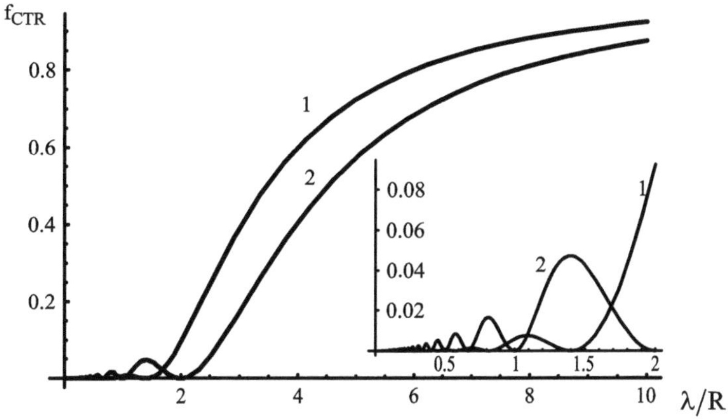

Fig. 7.2 Form factors versus the ratio λ/R for the (*line 1*) spherical and (*line 2*) cylindrical bunches. The inset shows the same plots in the region of small λ/R values

density along the z axis. As will be shown below (in Sect. 7.4), such maxima are completely absent for a Gaussian distribution of the charges in the bunch.

Bolotovskii and Serov [8] proposed another method for describing transition radiation generated by bunches of charged particles as a result of interference of radiations generated by thin layers of the bunch, which successively intersect the interface between the media (see Fig. 7.3).

The xy plane in this figure is an infinitely thin perfect conducting target. A circle of radius $a(t)$ is a bunch layer emitted at time t. Circle radius $a(t)$ is a function of time and the rate of its variation can be higher than the speed of light.

The following expression was obtained in [8] for the field of coherent transition radiation:

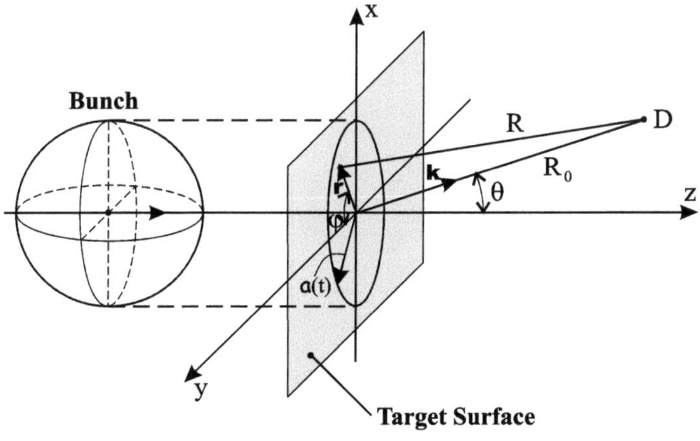

Fig. 7.3 Radiation generated by a thin layer of the bunch intersecting the interface between two media

$$E(\omega,\theta) = \frac{\sigma}{\pi \sin\theta}\left(\frac{\beta\sin\theta}{\omega\left(1-\beta^2\cos^2\theta\right)}\right)\times \\ \times \int_0^R a(t)\, J_1\left[\frac{\omega}{c}a(t)\sin\theta\right]\exp(i\omega t)\,dt, \tag{7.36}$$

where σ is the surface charge density, β is the particle velocity, and θ is the angle between the particle motion direction and wave vector.

The expression for the form factor for the target inclination angle $\psi = 0$ that is obtained from expression (7.22) is in agreement with expression (7.36) for perpendicular motion (up to an insignificant constant (see formula (7.34)).

To derive an expression for the form factor in the case of coherent diffraction radiation, we consider the scheme shown in Fig. 7.4. As mentioned above, coherent diffraction radiation in the far-field zone, as well as coherent transition radiation, can be considered as radiation generated by a point source. Then, the superposition of the fields of forward diffraction radiation generated by two electrons (the first of which is located in the center of the bunch with the coordinates $x = y = z = 0$ and the second has arbitrary coordinates x, y, and z) is represented in the form

$$\mathbf{E}_{1+2} = \mathbf{E}_1 + \mathbf{E}_2\exp(-i\Delta\phi), \tag{7.37}$$

where the phase shift $\Delta\varphi$ is determined only by the coordinates x, z

$$\Delta\varphi = \frac{2\pi}{\lambda}\left(x\sin\theta\sin\xi + \frac{z}{\beta}\right), \tag{7.38}$$

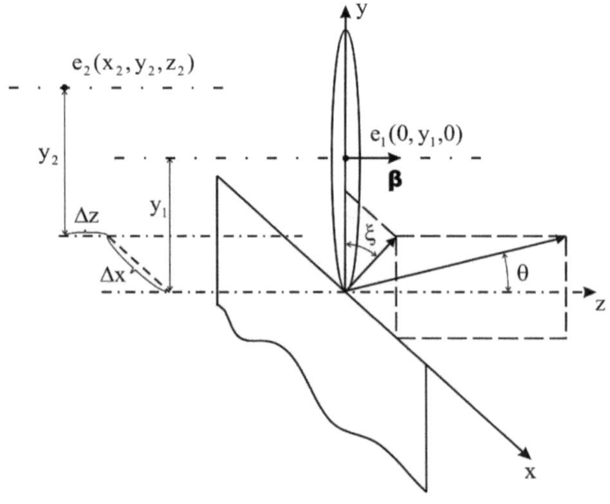

Fig. 7.4 Scheme of the generation of diffraction radiation by two charged particles

The strength of the diffraction radiation field depends on the y coordinate as

$$E(y) = E(0)\exp\left(-\frac{2\pi y}{\gamma\beta\lambda}\right)\sqrt{1+\gamma^2\beta^2\sin^2\theta\sin^2\xi}. \tag{7.39}$$

Thus, if the bunch is described by independent longitudinal, $S_L(z)$, and transverse, $S_T(x, y)$, distributions, the form factor of coherent diffraction radiation is represented in the form

$$f_{\mathrm{CDR}} = \left|\int dx\,dy\,dz\,S_L(z)\,S_T(x,y)\exp\left(-\frac{2\pi y\sqrt{1+\gamma^2\beta^2\sin^2\theta\sin^2\xi}}{\gamma\beta\lambda}\right)\times\right.$$
$$\left.\times\exp\left[-\frac{2\pi i}{\lambda}\left(x\sin\theta\sin\xi+\frac{z}{\beta}\right)\right]\right|^2 = f^T_{\mathrm{CDR}}f^L_{\mathrm{CDR}}, \tag{7.40}$$

where the transverse and longitudinal form factors are calculated by the formulas

$$f^T_{\mathrm{CDR}} = \left|\int dx\,dy\,S_T(x,y)\exp\left(-\frac{2\pi y}{\gamma\beta\lambda}\sqrt{1+\gamma^2\beta^2\sin^2\theta\sin^2\xi}\right)\times\right.$$
$$\left.\times\exp\left(-\frac{2\pi i\,x\sin\theta\sin\xi}{\lambda}\right)\right|^2. \tag{7.41}$$

$$f^L_{\mathrm{CDR}} = \left|\int dz S_L(z)\exp\left(-\frac{2\pi i}{\lambda}\frac{z}{\beta}\right)\right|^2. \tag{7.42}$$

If the transverse size of the bunch is much smaller than its longitudinal size (this condition is usually satisfied for ultrarelativistic electron beams), the transverse form factor (see expression (7.41)) coincides with unit with a good accuracy. In this case, the form factor of the bunch is completely determined by longitudinal form factor (7.42), which is independent of emission angles of a diffraction radiation photon.

7.2 Coherent Synchrotron Radiation in the Millimeter and Submillimeter Wavelength Ranges

In the first experiment [9] performed on a 180-MeV electron beam with bunch length $\ell_B \sim 1$ mm and a population (the number of electrons in a bunch) of $\sim 10^6$, the enhancement of the yield of coherent synchrotron radiation with wavelength $\lambda_{\mathrm{SR}} \approx 500\,\mu$m ($f = 0.6$ THz) by approximately five orders of magnitude as compared to incoherent synchrotron radiation was observed.

The beam of coherent synchrotron radiation usually acts on a finite object located at a given distance L from a synchrotron radiation source. Disregarding the effects of the near-field zone (associated with the distortion of the characteristics of synchrotron radiation, which are calculated for the wave ("far-field") zone [10], we consider the effect of a finite solid angle on the spectrum of coherent synchrotron radiation and on its power when the object under investigation is located at distances $L \gg \gamma^2\lambda$.

For quantitative analysis of the spectrum of coherent synchrotron radiation in this case, it is necessary to use the formulas of the theory of synchrotron radiation, which allow one to calculate the characteristics of radiation for the emission of a synchrotron radiation photon at arbitrary angles with respect to the instantaneous velocity of an electron (see, e.g., [11]). We use angular variable $t = \gamma\theta$ (θ is the polar angle of photon emission in the reference frame, where the instantaneous velocity of the electron is directed along the z axis) and spectral variable $y = \omega/\omega_c$, where ω_c is the typical frequency in the spectrum of synchrotron radiation given by the expression

$$\omega_c = \frac{3}{2}\frac{\gamma^3 c}{R}. \tag{7.43}$$

Here, R is the radius of an orbit. In terms of these variables, the spectral–angular distribution of the synchrotron radiation power (after the integration with respect to the azimuth angle measured from the instantaneous velocity) per electron is represented in the form

$$\frac{dP}{tdtdy} = P_0 \frac{27\sqrt{3}}{8\pi}\sqrt{1+t^2}\left(1+2t^2\right) y^2 K_{1/3}\left[y\left(1+t^2\right)^{3/2}\right], \tag{7.44}$$

where $P_0 = \frac{2}{3}\frac{e^2 c\gamma^4}{R^2} = \frac{2}{3}\alpha\,\hbar\, c^2 \frac{\gamma^4}{R^2}$ is the power of synchrotron radiation generated by the relativistic electron.

Figure 7.5 shows (line 1) the spectrum of synchrotron radiation after the integration of expression (7.44) over the entire t interval ($0 \le t \le \infty$) in comparison

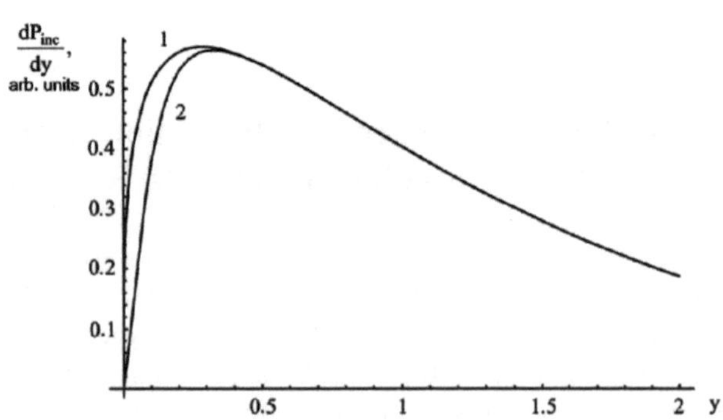

Fig. 7.5 Spectral distribution of the power of incoherent synchrotron radiation in (*line 1*) the total cone and (*line 2*) the angular cone $\theta \le 2.4\gamma^{-1}$ with respect to the instantaneous velocity of the electron

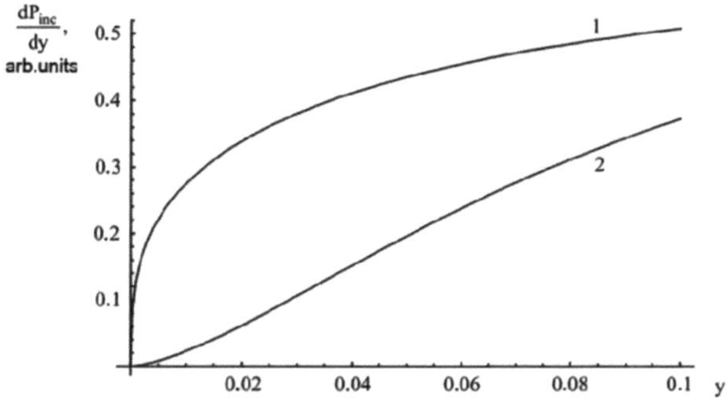

Fig. 7.6 Same as in Fig. 7.5, but for the softer part of the spectrum

with (line 2) the spectrum of synchrotron radiation obtained for the radiation cone $0 \leq t \leq t_{\max} = 2.4$. The last value is close to the instrumental angular acceptance in the experiment reported in [12] (± 30 mrad $\approx 2.4\gamma^{-1}$).

Figure 7.6 shows the spectra of synchrotron radiation in the region $y \ll 1$ ($\omega \ll \omega_c$). The total spectrum (line 1) in this region is described by the function $y^{1/3}$, whereas line 2 is almost a linear function of the variable y.

For the longitudinal form factor

$$f_{\mathrm{CSR}}^{L} = \exp\left(-\frac{2\pi^2\sigma^2}{\lambda^2}\right), \tag{7.45}$$

corresponding to the Gaussian longitudinal distribution of electrons in the bunch

$$S(z) = \exp\left(-\frac{z^2}{\sigma^2}\right), \tag{7.46}$$

the linear frequency spectrum of synchrotron radiation in a given cone makes it possible to obtain an analytical estimate of the power of coherent synchrotron radiation and, therefore, energy losses to radiation from the arc $\sim 2\pi R\gamma^{-1}$ (i.e., in time $\tau_0 \sim 2\pi R/\gamma c$). If the lower line in Fig. 7.6 is approximated by a linear function with the fitting constant $C_{\mathrm{inc}} = 3.92$:

$$\frac{dP_{\mathrm{inc}}}{dy} = P_0 C_{\mathrm{inc}} y, \tag{7.47}$$

the expression for the power of coherent synchrotron radiation per electron in the bunch is represented in the form

$$\frac{dP_{\mathrm{CSR}}}{dy} = N_{\mathrm{e}} P_0 C_{\mathrm{inc}} f_{\mathrm{CSR}}^{L} y = N_{\mathrm{e}} P_0 C_{\mathrm{inc}} y \exp\left[-\frac{9}{8}\left(\gamma^3\frac{\sigma}{R}\right)^2 y^2\right]. \tag{7.48}$$

After the integration over the spectrum (with respect to the variable y), we obtain the power lost by each electron in the bunch:

$$P_{\rm coh}^{\rm SR} = N_{\rm e} P_0 C_{\rm inc} \frac{4}{9} \frac{1}{\gamma^6} \left(\frac{R}{\sigma}\right)^2 = N_{\rm e} C_{\rm inc} \frac{8}{27} \alpha \hbar c^2 \frac{1}{\gamma^2 \sigma^2}. \tag{7.49}$$

According to this expression, the power of coherent synchrotron radiation is independent of the radius of the trajectory and is inverse proportional to the Lorentz factor squared. The dependence on the radius is absent, because the spectrum of coherent synchrotron radiation for reasonable lengths of the bunch $l_B \sim 2.36\sigma$ lies in the region $y \ll 1$, i.e., in the frequency region almost independent of R ($\lambda \sim \sigma \gg R/\gamma^3$).

Figure 7.7 shows (line 2) the spectrum of coherent synchrotron radiation calculated for the conditions close to the conditions of the described experiment [12]]:

$$\begin{aligned} N_{\rm e} &= 6 \cdot 10^8, & \gamma &= 80, & t_{\max} &= 2.4, \\ R &= 1\,{\rm m}, & \sigma &= 64\,\mu{\rm m}, & \ell_B &= 2.36\sigma. \end{aligned} \tag{7.50}$$

This figure also shows the spectrum of coherent synchrotron radiation after the integration with respect to the angles. Note that both spectra coincide for frequencies $\sim 100\,{\rm cm}^{-1}$, where coherent effect disappears.

For illustration, Fig. 7.8 shows the total spectrum of synchrotron radiation (coherent + incoherent). With the use of the power of coherent synchrotron radiation

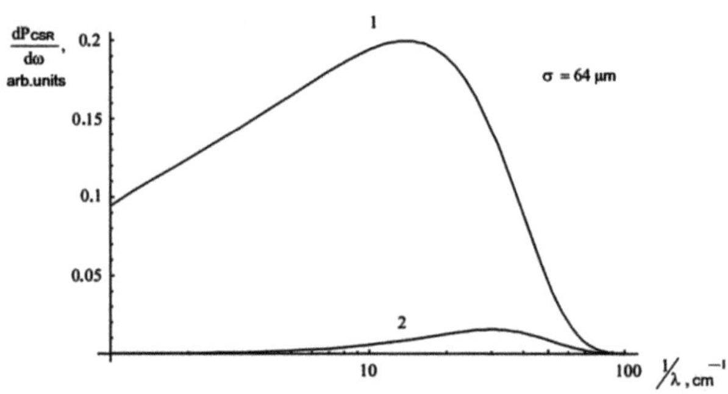

Fig. 7.7 Intensity of coherent synchrotron radiation for $\sigma = 64\,\mu$m in the (*line 1*) total cone and (*line 2*) cone $\theta \leq 2.4\gamma^{-1}$

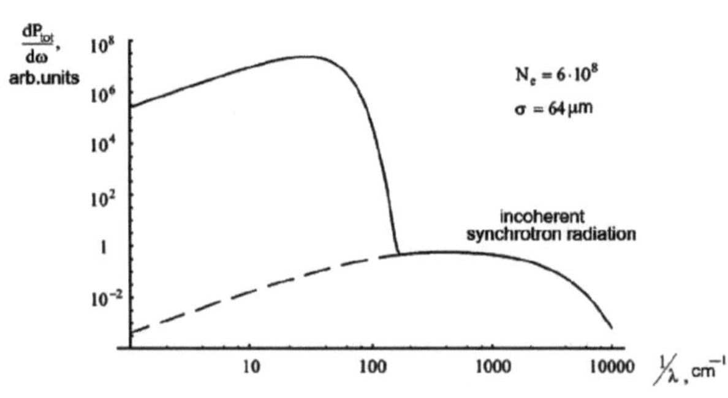

Fig. 7.8 Total spectrum of synchrotron radiation in the angular cone $\theta \leq 2.4\gamma^{-1}$ of an electron bunch with $N_{\rm e} = 6 \times 10^8$ and $\sigma = 64\,\mu$m. The upper and lower lines are coherent and incoherent radiations, respectively

generated by the beam given by expression (7.49) and "registration time" by the detector, which can be estimated for the experimental conditions as

$$t_2 = 2\frac{2\pi R}{c\gamma}, \tag{7.51}$$

we calculate radiation losses per electron (with the fitting parameter $C_{\rm inc} = 3.92$):

$$W_{\rm CSR} = P_{\rm coh} \cdot t_2 \approx 6 \times 10^3\,\text{eV}. \tag{7.52}$$

Then, for the bunch with population $N_{\rm e} = 6 \times 10^8$ and repetition rate $f_0 = 37.4 \cdot 10^6$ s^{-1}, we obtain the average power

$$\bar{P} = W_{\rm CSR} N_{\rm e} f_0 \approx 20\,\text{W}. \tag{7.53}$$

This value is in good agreement with the experimental data [12].

7.3 Coherent Diffraction Radiation

For comparison with the spectrum of coherent synchrotron radiation, we determine the spectrum of coherent diffraction radiation generated by a relativistic particle moving through the center of a circular hole of radius ρ [13] (see Sect. 3.1). In this case, the radiation power is not a universal characteristic, because the time scale of the generation of diffraction radiation is governed by two characteristic times:

$$\tau_{\rm DR} \approx \frac{1}{\omega_0} = \frac{\rho}{c\gamma}, \qquad \tau_B \approx \frac{\ell_B}{c}, \tag{7.54}$$

first of which determines the time of "burst" of diffraction radiation when a single charge moves through the hole and the second time is the time of the passage of the electron bunch of length ℓ_B through the hole. It is clear that the time scale in both cases depends on the parameters of the problem; for this reason, it is more natural to consider the spectral—angular distribution of energy lost to diffraction radiation. As shown in Chap. 3, radiation energy losses of a single particle are governed by the hole radius ρ and Lorentz factor and, e.g., for the "forward" diffraction radiation cone,

$$W_{\rm FDR} = \frac{1}{2} W \simeq \frac{1}{2} \alpha\, \hbar\, c \,\frac{\gamma}{\rho}. \tag{7.55}$$

To calculate the spectrum of forward diffraction radiation, it is convenient to use an expression for the diffraction radiation field, from which it is easy to obtain the spectral–angular distribution of the intensity:

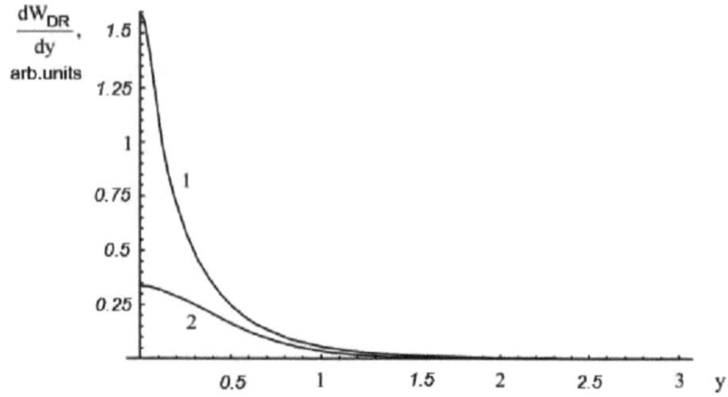

Fig. 7.9 Spectral distribution of the intensity of diffraction radiation generated by a relativistic charge moving through a hole of radius $\rho = 1$ mm (*lines 1 and 2* correspond to radiation in the total cone and in the angular cone $\theta \leq 2.4/\gamma$, respectively)

$$\frac{d^2 W_{DR}}{dy\, t\, dt} = \frac{4\alpha\hbar}{\pi}\omega_c \frac{t^2}{\left(1+t^2\right)^2}\left[y K_1(y)\, J_0(y\, t)\right]^2, \tag{7.56}$$

where $t = \gamma\theta$, $y = \omega/\omega_c$, and $\omega_c = c\gamma/\rho$ is the spectrum cutoff frequency of diffraction radiation. Similar to the case of synchrotron radiation, the spectral distribution of radiation depends on the emission angle of diffraction radiation photons.

Figure 7.9 shows the spectra of diffraction radiation in the (line 1) total cone and (line 2) cone with $t_{\max} = 2.4$. Similar to the case of synchrotron radiation, a significant decrease in the radiation intensity is observed in the "soft" part in the last case.

The frequency spectrum of coherent diffraction radiation is shown in Fig. 7.10 for the hole radius $\rho = 1$ mm and angular aperture $t_m = 2.4$ and for the parameters of the electron bunch for which the characteristics of coherent synchrotron radiation were calculated. As seen in Fig. 7.10, the spectrum of diffraction radiation in the coherence region $y \ll 1$ can be approximated by constant C_{DR}. In this case, the intensity spectrum of coherent diffraction radiation can be obtained by integrating expression (7.56) with respect to the angular variable:

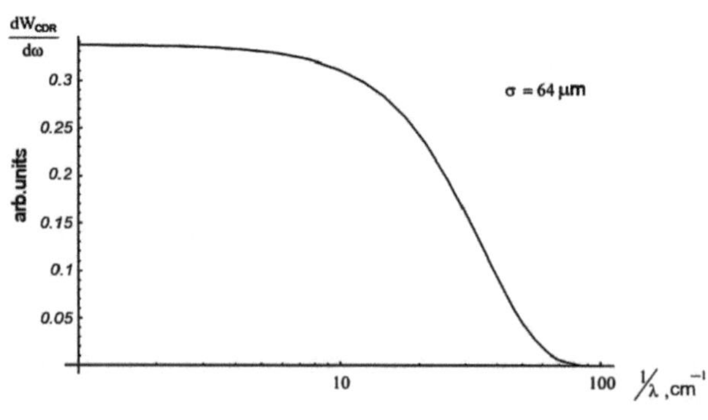

Fig. 7.10 Intensity of coherent diffraction radiation versus the wavenumber for the same parameters as in Fig. 7.9 and $\sigma = 64\,\mu$m

$$
\begin{aligned}
\frac{d^2 W_{\mathrm{CDR}}}{dy} &= N_{\mathrm{e}} \exp\left[-\frac{1}{2}\left(\frac{\gamma}{\rho}\right)^2 \sigma^2 y^2\right] \int \frac{d^2 W_{\mathrm{DR}}}{dytdt}\, tdt = N_{\mathrm{e}} \frac{4\alpha\hbar}{\pi}\, \omega_c \times \\
&\times \exp\left[-\frac{1}{2}\left(\frac{\gamma}{\rho}\right)^2 \sigma^2 y^2\right] \int_0^{t_m} tdt \frac{t^2}{\left(1+t^2\right)^2} \left[y K_1(y)\, J_0(y\,t)\right]^2 \\
&\approx N_{\mathrm{e}} \frac{4\alpha\hbar}{\pi}\, \omega_0 C_{\mathrm{DR}} \exp\left[-\frac{1}{2}\left(\frac{\gamma}{\rho}\right)^2 \sigma^2 y^2\right].
\end{aligned}
\tag{7.57}
$$

After the integration over the spectrum in the interval $0 \le y < \infty$ (this integration does not give large errors owing to a rapid decrease in the Gaussian for large y values for which the approximation of the spectral density by a constant is invalid), the expression for radiation losses of each electron in the bunch to coherent diffraction radiation is obtained in the form

$$
W_{\mathrm{CDR}} \approx N_{\mathrm{e}} \frac{4\alpha\hbar}{\pi}\, \omega_c C_{\mathrm{DR}} \sqrt{\frac{\pi}{2}} \frac{\rho}{\gamma\sigma} = N_{\mathrm{e}} \frac{4\alpha\,\hbar c}{\sqrt{2\pi}}\, C_{\mathrm{DR}} \frac{1}{\sigma}. \tag{7.58}
$$

Similarly to the mechanism of coherent synchrotron radiation considered above, this expression is independent of the hole radius, which is one of the main size parameters. This is due to the fact that the quantity $\gamma\lambda_{\mathrm{coh}} \sim \gamma\sigma$ in the case under consideration is much larger than the hole radius and this relation corresponds to transition radiation with a constant intensity spectrum. For the parameters used above, estimate (7.58) yields the value $W_{\mathrm{CDR}} \sim 6 \cdot 10^4$ eV, which is an order of magnitude larger than energy losses through the mechanism of coherent synchrotron radiation.

Figure 7.11 shows the total spectrum of diffraction radiation generated by the electron bunch with the above parameters in the logarithmic scale. Comparison of this figure with Fig. 7.8 shows a noticeable excess of the spectral density of coherent

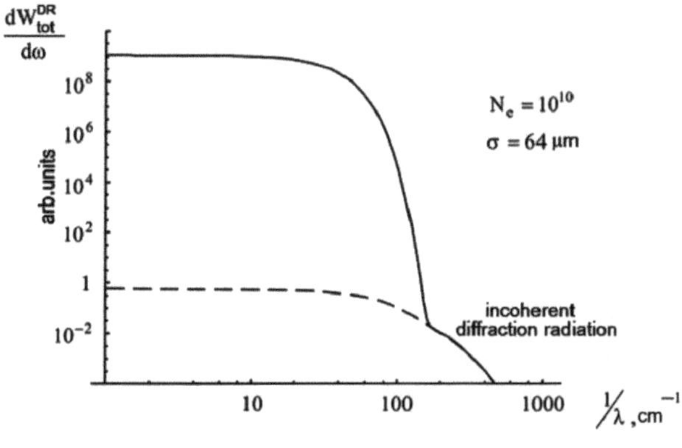

Fig. 7.11 Spectrum of diffraction radiation generated by an electron bunch with $N_{\mathrm{e}} = 10^{10}$; the other parameters are the same as in Fig. 7.10

diffraction radiation in the frequency range $< 100\,\mathrm{cm}^{-1}$ as compared to coherent synchrotron radiation. This excess is responsible for the excess of radiation losses in the case of coherent diffraction radiation.

In recent years, the wavelength range between the traditional microwave electronics (with frequencies below 100 GHz) and photonics (with frequencies higher than 100 THz, i.e., infrared and optical ranges) has attracted large attention [14].

The existing laboratory sources ensure the yield of radiation in the terahertz range with a power of less than 1 mW (radiation from a mercury-vapor lamp in its characteristics is close to radiation from a black body with a temperature of $\sim 20,000$ K). Note that a frequency of 1 THz corresponds to the wavelength $\lambda = 300\,\mu\mathrm{m}$; hence, coherent radiation generated by an electron bunch with length $\ell_B \sim 300\,\mu\mathrm{m}$ contains a significant terahertz component. Additional advantages of such a source (independently of the radiation mechanism) are the directivity of radiation and short duration ($\sim \frac{\ell_B}{c}$) [15].

The first real source of terahertz radiation on the basis of an electron accelerator whose average power is three orders of magnitude higher than that of the standard sources was created in 2002 [16]. In that work, the terahertz source was created on the basis of a 40-MeV electron beam, which is a sequence of short bunches with a charge of 0.1 nC, a duration of 500 fs, and a frequency of 37.4 MHz, which pass through a magnetic field with the trajectory radius $R = 1$ m.

Generation of terahertz radiation by short electron bunches through the mechanism of coherent diffraction radiation can be sometimes more efficient than that through the mechanism of coherent synchrotron radiation. Backward coherent diffraction radiation has an additional advantage; namely, by choosing the target inclination angle, the beam of coherent diffraction radiation can be directed at arbitrary angle to the electron momentum up to the right angle, which significantly reduces background loads.

This last characteristic of backward coherent diffraction radiation is very promising for non-invasive diagnostics of the length of an electron bunch. The term "non-invasive diagnostics" usually means the absence of interaction of the beam under investigation with a solid target, i.e., the complete absence of ionization losses and multiple scattering. Nevertheless, the energy of the initial particle decreases owing to radiation losses. Further, if the diffraction target is, for example, inclined at an angle of 45° to the initial beam, the backward diffraction radiation cone is directed along the mirror reflection direction, i.e., at an angle of 90°. In this case, the photons in the backward diffraction radiation cone "carry" not only energy, but also the transverse momentum. This fact, owing to the recoil effect, gives rise to the transfer of the transverse momentum to the initial particle, i.e., to the deflection of the particle trajectory from the initial direction. The angle of deflection for the most interesting case of backward coherent diffraction radiation generated by a target inclined at an angle of 45° can be estimated with the use of the simple model proposed in [17].

If the length of the relativistic bunch, σ, satisfies the condition $\gamma\sigma/2\pi \leq h$, where h is the impact parameter, radiation losses in the backward cone of coherent diffraction radiation of each electron (which are equal to losses in the forward coherent diffraction radiation cone) in the bunch with the number of electrons N_e can be estimated as (see, e.g., formula (7.58)):

$$\Delta W_{\mathrm{CDR}} \sim N_{\mathrm{e}} \frac{\alpha \hbar c}{\sigma}. \tag{7.59}$$

Then the momentum carried by photons of backward coherent diffraction radiation in the direction perpendicular to the electron beam is estimated from expression (7.59) as

$$K_{\mathrm{CDR}} \sim \frac{\Delta W_{\mathrm{CDR}}}{c}. \tag{7.60}$$

Therefore, the transverse momentum acquired by the initial electron is governed by the bunch length and the number of electrons in the bunch:

$$\Delta p_{\perp} \sim K_{\mathrm{CDR}} \sim N_{\mathrm{e}} \frac{\alpha \hbar}{\sigma}. \tag{7.61}$$

This expression allows one to estimate the angle of deflection of the initial electron:

$$\Delta \theta \sim \frac{\Delta p_{\perp}}{p_0} \sim \frac{N_{\mathrm{e}} \alpha \hbar / \sigma}{\gamma m c} = \frac{N_{\mathrm{e}} r_0}{\gamma \sigma}, \tag{7.62}$$

where $r_0 = 2.82 \cdot 10^{-13}$ cm is the classical electron radius.

In the accelerator physics, the processes of passage of short bunches of charged particles through conducting structures of various geometries are considered. In this case, the particles in a bunch are subjected to the fields generated by the currents induced in the elements of a structure by a passing beam. This action is calculated in terms of the wakefields [18].

Stupakov [19] considered the passage of an electron bunch through a "pyramidal" collimator. If the angle at the vertex of the pyramid is equal to 45° and the electron bunch moves near the wall of the collimator rather than along its axis, the angle of deflection is estimated by the formula

$$\Delta \theta \sim \frac{\sqrt{\pi}}{4} \frac{N_{\mathrm{e}} r_0}{\gamma \ell_B}, \tag{7.63}$$

where ℓ_B is the length of the bunch. Comparison expressions (7.62) and (7.63) shows that they are in good agreement; hence, formulas similar to formula (7.62) can be used to estimate the "perturbation" of the initial beam by a diagnostic diffraction target.

Concluding, we note that, for real experimental conditions, the spectra shown in Figs. 7.8 and 7.11 should be corrected in the "soft" part of the spectrum. In the first case (coherent synchrotron radiation), this correction is due to the suppression of the yield of radiation caused by the "screening" of the field of an electron by the poles of a magnet that are located at a given distance [20]. In the second case (coherent diffraction radiation), this correction is attributed to the finiteness of the outer diameter of the target [21].

7.4 Coherent Smith—Purcell Radiation

Note that form factor (7.40) also describes coherent Smith—Purcell radiation, because Smith—Purcell radiation can be considered as resonant diffraction radiation as mentioned above. In [22, 23], expressions for f_{CSPR} were derived by another method, but the resulting expression coincides with formula (7.40).

For relativistic energies, which are of most interest, the effective transverse size of the beam,

$$\langle y\rangle = \frac{\int y\, S_T(x,y)\, dx\, dy}{\int S_T(x,y)\, dx\, dy}, \tag{7.64}$$

usually satisfies the relation

$$\langle y\rangle \ll \gamma\lambda. \tag{7.65}$$

For this reason, the first exponential factor in the integrand in expression (7.41) can be changed to unit. Then, the expression for the transverse form factor is significantly simplified:

$$f_{\mathrm{CSPR}}^{T} = \left|\int d\mathbf{r}_T\, S_T\left(\mathbf{r}_T\right) \exp\left(\frac{2\pi i}{\lambda} x \sin\theta \sin\xi\right)\right|^2. \tag{7.66}$$

For the simplest transverse distribution (a cylindrical bunch with uniform filling, as above), integral (7.41 is calculated analytically:

$$f_{\mathrm{CDR}}^{T} = \left[\frac{\lambda}{\pi r\, \sin\theta \sin\xi} J_1\left(\frac{2\pi}{\lambda} r_0 \sin\theta \sin\xi\right)\right]^2. \tag{7.67}$$

In this case, the longitudinal form factor for coherent Smith–Purcell radiation coincides with the second of formulas (7.24) and is independent of the radiation emission angle θ.

In some works (see, e.g., [23]), the form factor for the case of coherent Smith—Purcell radiation is represented by analogy with the case of coherent synchrotron radiation, i.e., with the use of expression (7.6). In this model, when calculating the longitudinal form factor, the factor $\exp\left(-\frac{2\pi i}{\lambda} z \cos\theta\right)$ remains in the integrand. As a result, the longitudinal form factor in [23] had the form

$$f_{\text{CSPR}}^{L} \sim \exp\left(-\frac{4\pi^2\sigma^2}{\lambda^2}\cos^2\theta\right) \tag{7.68}$$

if the longitudinal distribution of electrons in the bunch is described by a Gaussian with variance σ^2. Using this formula, one can erroneously conclude that coherence for Smith—Purcell radiation at angle $\theta = \pi/2$ is observed for any wavelength and for arbitrary length of the bunch.

If both the longitudinal and transverse distributions of particles in the electron bunch are Gaussians, i.e.,

$$\begin{aligned} S_L(z) &= \frac{1}{\sqrt{\pi}\,\sigma_L}\exp\left(-\frac{z^2}{\sigma_L^2}\right), \\ S_T(r_T) &= \frac{1}{\pi\,\sigma_T^2}\exp\left(-\frac{r_T^2}{\sigma_T^2}\right), \end{aligned} \tag{7.69}$$

the form factor is given by the expression

$$\begin{aligned} f_{\text{CSPR}} &= f_{\text{CSPR}}^{L}\cdot f_{\text{CSPR}}^{T} = \\ &= \exp\left\{-\frac{2\pi^2\sigma_L^2}{\lambda^2}\right\}\exp\left\{-\frac{2\pi^2\sigma_T^2}{\lambda^2}\sin^2\theta\sin^2\xi\right\}. \end{aligned} \tag{7.70}$$

According to this formula, the transverse form factor for radiation in the plane perpendicular to the grating ($\xi = 0$) is equal to unit for any wavelength. However, the situation is more complicated outside of this plane.

Under the condition

$$\lambda \gg \pi\sigma_T\sin\theta\sin\xi, \tag{7.71}$$

the transverse form factor is close to unit (spatial coherence is not violated). However, in contrast to radiation mechanisms considered in Sect. 7.1, the mechanism of coherent Smith—Purcell radiation exhibits a strict relation between the wavelength and emission angle. Therefore, condition (7.71) can be written in the form

$$\sigma_T \ll \frac{d\tan(\theta/2)}{n\pi\sin\xi}, \qquad n = 1, 2, 3, \ldots, \tag{7.72}$$

which indicates that the spatial coherence can be violated for sufficiently small polar angles θ and nonzero azimuth angles. Figure 7.12 shows the effect of the transverse size of the bunch on the intensity of coherent Smith—Purcell radiation for the "transverse" grating in which strips are perpendicular to the electron trajectory.

Criterion (7.72) can be important for nonrelativistic and moderately relativistic electrons (e.g., if $\gamma \leq 10$). In the ultrarelativistic case, where $\sin\xi \sim \gamma^{-1}$,

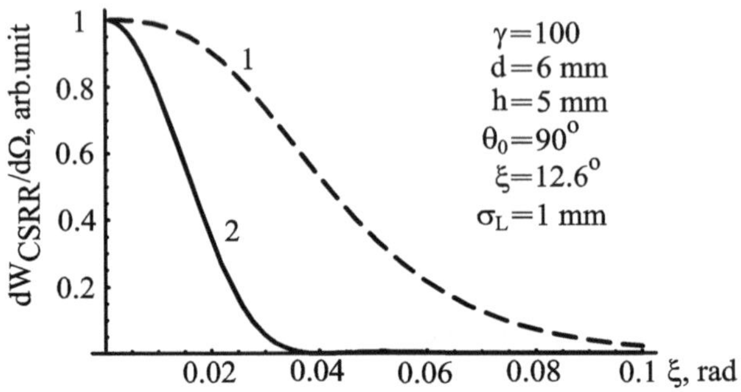

Fig. 7.12 Azimuthal distribution of coherent Smith—Purcell radiation for electron bunches of the diameters (*line 1*) $\sigma_T = 10\,\text{mm}$ and (*line 2*) $\sigma_T = 1\,\text{mm}$

condition (7.72) is almost always satisfied and the main criterion of coherence is the closeness of the longitudinal form factor to one:

$$\lambda \gg \pi\sigma_L \tag{7.73}$$

or, with the use of the Smith—Purcell dispersion relation:

$$\sigma_L \ll \frac{d\left(\beta^{-1} - \cos\theta\right)}{\pi n}, \quad n = 1, 2, 3, \ldots . \tag{7.74}$$

The authors of [23–25] proposed to use condition (7.74) for measuring the length of an ultrashort electron bunch ($\sigma_L \leq 1\,\text{mm}$), for which the traditional methods are inapplicable.

Figure 7.13 shows the yield of coherent Smith—Purcell radiation as a function of the polar observation angle for various lengths of the electron bunch. As seen in this figure, the angle corresponding to the peak of coherent Smith—Purcell radiation depends on the relation between the length of the bunch, σ_L, and the grating period [26]. Using a detector with a sufficiently broad passband, one can determine the length of the short electron bunch from measurement results for the angular dependence of coherent Smith—Purcell radiation from a grating with a chosen period (Figure 7.13 shows the results of calculation for $\sigma_L = 40$, 60, and $80\,\mu\text{m}$ and the grating with a period of 0.6 mm).

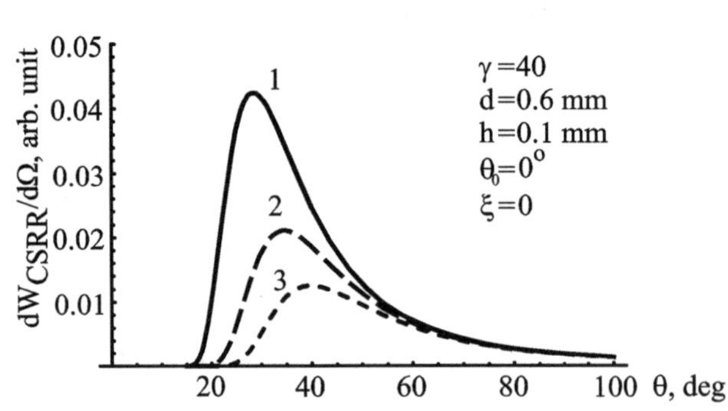

Fig. 7.13 Angular distribution of coherent Smith—Purcell radiation generated by electron bunches of the lengths $\sigma_L =$ (*line 1*) 40, (*line 2*) 60, and (*line 3*) $80\,\mu\text{m}$

According to this figure, with the use of the yield of coherent Smith—Purcell radiation measured in the angular interval from $\theta_{\min} = 20°$ to $\theta_{\max} = 50°$ by the detector with the sensitivity band from $\lambda_{\min}(\theta = 20°) = 36\,\mu\text{m}$ to $\lambda_{\max}(\theta = 50°) = 214\,\mu\text{m}$, it is easy to determine the position of the maximum with an accuracy of several arc degrees, which makes it possible to measure the length of the electron bunch with the required accuracy. The similar technique to measure a bunch length was developed in papers [27, 28].

References

1. Michel, C.F.: Intense coherent submillimeter radiation in electron storage ring. Phys. Rev. Lett. **48**, 580 (1982)
2. Nodvick, J.S., Saxon, D.S.: Suppression of coherent radiation by electrons in a synchrotron. Phys. Rev. **96**, 180 (1954)
3. Blum, E.B., Happek, U., Sievers, A.J.: Observation of coherent synchrotron radiation at the Cornell linac. Nucl. Instrum. Methods Phys. Res. A **307**, 568 (1991)
4. Shibata, Y., Ishi, K., Takahashi, T. et al.: Observation of coherent synchrotron, Cherenkov and wake-field radiation at millimeter wave lengths using an L-band linear accelerator. Phys. Rev. A. **44**, R3449 (1991)
5. Alferov, D.F., Bashmakov, Yu.A., Bessonov, E.G.: Theory of coherent undulator radiation. Sov. Zhurnal Tekhnicheskoi Fiziki **48**, 1598 (1978)
6. Garibyan, G.M., Yang, C.: X-ray Transition Radiation. Yerevan, Armenia (1983)
7. Op't, W.P.E.M., Smorenburg, P.W., van Oudheusden, T. et al.: Theory of coherent radiation generated by ellipsoidal electron bunches. Phys. Rev. ST-AB **10**, 012802 (2007)
8. Bolotovskiy, B.M., Serov, A.V.: Transition radiation of lengthy system of charges. Tech. Phys. **47**, 1 (2002)
9. Nakazato, T., Oyamada, M., Niimura, N. et al.: Observation of coherent synchrotron radiation. Phys. Rev. Lett. **63**, 1245 (1989)
10. Bosch, R.A.: Extraction of edge radiation within a straight section of Aladdin. Rev. Sci. Instrum. Rev. Sci. Instrum. **73**, 1423 (2002)
11. Baier, V.N., Katkov, V.M., Fadin, V.N.: Radiation from Relativistic Electrons. Moscow, Atomizdat (in Russian) (1973)
12. Neil, G.P., Carr, G.L., Gubeli Joseph, F. et al.: Production of high power femtosecond terahertz radiation. Nucl. Instrum. Methods Phys. Res. A **507**, 537 (2003)
13. Potylitsyn, A.P.: Transition radiation and diffraction radiation. Similarities and differences. Nucl. Instrum. Methods Phys. Res. B **145**, 169 (1998)
14. Williams, G.P.: Filling the THz gap-high power sources and applications. Rep. Prog. Phys. **69**, 301 (2006)
15. Neil, G.P.: FEL oscillators. Proceedings of the 2003 Particle Accelerator Conference, p. 181 (2003)
16. Carr, G.L., Martin, M.C., McKinney, W.R. et al.: High-power terahertz radiation from relativistic electrons. Nature **420**, 153 (2002)
17. Potylitsyn, A.P.: Thomson scattering of coherent diffraction radiation by an electron bunch. Phys. Rev. E **60**, 2272 (1999)
18. Xiang, D., Huang, W-H.: Ultrashort electron bunch length measurement with diffraction radiation deflector. Phys. Rev. ST Accel. Beams **10**(1), 012801 (2007)
19. Stupakov, G.V.: Geometrical Wake of a Smooth Flat Collimator. SLAC-PUB-96-7167 (1996)
20. Bosch, R.A.: Shielding of infrared edge and synchrotron radiation. Nucl. Instrum. Methods Phys. Res. A **482**, 789 (2002)

21. Potylitsyn, A.P.: Scattering of coherent diffraction radiation by a short electron bunch. Nucl. Instrum. Methods Phys. Res. A **455**, 213 (2000)
22. Brownell, J.H., Walsh, J., Doucas, G.: Spontaneous Smith-Purcell radiation described through induced surface currents. Phys. Rev. E **57**, 1075 (1998)
23. Shibata, Y., Hasebe, S., Ishi, K. et al.: Coherent Smith-Purcell radiation in the millimeter-wave region from a short-bunch beam of relativistic electrons. Phys. Rev. E **57**, 1061 (1998)
24. Nguyen, D.C.: Design of a Smith-Purcell radiation bunch length diagnostic. Nucl. Instrum. Methods Phys. Res. A **393**, 514 (1997)
25. Lampel, M.C.: Advanced Accelerator Concepts, 8th Workshop., CP 472, 785, (1999)
26. Korbly, S.E., Brown, W.J., Shapiro, M.A. et al.: Proceedings of the 2001 Particle Accelerator Conference, p. 2347 (2001)
27. Doria, A., Gallerano, G.P., Giovenale, E. et al.: Can coherent Smith–Purcell radiation be used to determine the shape of an electron bunch? Nucl. Instrum. Methods Phys. Res. A **483**, 263 (2002)
28. Blackmore, V., Doucas, G., Perr, C. et al.: First measurements of the longitudinal bunch profile of a 28.5 GeV beam using coherent Smith–Purcell radiation. Phys. Rev. ST-AB **12**, 032803 (2009)

Chapter 8
Diffraction Radiation in the Pre-wave (Fresnel) Zone

8.1 Transition Radiation in the Pre-wave (Fresnel) Zone

The notion of the formation (coherence) length introduced in the early 1950s (see detailed references in [1, 2]) was described in the framework of classical electrodynamics in Sects. 1.2 and 2.2 for bremsstrahlung and polarization radiation mechanisms, respectively. The quantum-mechanical consideration of this problem provides the identical result [2]. The notion of the formation length appeared to be useful for describing not only electromagnetic, but also strong interactions [3, 4]. The main physical idea underlying the notion of the formation length for polarization emission mechanisms (including transition radiation and diffraction radiation) is the interference of the radiation fields emitted from different points of the emitting substance along particle trajectory (see Sect. 2.2). This concept is very close to the Fresnel zone widely known in optics. In Sects. 1.2 and 2.2, expression (2.11) for the formation length was obtained in the form

$$l_c \sim \frac{2\pi}{\omega/v - k_z} = \frac{\lambda}{\beta^{-1} - n_z}.$$

The maximum formation length for forward radiation reaches $2\gamma^2$, whereas the formation length for emission angles much larger than γ^{-1} is about λ. The latter case corresponds to backward transition (diffraction) radiation, i.e., the case where radiation is generated in the mirror reflection direction with respect to the target surface. An additional reason is associated with not the notion of the formation length, but with the transverse sizes of a source and manifests the same process of the interference of radiations emitted from different points of the target. Indeed, let us consider backward transition radiation (as well as diffraction radiation) as a process induced by the radiation from polarization currents. Then, as shown above, the region on the target where currents are induced has the typical size d $\gamma\lambda$. With an increase in the Lorentz factor or the emitted wavelength the size of the radiation source can obviously become macroscopic. In optics, the initial transverse size of the source d, is used to define the distances characterizing the so-called near (Fresnel) zone and far (Fraunhofer) zone [5]:

A.P. Potylitsyn et al., *Diffraction Radiation from Relativistic Particles*, STMP 239, 221–250, DOI 10.1007/978-3-642-12513-3_8,

$$l_{\mathrm{nf}} \ll \frac{\pi d^2}{\lambda}, \quad l_{ff} \gg \frac{\pi d^2}{\lambda}.$$

The distribution of the scattered-beam intensity in the far zone coincides with the angular distribution of the initial beam. As shown in [6], the description of the backward transition radiation taking into account the finite size of the radiation source on the target leads to the notion similar to the Fresnel zone in optics, the so-called "pre-wave zone." Considering instead of the 'optical' size of the beam, we can estimate the distance corresponding to the Fresnel zone:

$$a \ll \pi \gamma^2 \lambda,$$

When a detector for forward transition radiation is placed at distances smaller than the formation length, the intensity of the detected radiation is smaller than the Ginzburg-Frank result [7] obtained for the far zone. For this reason, the inclusion of the effect of the pre-wave (Fresnel) zone even for well-studied transition radiation is a nontrivial problem. The calculation of the effects of the pre-wave zone is even more nontrivial for diffraction radiation, which differs from transition radiation in the necessity to solve extremely complicated boundary value problems. Boundary value problems for diffraction radiation with the dielectric properties of the substance usually have no exact solution; for this reason, we use some semi-quantitative concepts in this chapter. In addition, we note that the detector for backward transition radiation is usually placed at finite distances. In this case, the radiation intensity does not decrease, but the radiation cone strongly broadens under the condition

$$a \le \gamma^2 \lambda$$

The pre-wave zone effect was observed experimentally for backward transition radiation [8] and backward diffraction radiation [9].

Before considering diffraction radiation in the pre-wave zone, let us consider this effect for transition radiation. The essence of this effect is that the approximation of a point source of transition radiation (valid for large distances a, when the size of the source of transition radiation on the surface of a target on the order of $\gamma\lambda$ can be disregarded) is inapplicable for comparatively small distances a between the detector and target. This point-source approximation (wave or far-field zone approximation) was used in most works devoted to transition radiation beginning with [1, 10]. The criterion of the violation of the applicability of the far-field zone approximation can be obtained from a simple geometrical consideration (see Fig. 8.1).

When an ultrarelativistic particle intersects the target inclined at angle θ_0 (for the sake of simplicity, we take $\theta_0 = 45°$, see Fig. 8.1), the effective target-region section excited by the field of the particle and emitting electromagnetic radiation with wavelength λ is an ellipse with the major axis $\gamma\lambda/\sin\theta_0$ along the Y_T axis and the minor axis $\gamma\lambda$ along the X_T axis. The coordinates of a point on the target and detector are denoted as X_T, Y_T and X_D, Y_D, respectively (the X_T, X_D axes are perpendicular to the figure plane).

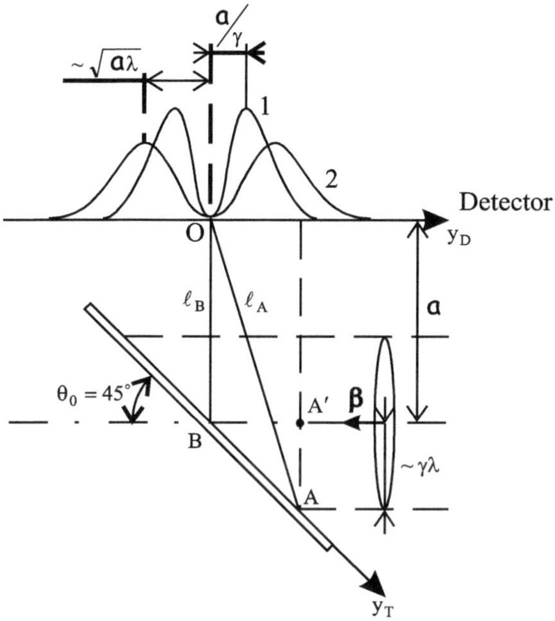

Fig. 8.1 Formation of the "image" of the angular distribution of transition radiation on the plane of the detector placed in the (1) far-field zone ($a \gg \gamma^2\lambda$) and (2) pre-wave zone ($a < \gamma^2\lambda$)

Let us consider the interference of waves emitted from points A and B in the target, which are excited when the ultrarelativistic particle passes through point B, at point O of the detector (see Fig. 8.1).

The waves from points A and B interfere when the following condition on the phase shift is satisfied:

$$\Delta\varphi = k\Delta r - \omega\Delta t \ll \pi. \tag{8.1}$$

Here, $k = 2\pi/\lambda$ is the length of the wave vector, $\Delta r = \ell_A - \ell_B$, $\Delta t = \gamma\lambda/\beta c$ is the time in which the particle passes the distance $A'B$ (see Fig. 8.1). Assuming that $\gamma\lambda << a$, we expand Δr in the small parameter $\gamma\lambda/a$ up to the second order terms:

$$\Delta r = \sqrt{(a+\gamma\lambda)^2 + (\gamma\lambda)^2} - a \approx \gamma\lambda + \frac{\gamma^2\lambda^2}{2a}. \tag{8.2}$$

Condition (8.1) is written for the central point B of the emitting region and a point A at the periphery of this region. Thus, when condition (8.1) is satisfied, all points in the target excitation region ("illumination" region) coherently emit waves with the wavelength λ, i.e., the size of the luminous spot can be disregarded (far-field-zone approximation). The far-field-zone condition that relates the Lorentz factor of the particle, radiation wavelength, and the distance between the detector and target follows from Eq. (8.1):

$$a \gg \gamma^2\lambda. \tag{8.3}$$

Note that the wave zone in optics, where radiation from nonrelativistic charges is primarily considered, is defined as the distance much larger than the wavelength and the size of the region of the motion of the charges emitting radiation [11]. Condition (8.3) characterizes the distance a that is $\gamma \gg 1$ times larger than the typical size of the emitting region, $\gamma\lambda$. Therefore, the opposite condition

$$a < \gamma^2\lambda \tag{8.4}$$

characterizes the so-called "pre-wave" zone. In contrast to the wave zone, the size of the source should be taken into account in the pre-wave zone.

Note that the length of the pre-wave zone, $\gamma^2\lambda$, formally coincides with the radiation formation length (see Sect. 2.2). These two quantities have both differences and similarities. The differences are that the radiation formation length is a distance along the trajectory of the charged particle and that, for backward radiation considered in this section (the other terms: backward radiation and radiation at the mirror reflection angle from the surface), the coherence length is as small as about λ (except for the case of grazing incidence); i.e., the formation length in the ultrarelativistic case strongly differs from the pre-wave-zone size. However, these quantities are similar, because both quantities are obtained from the condition of the coherence of radiation from emitting-region points and are the zones of the formation of radiation from elementary emitters (electrons bound in the target substance) excited by the field of the charged particle.

Let us write formula (1.34) in terms of the wavelength:

$$l_c(\theta) = \frac{\beta\lambda}{1 - \beta\cos\theta}. \tag{8.5}$$

For forward radiation, this formula provides

$$l_c(0) \approx 2\gamma^2\lambda, \tag{8.6}$$

whereas the formation length for radiation angle $\theta = 2\theta_0 = 90^\circ \gg \gamma^{-1}$ (as for the geometry shown in Fig. 8.1) is comparable with the wavelength:

$$l_c\left(90^\circ\right) \approx \lambda. \tag{8.7}$$

Thus, the detector for transition radiation is almost always placed at a distance from the target that is much larger than the formation length [12]. However, in dependence on the Lorentz factor and wavelength of detected radiation, the detector can be placed in the pre-wave zone (i.e., when condition (8.4) is satisfied). In other words, when condition (8.4) is satisfied, the detector is placed in the region, where a wave emitted from the target is not spherical.

In the experiment [8] with a 900-MeV electron beam ($\gamma = 1760$), the detecting system was placed at the distance $a = 100\,\text{cm}$ from the target. For the wavelength $\lambda = 1.6\,\mu\text{m}$, the quantity $\gamma^2\lambda \approx 500\,\text{cm}$ is much larger than this distance. The broadening of the angular distribution of backward transition radiation on the detector plane that was observed in the experiment (i.e., the shift of the distribution peak at an angle exceeding γ^{-1}) is conclusive evidence of the effect of the pre-wave zone of transition radiation. Below, the pre-wave-zone effect will be characterized by the dimensionless parameter R:

$$R = \frac{a}{\gamma^2\lambda}, \tag{8.8}$$

i.e., $R \gg 1$ for the far-field zone.

In order to illustrate the procedure of the calculation of the components of the transition-radiation field in the pre-wave zone for obtaining the distribution of the transition-radiation intensity on the detector surface, we first consider the geometry of backward transition radiation when the particle perpendicularly intersects the infinite perfectly conducting interface ($\theta_0 = 90°$); i.e., the target and detector are placed perpendicularly to the particle trajectory. In the simplest approximation, where the longitudinal component of the field of the relativistic electron is disregarded and the transverse field components $E^0_{x,y}$ are described by a packet of electromagnetic waves, expressions for the components of the field reflected from the detector surface are written according to the known optical laws [13]:

$$\begin{Bmatrix} E^D_X(X_D, Y_D) \\ E^D_Y(X_D, Y_D) \end{Bmatrix} = \text{const} \int dX_T dY_T \begin{Bmatrix} E^0_X(X_T, Y_T) \\ E^0_Y(X_T, Y_T) \end{Bmatrix} \frac{e^{ikr}}{r}. \tag{8.9}$$

The coordinates X and Y in expression (8.9) on the target and detector are denoted by subscripts T and D, respectively. The coordinates X_D, Y_D enter into the phase through the distance r, where

$$r = \sqrt{a^2 + (X_T - X_D)^2 + (Y_T - Y_D)^2}, \quad k = \frac{2\pi}{\lambda}. \tag{8.10}$$

Of applied interest is the case where the following condition is satisfied:

$$a \gg \gamma\lambda. \tag{8.11}$$

In this case, $X_{T,D} \ll a$ and $Y_{T,D} \ll a$, and these relations allow the expansion of r in the powers of $\frac{X_{T,D}}{a}$ and $\frac{Y_{T,D}}{a}$. Retaining the terms of the first and second orders in the argument of the exponential and the leading term ($r \approx a$) in the denominator, one can obtain [14]

$$\begin{Bmatrix} E_X^D(X_D, Y_D) \\ E_Y^D(X_D, Y_D) \end{Bmatrix} = \text{const} \exp\left[i\frac{k}{2a}\left(X_D^2 + Y_D^2\right)\right]$$

$$\int dX_T dY_T \begin{Bmatrix} X_D \\ Y_D \end{Bmatrix} \frac{K_1\left(k\sqrt{X_T^2+Y_T^2}/\gamma\beta\right)}{\gamma\beta} \times$$

$$\times \exp\left[i\frac{k}{2a}\left(X_T^2 + Y_T^2\right)\right] \exp\left[-i\frac{k}{a}\left(X_T X_D + Y_T Y_D\right)\right]. \tag{8.12}$$

Here, expression (1.51) for the transverse field of the electron $E^0_{x,y}$ is substituted. In view of the symmetry of the problem, the further calculation will be performed in the polar coordinate system:

$$\begin{Bmatrix} X_{T,D} \\ Y_{T,D} \end{Bmatrix} = R_{T,D} \begin{Bmatrix} \cos\varphi_{T,D} \\ \sin\varphi_{T,D} \end{Bmatrix}, \quad dX_T dY_T = R_T dR_T d\varphi_T. \tag{8.13}$$

In addition, we use the dimensionless variables

$$r_T = \frac{2\pi R_T}{\gamma\lambda}, \quad r_D = \frac{\gamma}{a} R_D. \tag{8.14}$$

With the use of the known representations of the Bessel functions, the azimuthal integral in Eq. (8.12) is calculated analytically:

$$X_T X_D + Y_T Y_D = R_T R_D \cos(\varphi_T - \varphi_D),$$

$$\int d\varphi_T \begin{Bmatrix} \cos\varphi_T \\ \sin\varphi_T \end{Bmatrix} \exp\left\{-\tfrac{k}{a} R_T R_D \cos\left(\varphi_T - \varphi_D\right)\right\} =$$

$$= \int d\varphi_T \begin{Bmatrix} \cos\varphi_T \\ \sin\varphi_T \end{Bmatrix} \exp\left\{-r_T r_D \cos\left(\varphi_T - \varphi_D\right)\right\} = \begin{Bmatrix} \cos\varphi_D \\ \sin\varphi_D \end{Bmatrix} J_1\left(r_T r_D\right). \tag{8.15}$$

Then, omitting the insignificant phase factor in front of the integral in Eq. (8.12), we arrive at the following expression for the radial component of the transition-radiation field:

$$E_x^D = E_D \cos\varphi_D, \quad E_y^D = E_D \sin\varphi_D,$$

$$E_D\left(r_D, R\right) = \text{const} \int r_T dr_T K_1\left(r_T\right) J_1\left(r_T r_D\right) \exp\left(i\frac{r_T^2}{4\pi R}\right). \tag{8.16}$$

In the wave zone ($R \gg 1$), the exponential in the integrand can be replaced by one. In this case, the integration with respect to r_T is performed analytically:

$$E_D(r_D, \infty) = \int_0^\infty r_T\, dr_T\, K_1\left(r_T\right) J_1\left(r_T r_D\right) = \frac{r_D}{1 + r_D^2}, \tag{8.17}$$

i.e., the radial distribution of the transition radiation intensity on the detector is described by the expression

$$I\left(r_D\right) = \text{const}\left|E_D\left(r_D\right)\right|^2 = \text{const}\frac{r_D^2}{\left(1+r_D^2\right)^2}. \tag{8.18}$$

After the inverse transition to the dimensional variable R_D (see formula (8.14)), expressing the radius R_D in terms of the emission angle of a transition radiation photon, θ, we arrive at the relation

$$r_D = \gamma\theta, \tag{8.19}$$

and, correspondingly, at the known angular distribution of transition radiation photons recorded by the detector:

$$I\left(\theta\right) = \text{const}\frac{\gamma^2\theta^2}{\left(1+\gamma^2\theta^2\right)^2} = \text{const}\frac{\gamma^2\theta_x^2+\gamma^2\theta_y^2}{\left(1+\gamma^2\theta^2\right)^2}. \tag{8.20}$$

Here, Cartesian angular variables $\theta_x = \frac{X_D}{a}$ and $\theta_y = \frac{Y_D}{a}$ are used. Let us calculate the transition-radiation field in the pre-wave zone in order to compare with result (8.18). To simplify the calculations, we use the approximation

$$r_T K_1\left(r_T\right) = \left(1 + 0,57\,r_T - 0,04 r_T^2\right)e^{-r_T}, \tag{8.21}$$

which provides an accuracy of several percent in the interval $0 \le r_T \le 5$ (see Fig. 8.2).

When calculating integral (8.16), the integration with respect to the variable r_T can be performed in the limits

$$0 \le r_T \le r_{T\,\max} = 5, \tag{8.22}$$

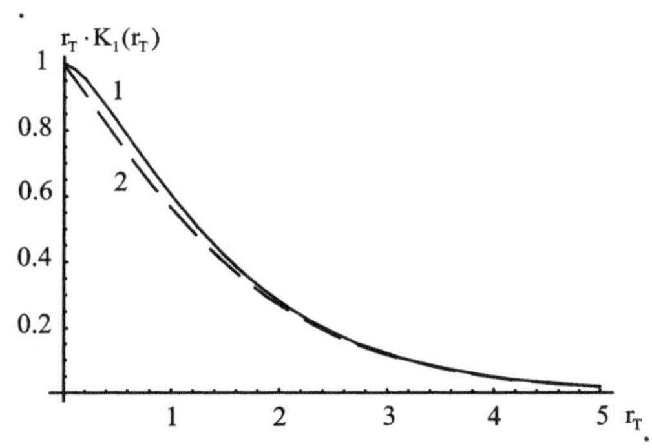

Fig. 8.2 Approximation of (*upper curve*) the function $r_T K_1(r_T)$ by (*lower curve*) formula (8.21)

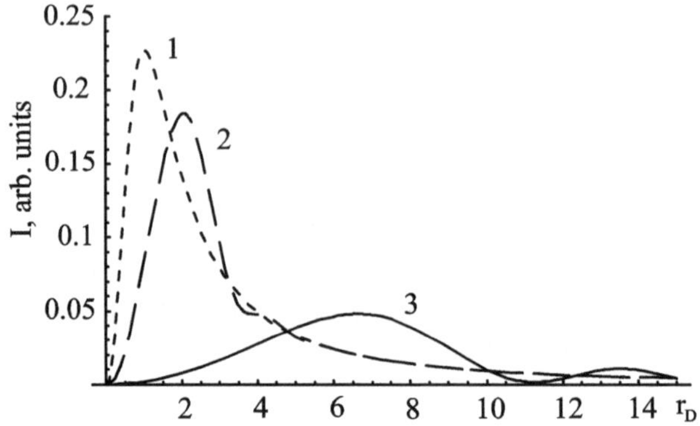

Fig. 8.3 Radial distribution of the transition radiation intensity on the detector surface for various distances between the detector and target, curve 1-$R = 10/2\pi$, curve 2-$R = 1/2\pi$, and curve 3-$R = 0.1/2\pi$

because an increase in the upper integration limit, e.g., to $r_{T\ \max} = 10$ changes the integrals by less than 1%. Figure 8.3 shows the radial dependences of the intensity $I \sim |E_D(r_D, R)|^2$ for various R values, which are in good agreement with the results of the calculations by Verzilov (see [6, Figure 3]).

The transition radiation intensity on the detector, $|E_D(r_D, R)|^2$, is maximal at the radius $r_D = r_0$ that ensures the maximum overlapping of oscillating integrand functions $J_1(r_T r_D)$ and $\mathrm{Re}\left[\exp\left(i\frac{r_T^2}{4\pi R}\right)\right]$. Since the leading contribution to integral (8.16) comes from the region $r_T \le 1$, this requirement reduces to the requirement of the coincidence of the first zeros of the both functions. Thus, there are two relations

$$r_T r_0 = 3.832,$$
$$\frac{r_T}{4\pi R} = \frac{\pi}{2}, \tag{8.23}$$

from which the relation between r_0 and R is obtained in the form

$$r_0 = \frac{3.832}{\sqrt{2}\pi\sqrt{R}} = \frac{0.863}{\sqrt{R}}. \tag{8.24}$$

In the dimensional variables,

$$R_D^0 = 0.863\sqrt{a\lambda}. \tag{8.25}$$

Estimate (8.25) provides the values $r_0 = 2.16$ and 6.84 for $R = \frac{1}{2\pi}; \frac{0.1}{2\pi}$, respectively, which are in satisfactory agreement with the respective exact values $r_0 = 2.05$ and 6.65 (see Fig. 8.3). According to the first of equations (8.23), the leading contribution to the transition radiation intensity in the pre-wave zone comes from the target region

$$R_T \leq \sqrt{\frac{a\lambda}{2}}, \tag{8.26}$$

i.e., estimate (8.25) is independent of the Lorentz factor in the case under consideration.

As seen in Fig. 8.3, first, the distribution of the transition radiation intensity in the pre-wave zone ($R \sim 0.01$) is broadened as compared to the wave zone ($R \sim 1$) and, second, the position of the peak in the intensity distribution in the pre-wave zone depends on the wavelength.

Bosch [15, 16], who developed a somewhat different approach to the problem of the pre-wave zone, obtained the close result

$$R_D^0 \approx \sqrt{a\lambda}. \tag{8.27}$$

In addition to cited work [12], transition radiation in the pre-wave zone was experimentally investigated in [17] with a 6-MeV electron beam ($\gamma \approx 12$). The angular distribution of optical transition radiation was studied with the same beam in the experiment reported in [18]. In the optical range ($\lambda \sim 0.5\,\mu$m), the parameter value $\gamma^2\lambda \approx 0.07$ mm allows the use of the formulas for the far-field zone to calculate the angular dependences of optical transition radiation and their comparison with experimental data (see Fig. 8.4).

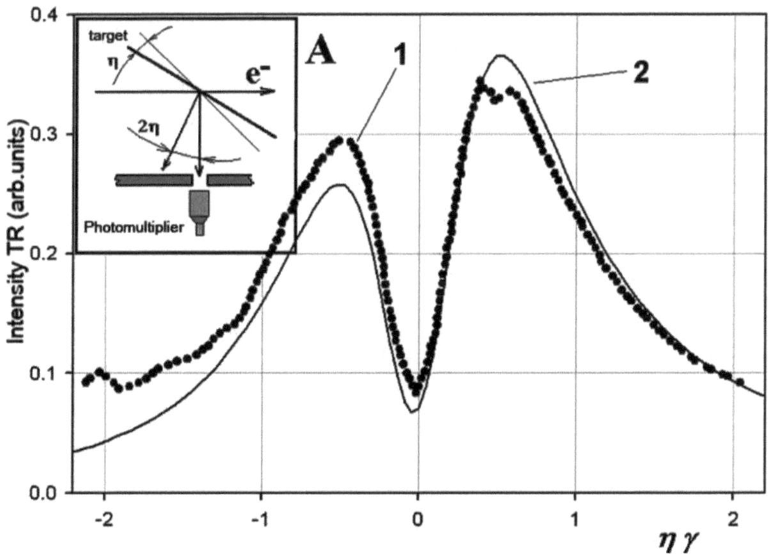

Fig. 8.4 Intensity of backward optical transition radiation generated by 6.1-MeV electrons versus the target inclination angle η: curve 1-experimental points in comparison with the theoretical curve 2-calculated in the model from [19]

As follows from this figure, the experimental data are satisfactorily described by the known theory with allowance for the divergence of an electron beam [19].

However, in the millimeter range, which corresponds to coherent transition radiation (in the described experiment, the length of an electron bunch was equal to 6–8 mm; therefore, coherent effects were manifested in the wavelength region $\lambda > 10$ mm), the criterion of the far-field zone corresponds to the following distance between the detector and target:

$$a \gg \gamma^2\lambda = 1440\,\text{mm}. \tag{8.28}$$

In the described experiment, this distance was equal to 360 mm$= 0.25\gamma^2\lambda$.

Figure 8.5 shows the experimental angular distribution of coherent transition radiation for the distance $a \approx 0.25\,\gamma^2\lambda$ in comparison with the theoretical dependence calculated by formula (8.16) with allowance for a finite angular aperture of the detector. As follows from the figure, the angular distribution of transition radiation in the pre-wave zone is much wider than the characteristic radiation angle γ^{-1}.

The use of additional optical elements in the detection system proposed in [17] (a parabolic mirror at the focus of which a detector is placed) and in [20] (a focusing lens) makes it possible to eliminate the effect of the near-field zone and to obtain angular distribution (8.20), which is typical for the far-field zone, for any distance between the detector and target.

Figure 8.6 shows the angular distribution of the intensity of backward coherent transition radiation measured with the use of the focusing parabolic mirror placed in the pre-wave zone and a detector placed at the focus of the mirror [17]. As follows from the figure, the effect of the pre-wave zone is almost completely eliminated in this geometry.

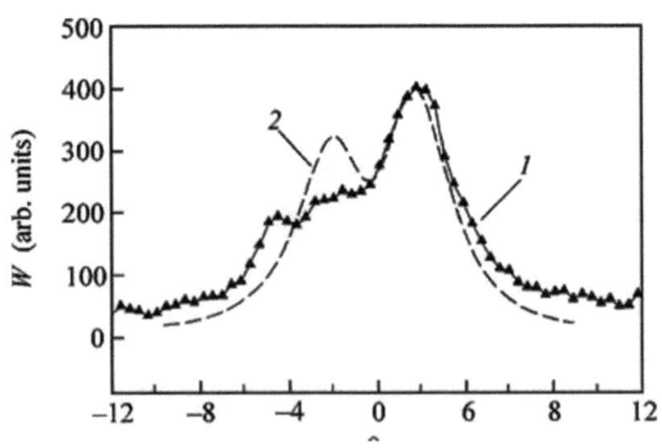

Fig. 8.5 Angular distribution of the intensity of coherent transition radiation in the near-field zone ($a \approx 0.25\,\gamma^2\lambda$).: experimental points 1 in comparison with theoretical curve 2 calculated for the experimental conditions

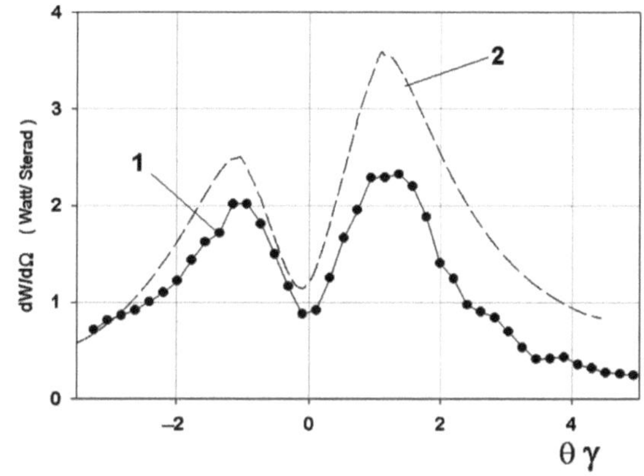

Fig. 8.6 Angular distribution of the power of coherent transition radiation per unit solid angle as (1) measured with the use of the parabolic optics and (2) calculated in the far-field zone approximation

8.2 Diffraction Radiation in the Pre-wave (Fresnel) Zone as a Tool for Beam Diagnostics

The field of backward diffraction radiation generated by a charged particle moving through the center of a circular hole of radius ρ in a perpendicular perfectly conducting target is written in the complete analogy with formula (8.16):

$$E_D^{DR}(r_D, R, \rho_0) = \text{const} \int_{\rho_0}^{\infty} r_T dr_T K_1(r_T) J_1(r_T r_D) \exp\left(i \frac{r_T^2}{4\pi R}\right). \tag{8.29}$$

Here, the dimensionless radius of the hole is given by the expression

$$\rho_0 = \frac{2\pi\rho}{\gamma\lambda}. \tag{8.30}$$

Figure 8.7 shows the distribution of the intensity of diffraction radiation on the surface of the detector for various R values and the fixed radius of the hole $\rho_0 = 0.33$. As in the case of transition radiation, the distribution is broadened as the parameter R decreases. Note that the intensity of diffraction radiation (as well as transition radiation) with a given wavelength that is registered by the detector,

$$I(\lambda, \rho_0) = \text{const} \int_{\rho_0}^{\infty} \left|E_D^{DR}(r_D, R, \rho_0)\right|^2 r_D dr_D, \tag{8.31}$$

is independent of the parameter R, i.e., of the detector position (in the wave or pre-wave zone), and is determined only by the hole radius ρ_0. In particular, for

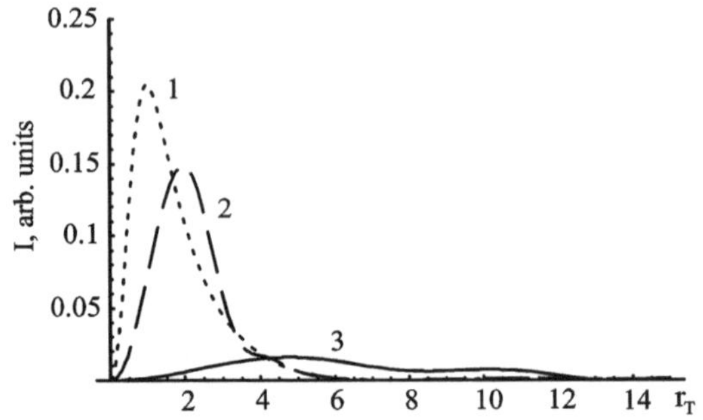

Fig. 8.7 Radial distribution of the intensity of diffraction radiation generated by a charge passing through a hole with the radius $\rho_0 = 0.33$ for (1) $R = 10/2\pi$, (2) $R = 1/2\pi$, and (3) $R = 0.1/2\pi$

$\rho_0 = 0.33$, the yield of diffraction radiation is $\sim 32\%$ of the intensity of transition radiation.

As mentioned above (see Sect. 3.1), Dnestrovskii and Kostomarov [21] considered for the first time the characteristics of diffraction radiation generated by a relativistic particle passing through the center of a circular hole in an infinite conducting screen. In that paper, diffraction radiation in the far-field zone was considered as a function of the parameter $x = 2\pi\rho/(\beta\gamma\lambda)$, i.e., of the ratio of the hole radius ρ to the "transverse" size of the field, $\gamma\lambda$.

In the ultrarelativistic case, the parameter x coincides with the notation $x = \rho_0$ used above for the dimensionless hole radius (see Eq. (8.30)). Further, with the use of the $\sin\theta = \theta$ approximation, the spectral–angular distribution of diffraction radiation obtained in the cited work can be represented in the form (see also formula (3.24))

$$\frac{d^2W}{\hbar d\omega d\Omega} = \frac{2\alpha}{\pi^2}\frac{\theta^2}{\left(\gamma^{-2}+\theta^2\right)^2}\left[\rho_0 K_1(\rho_0)\, J_0(\rho_0\gamma\,\theta)\right]^2. \tag{8.32}$$

In the far-field zone ($R \geq 1$), this distribution can be compared with the angular distribution of diffraction radiation for the field given by Eq. (8.29). In this case, as mentioned above, the polar angle is changed to the expression $\theta = r_d/\gamma$. Figure 8.8a shows the angular distribution of diffraction radiation calculated within the framework of both models in the far-field zone ($R = 1$) for the hole with the radius $\rho_0 = 0.33$. The models are in good agreement with each other in the far-field zone, whereas the model proposed in [21] is inapplicable in the pre-wave zone ($R = 0.1$, see Fig. 8.8b). A comparison of the distributions of transition radiation and diffraction radiation from azimuthally symmetric targets, which are presented in Figs. 8.3 and 8.7, shows that the presence of the hole in the target does not noticeably change the shape of the distributions, but slightly reduces the detected radiation intensity.

In both cases, an azimuthally symmetric funnel-shaped distribution with zero minimum in the mirror reflection direction is observed. However, azimuthal

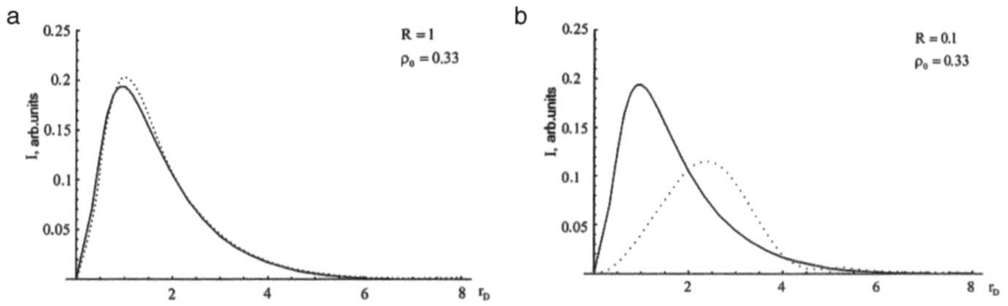

Fig. 8.8 Angular distribution of diffraction radiation in the (**a**) far-field zone ($R = 1$) and (**b**) pre-wave zone ($R = 0.1$). Radiation is generated by a particle passing through the center of the hole (with the radius $\rho_0 = 0.33$). The solid line is the Dnestrovskii—Kostomarov model and the points are obtained by formula (8.29)

symmetry is strongly broken in the case of the generation of diffraction radiation by the particle passing with the impact parameter h near a perpendicular half-plane. In the wave zone, where the source of radiation can be considered as point-like, the angular distribution of diffraction radiation nevertheless looses azimuthal symmetry (see formula (3.59)):

$$\frac{d^2W}{\hbar d\omega d\Omega} = \frac{\alpha}{4\pi^2}\exp\left(-z\sqrt{1+\gamma^2\theta_x^2}\right)\frac{\gamma^{-2}+2\theta_x^2}{\left(\gamma^{-2}+\theta_x^2\right)\left(\gamma^{-2}+\theta_x^2+\theta_y^2\right)} =$$
$$= \frac{\alpha}{4\pi^2}\exp\left(-z\sqrt{1+\gamma^2\theta^2\cos^2\varphi}\right)\frac{\gamma^{-2}+2\theta^2\cos^2\varphi}{\left(\gamma^{-2}+\theta^2\cos^2\varphi\right)\left(\gamma^{-2}+\theta^2\right)}. \tag{8.33}$$

In the wave zone, the maximum of radiation corresponds to the mirror reflection direction and distribution (8.33) is symmetric with respect to reflections in the vertical and horizontal planes (i.e., is invariant under the change $\theta_x \rightleftarrows -\theta_x$, $\theta_y \rightleftarrows -\theta_y$).

In the pre-wave zone, there is only "natural" symmetry with respect to the change $\theta_x \rightleftarrows -\theta_x$ (i.e., with respect to reflection from the plane perpendicular to the edge of the half-plane). The components of the diffraction-radiation field in the pre-wave zone for the semi-infinite perfectly conducting target are calculated by formula (8.12), where the integration is performed over the surface of the target whose edge is spaced from the electron trajectory by the impact parameter h (i.e., in the range $h \le Y_T \le \infty$ of the Y_T coordinate and in the range $-\infty < X_T < \infty$ of the X_T coordinate).

Figure 8.9 shows the distribution of the intensity of diffraction radiation in the X_D coordinate for the parameter (a) $R = 10/2\pi$, (b, line 1) $R = 1/2\pi$, and (b, line 2) $R = 0.1/2\pi$. It can be shown that the distribution in the first case coincides with an accuracy better than 1% with the angular distribution of diffraction radiation in the far-field zone given by Eq. (8.33) at $\theta_y = 0$. As in the case of transition radiation, when the distance between the target and detector decreases (when entering the pre-wave zone), the "image" of the angular distribution of diffraction

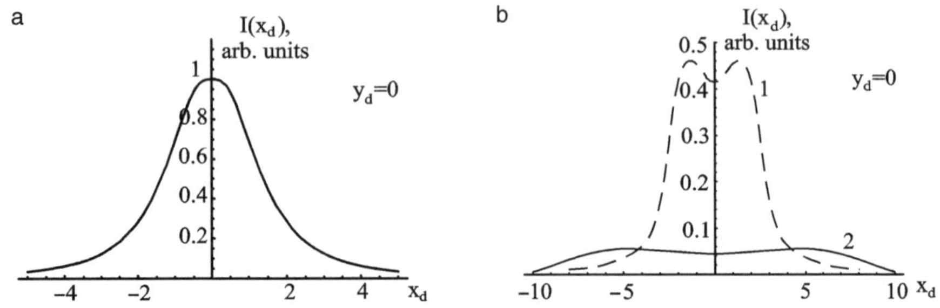

Fig. 8.9 Distribution of diffraction radiation on the surface of the detector in the plane parallel to the edge of the target for the impact parameter $h = 0.1\gamma\lambda/2\pi$ and (**a**) $R = 10/2\pi$, (**b**, *line 1*) $R = 1/2\pi$, and (**b**, *line 2*) $R = 0.1/2\pi$

radiation in the plane parallel to the edge of the half-plane is broadened with a decrease in the parameter R, remaining symmetric with respect to $x_D = 0$.

In contrast to the distribution of diffraction radiation in the far-field zone (see formula (8.33)), the distribution of diffraction radiation in the plane perpendicular to the edge of the target (distribution in the angular variable Y_D) changes significantly in the pre-wave zone.

Figure 8.10a shows the image of the angular distribution in the far-field zone in comparison with that for $R = 10/2\pi = 1.6$. We point to asymmetry in the distribution, which increases with a decrease in the parameter R (see Fig. 8.10b). The peak in the distribution is shifted towards positive y_T values, i.e., towards the half-space in which the diffraction target is placed. Nevertheless, variation in the impact parameter almost does not affect the image of the angular distribution of diffraction radiation in the pre-wave zone.

Fundamentally different information can be acquired by analyzing the shape of the "luminous region" on the surface of targets. This analysis can be performed by placing the detector at the focus of the optical lens, i.e., when measuring the size of the radiation source (see Fig. 8.11).

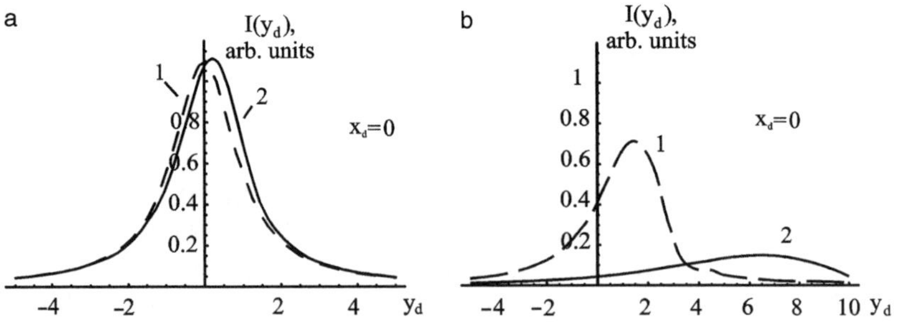

Fig. 8.10 Distribution of diffraction radiation on the surface of the detector in the plane perpendicular to the edge of the target for the impact parameter $h = 0.1\gamma\lambda/2\pi$ (**a**, *line 1*) in the far-field zone and for (**a**, *line 2*) $R = 10/2\pi$, (**b**, *line 1*) $R = 1/2\pi$, and (**b**, *line 2*) $R = 0.1/2\pi$

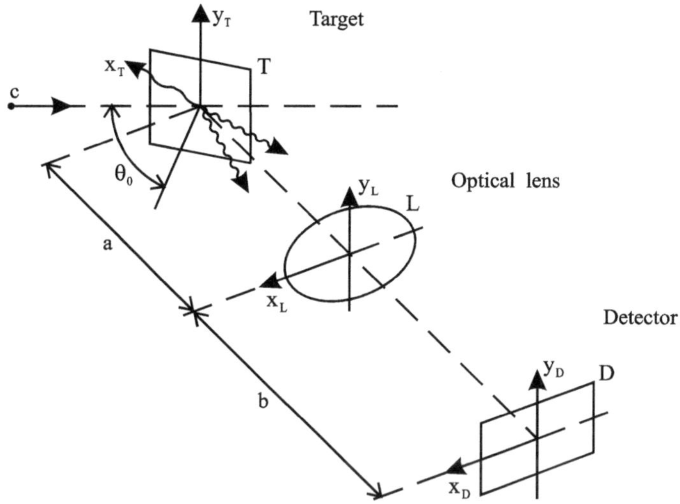

Fig. 8.11 Optical scheme for measuring the size of the radiation source

We again begin the consideration of the features of the image of the source on the detector with the case of backward transition radiation. To simplify the calculations, we consider the geometry of the perpendicular passage of the particle through the target. In actual cases, the inclination angle of the target, θ_0, with respect to the particle trajectory differs from 90°; however, was shown in [19], the angular characteristics of the backward transition radiation of ultrarelativistic particles in the wave zone (as well as the characteristics of backward diffraction radiation) are determined with respect to the mirror reflection direction and are independent of the target inclination angle θ_0 if $\theta_0 \gg \gamma^{-1}$. Following [14], the components of the transition-radiation field that is generated by the ultrarelativistic particle in the infinite perfectly conducting screen and is focused by the lens with the focus distance f on the detector are represented in the form

$$E^D_{x,y}(\mathbf{R}_D) = \text{const} \int dR_T \int dR_L \begin{Bmatrix} \cos\varphi_T \\ \sin\varphi_T \end{Bmatrix} \times \frac{K_1(kR_T/\beta\gamma)}{\beta\gamma} \times$$
$$\times \exp\left(i\frac{k}{2a}R_T^2\right) \exp\left(i\frac{k}{2a}R_L^2\right) \exp\left(-i\frac{k}{a}R_L R_T \cos\left(\varphi_T - \varphi_L\right)\right) \times \quad (8.34)$$
$$\times \exp\left(-i\frac{k}{2f}R_L^2\right) \exp\left(i\frac{k}{2b}\left(\mathbf{R}_L - \mathbf{R}_D\right)^2\right).$$

Here, b is the distance between the target and lens, a is the distance between the lens and detector, the condition $\frac{1}{a} + \frac{1}{b} = \frac{1}{f}$ is assumed, and subscript L denotes the coordinates on the surface of the lens. Below, we will use the dimensionless variables defined by analogy with relations (8.14):

$$r_T = \frac{2\pi R_T}{\gamma\lambda},\ r_L = \frac{\gamma}{a} R_L,\ r_D = \frac{2\pi R_D}{\gamma\lambda}. \tag{8.35}$$

Then, Eq. (8.34) can provide the following expression for the radial components of the transition-radiation field on the surface of the detector for the infinite interface:

$$\begin{aligned} E_D(r_D, R) = \text{const} \int_0^{\infty} r_T dr_T \int_0^{r_m} r_L dr_L K_1(r_T)\, J_1(r_T r_L) \times \\ \times \exp\left(i \frac{r_T^2}{4\pi R}\right) J_1\left(r_L \frac{r_D}{M}\right) \end{aligned} \tag{8.36}$$

Here, $M = b/a$ is the magnification factor of the lens with the radius r_m over the aperture of which integration is performed. Further calculations will be carried out for the case of $a = b = 2f$, i.e., for the magnification factor of the lens $M = 1$. In the wave zone ($R \gg 1$), with the use of the result given by Eq. (8.18), one can obtain the simple integral

$$E_D(r_D, \infty) = \text{const} \int_0^{r_m} r_L dr_L \frac{r_L}{1 + r_L^2} J_1(r_L r_D), \tag{8.37}$$

where $r_m = \frac{\gamma}{a} R_{\max}$ is the angular aperture of the lens at angles γ^{-1}.

The resulting expression up to the notation coincides with the result obtained in [22], which was based on another approach. To obtain field (8.36) on the surface of the detector, it is necessary to calculate the double integral. Note that the "inner" integral in this expression is calculated analytically:

$$\begin{aligned} G(r_T, r_D, r_m) = \int_0^{r_m} r_L dr_L\, J_1(r_T r_L)\, J_1(r_L r_D) = \\ = \frac{r_m}{r_D^2 - r_T^2} \left\{ r_T J_0(r_m r_T)\, J_1(r_m r_D) - r_D J_0(r_m r_D)\, J_1(r_m r_T) \right\}. \end{aligned} \tag{8.38}$$

Then, the field on the detector is expressed in terms of the integral:

$$E_D(r_D, R) = \int r_T dr_T\, K_1(r_T)\, G(r_T, r_D, r_m) \exp\left(i \frac{r_T^2}{4\pi R}\right) \tag{8.39}$$

Figure 8.12 shows the normalized shape of the image of the transition-radiation source on the detector $|E_D(r, R)|^2$ for the lens apertures $r_m = 50,\ 100$ in the wave and pre-wave zones.

For the Lorentz factor $\gamma = 1000$, these lens apertures correspond to the angles $\theta_m = 0.1$ and $\theta_m = 0.05$ rad. When the aperture is halved, the radius $r_D^{\max}$ is doubled. The images calculated by formula (8.39) in the "extremely near-field" zone (for $R = 0.001$) are shown in Fig. 8.12b. Note that the normalized shape of the distribution of the transition radiation intensity on the detector depends only slightly on the parameter R if the lens aperture expressed in terms of the angles $1/\gamma$ is

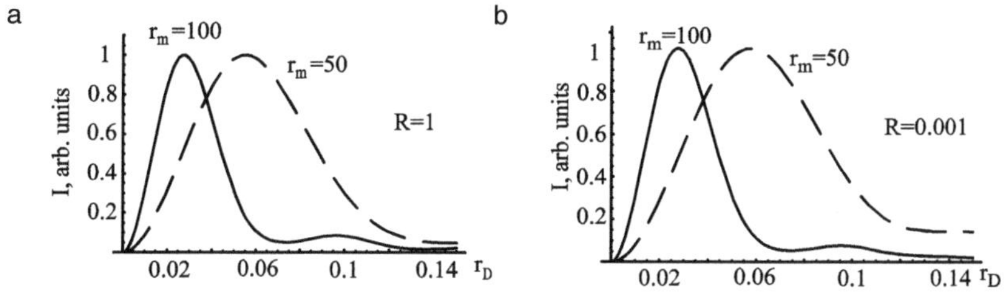

Fig. 8.12 Normalized intensity of the image of the transition-radiation source on the detector plane for the lens aperture (*left curve*) $r_m = 100$ and (*right curve*) $r_m = 50$ in the (**a**) wave zone ($R = 1$) and (**b**) pre-wave zone ($R = 0.1$)

unchanged. In the actual case, variation in the distance a between the lens and target at unchanged other parameters of the problem is accompanied by change not only in the parameter R, but also in the lens aperture r_m at its fixed diameter.

Figure 8.13 shows similar distributions (for γ=1000) calculated with the use of the expressions derived in [22]. The results obtained by different methods coincide with good accuracy. In particular, for the Lorentz factor $\gamma = 1000$ and aperture $\theta_m = 0.1$ rad, the maximum in the distribution corresponds to the radius $\rho_L^{\max} = 4.4\lambda$ (see Fig. 8.13). As follows from Fig. 8.12a (left curve), the maximum at $r_m = 100$ corresponds to the dimensionless variable $r_D = 0.0275$, which corresponds to the radius $r_D^{\max} = \frac{1000}{2\pi} \cdot 0.0275\,\lambda = 4,38\ \lambda$ at $\gamma = 1000$.

Figure 8.14 shows the image of the transition-radiation "spot" for $R = 0.1$, $r_m = 25$ and $R = 0.01$, $r_m = 250$, which corresponds to a fixed lens diameter with an order of magnitude change in the distance a. For convenient comparison, the right curve is increased by a factor of 100. As follows from the figure, the effective

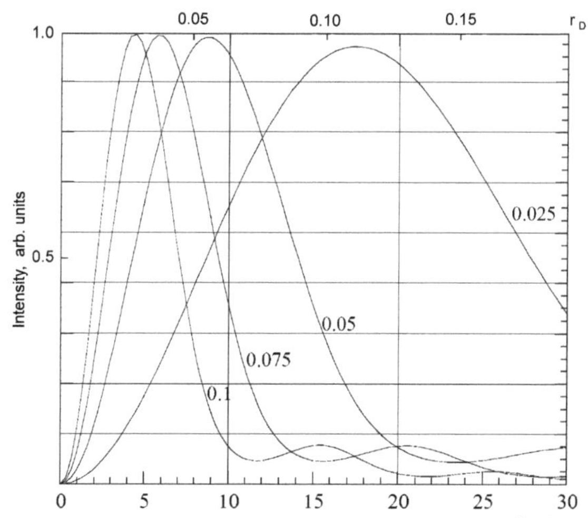

Fig. 8.13 Normalized intensity of the image of the transition-radiation source for $\gamma = 1000$. The abscissa axis shows the coordinate on the detector plane divided by $\lambda\, M$ (see [22])

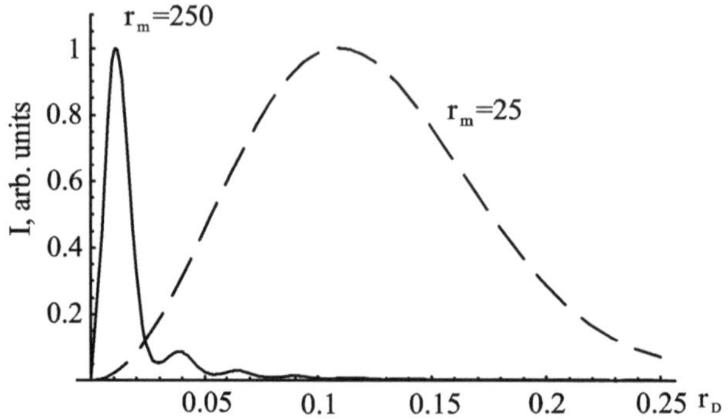

Fig. 8.14 Image of the transition-radiation source at a given diameter of the lens for various distances a between the target and lens for (*left curve*) $R = 0.01$ and $r_m = 250$ and (*right curve*) $R = 0.1$ and $r_m = 25$

"size of the spot" (detector resolution) decreases by a factor of 10 when the lens approaches the target (this decrease factor coincides with the increase factor of the lens angular aperture).

The spatial resolution can be estimated from the plots shown in Fig. 8.14. For the lens aperture $r_m = 250$, the maximum in the image of the transition-radiation source corresponds to the value $r_0 = 0.012$, or, in the dimensional variables, to

$$R_0 = \frac{\lambda}{2\pi/\gamma r_0} = \frac{\lambda}{523/\gamma} \sim \frac{\lambda\gamma}{2 \cdot 250} = \frac{\lambda}{2\theta_m}. \tag{8.40}$$

A similar estimate is obtained for all other cases.

The approach developed in the preceding section allows one to obtain the image of the circular hole with the radius ρ through the center of which the charged particle passes and generates optical diffraction radiation. The field of diffraction radiation on the detector is calculated by a formula similar to Eq. (8.36), where zero lower limit is changed to the hole radius ρ_0:

$$E_D^{DR}(r_D, R, \rho_0) = \int_{\rho_0}^{\infty} r_T dr_T K_1(r_T)\, G(r_T, r_D, r_m) \exp\left(i\frac{r_T^2}{4\pi R}\right). \tag{8.41}$$

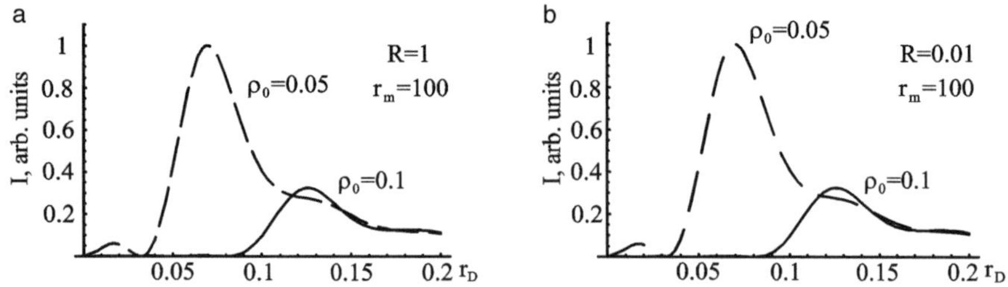

Fig. 8.15 Image of the circular hole for aperture $r_m = 100$ at (*left curve*) $\rho_0 = 0.05$ and (*right curve*) $\rho_0 = 0.1$ in the (**a**) wave zone ($R = 1$) and (**b**) pre-wave zone ($R = 0.01$)

Figure 8.15 shows such images for $\rho_0 = 0.05$ and $\rho_0 = 0.1$ in the far-field zone ($R = 1$) in comparison with similar images in the pre-wave zone. If the spatial resolution of the transition-radiation source (in other words, the size of the diffraction-radiation image from the hole with infinitely small radius) is determined by the quantity $\lambda/2\theta_m$ (or, in the dimensionless variables, $\Delta r_D \sim 0.012$ for $r_m = 250$), then it should be expected that the image of the diffraction-radiation spot almost coincides with transition radiation when $r_0 < \Delta r_D$, whereas the image of the hole can be obtained on the detector when $r_0 > \Delta r_D$. As follows from the figure, the radiation intensity on the detector almost vanishes when $r_D < r_0$ and reaches a maximum when $r_D^m \approx r_0 + \Delta r_D^{\mathrm{DR}}$ and the width of the peak is determined by the quantity $\Delta r_D^{\mathrm{DR}} \approx 0.04$ in both cases.

This result indicates that the lens with a sufficiently large aperture makes it possible, first, to compensate the effect of the broadening of the distribution of transition radiation and diffraction radiation, which is associated with the pre-wave zone, and, second, to obtain the image of the diffraction-radiation source, which, in the case under consideration, corresponds to a ring whose inner radius is equal to the hole radius and the width is determined by the quantity $\lambda/2\theta_m$. In other words, the effective size of the luminous region is governed by the lens aperture θ_m rather than by the Lorentz factor. A similar effect for the case of transition radiation was physically interpreted in [24, 23]. When the aperture is halved (to $r_m = 50$), the distribution width is more than doubled as compared to the case of $r_m = 100$ (see Fig. 8.16). In this case, the image of the small hole (with the radius $\rho_0 = 0.05$) appears to be sufficiently "smeared".

The developed approach makes it possible to take into account the effect of the outer size of the target on the characteristics of transition radiation and diffraction radiation. This effect should be taken into account if the outer diameter of the target is comparable with the target excitation region $\sim \gamma\lambda$.

For the azimuthally symmetric geometry under consideration (i.e., for the circular target with the radius $\rho_{\max}$ in the case of transition radiation and for the ring target with the same outer radius in the case of diffraction radiation), the upper limit

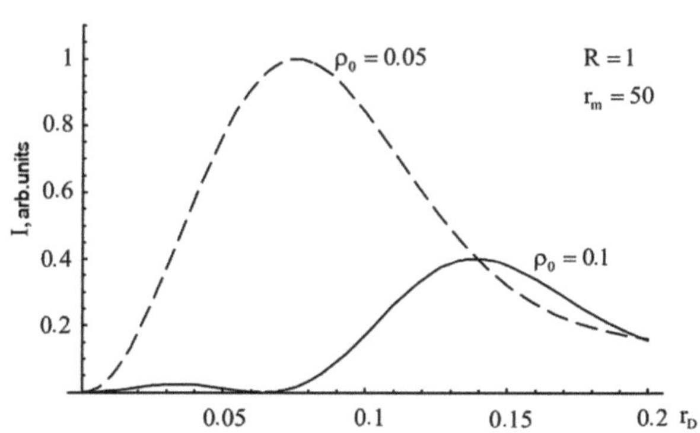

Fig. 8.16 Image of the hole in the wave zone ($R = 1$) with low resolution ($r_m = 50$)

in formulas and (8.41) is changed to the quantity $\rho_m = \frac{2\pi\rho_{\max}}{\gamma\lambda}$. In the wave zone ($R \gg 1$), a simpler method is the direct use of formula, which can be reduced to the simpler expression

$$\begin{aligned} E^{\mathrm{DR}}(r_L) &= \int_{\rho_0}^{\rho_m} r_T dr_T\, K_1(r_T)\, J_1(r_T r_L) = \\ &= -\frac{r_T}{1+r_L^2}\left[K_0(r_T)\, J_1(r_T r_L) + r_L\, K_1(r_T)\, J_0(r_T\, r_L)\right]\Bigg|_{\rho_0}^{\rho_{\max}} . \end{aligned} \tag{8.42}$$

For the hole radius $\rho_0 \ll 1$, $\rho_0\, K_0(\rho_0) \approx 0$ and $\rho_0 K_1(\rho_0) \approx 1$ with good accuracy and, therefore, in the limit $r_{\max} \to \infty$,

$$E_\infty^{\mathrm{DR}}(r_L) \approx \frac{r_L}{1+r_L^2} J_0(\rho_0 r_L)\,. \tag{8.43}$$

Result (8.43) agrees well with the previous result [25]. However, when the axial symmetry of the problem is broken (e.g., when the point of the intersection of the target by the particle does not coincide with the target center), the above formulas are inapplicable.

It is of interest to estimate whether information on the transverse coordinate of the particle passing through the hole can be acquired from the image of diffraction radiation from the hole in the presence of focusing on the detector. In this case, it is necessary to pass to the Cartesian coordinates in the initial formulas, because the azimuthal symmetry of the problem is absent. As above, subscripts T, L and D denote the Cartesian coordinates on the target, lens, and detector, respectively. The dimensionless coordinates are introduced similarly to Eqs. (8.35):

$$\begin{Bmatrix} x_T \\ y_T \end{Bmatrix} = \frac{2\pi}{\gamma\lambda}\begin{Bmatrix} X_T \\ Y_T \end{Bmatrix}, \quad \begin{Bmatrix} x_L \\ y_L \end{Bmatrix} = \frac{\gamma}{a}\begin{Bmatrix} X_L \\ Y_L \end{Bmatrix}, \quad \begin{Bmatrix} x_D \\ y_D \end{Bmatrix} = \frac{2\pi}{\gamma\lambda}\begin{Bmatrix} X_D \\ Y_D \end{Bmatrix}. \tag{8.44}$$

Then, the calculation of the field components on the detector reduces to a quadruple integral over the surface of the target and over the lens aperture:

$$\begin{aligned} \begin{Bmatrix} E_x^D(x_D, y_D) \\ E_y^D(x_D, y_D) \end{Bmatrix} &= \mathrm{const} \int dx_T dy_T \int dx_L\, dy_L \begin{Bmatrix} x_T \\ y_T \end{Bmatrix} \frac{K_1\left(\sqrt{x_T^2+y_T^2}\right)}{\sqrt{x_T^2+y_T^2}} \times \\ &\times \exp\left[i\frac{x_T^2+y_T^2}{4\pi R}\right] \exp\left[-i\left(x_T x_L + y_T x_L\right)\right] \exp\left[-i\left(x_L\frac{x_D}{M} + y_L\frac{y_D}{M}\right)\right]. \end{aligned} \tag{8.45}$$

Here, $R = \frac{a}{\gamma^2\lambda}$ and M is the magnification factor. This integral is most simply calculated for a rectangular lens:

$$-x_m \le x_L \le x_m; \quad -y_m \le y_L \le y_m. \tag{8.46}$$

In this case, the inner double integral (over the surface of the lens) is again calculated analytically:

$$\begin{aligned}
&\int_{-x_m}^{x_m} dx_L \int_{-y_m}^{y_m} dy_L \exp\left[-ix_L\left(x_T + \frac{x_D}{M}\right)\right] \exp\left[-iy_D\left(y_T + \frac{y_D}{M}\right)\right] = \\
&= 4\frac{\sin\left[x_m\left(x_T + \frac{x_D}{M}\right)\right]}{x_T + \frac{x_D}{M}} \frac{\sin\left[y_m\left(y_T + \frac{y_D}{M}\right)\right]}{y_T + \frac{y_D}{M}} = \\
&= G_x\left(x_T, x_D, x_m\right) G_y\left(y_T, y_D, y_m\right).
\end{aligned} \tag{8.47}$$

Below, we again consider the case $M = 1$. In this case, expression (8.45) reduces to the double integral:

$$\begin{aligned}
&\begin{Bmatrix} E_x^D\left(x_D, y_D\right) \\ E_y^D\left(x_D, y_D\right) \end{Bmatrix} = \text{const} \int dx_T dy_T \begin{Bmatrix} x_T \\ y_T \end{Bmatrix} \frac{K_1\left(\sqrt{x_T^2 + y_T^2}\right)}{\sqrt{x_T^2 + y_T^2}} \times \\
&\times \exp\left[i\frac{x_T^2 + y_T^2}{4\pi R}\right] G_x\left(x_T, x_D, x_m\right) G_y\left(y_T, y_D, y_m\right).
\end{aligned} \tag{8.48}$$

The integration of expressions (8.48) in the limits

$$-5 \le x_T, y_T \le 5 \tag{8.49}$$

for $R \ge 1$ provides the standard images of the components of the transition-radiation field (see Fig. 8.17).

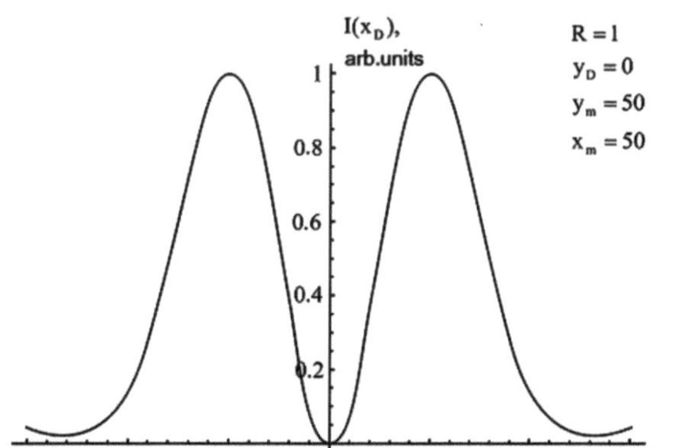

Fig. 8.17 Image of the transition-radiation source in the wave zone with the parameters $x_m = y_m = 50$, $y_D = 0$, and $R = 1$

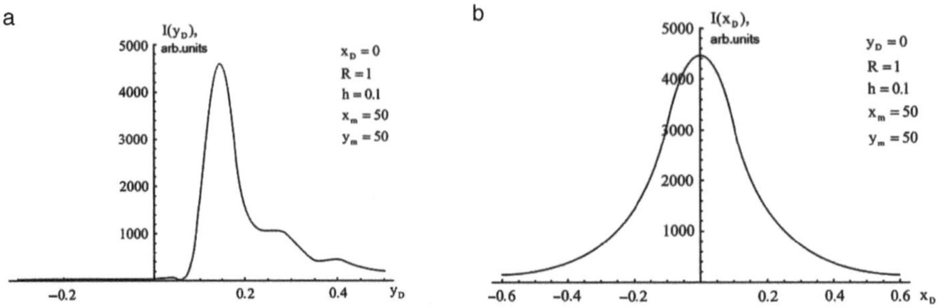

Fig. 8.18 Image of the diffraction radiation source in the plane (**a**) perpendicular and (**b**) parallel to the edge of the semi-infinite target

Figure 8.18 shows the image of the diffraction-radiation source under the same conditions when the particle passes over the half-plane at the distance $H = \frac{0.1}{2\pi}\gamma\lambda$. As in Fig. 8.18, a symmetric distribution in the x_d coordinate is seen (Fig. 8.18b), whereas the distribution in the y_d coordinate carries information on the impact parameter. Indeed, the characteristic size obtained with the use of the half-maximum of the right slope is $y_d \approx 0.1$, which corresponds to the dimensionless parameter $h = \frac{2\pi}{\gamma\lambda}H = 0.1$.

It is obvious that, in the case of the passage through the center of the slit with the width $2H = \frac{0.2}{2\pi}\gamma\,\lambda$ in the plane perpendicular to the edge of the slit, the symmetric image of the slit is observed (see Fig. 8.19a) and the spatial resolution of the optical scheme is determined by the lens aperture (curves 1 and 2 correspond to $x_m = y_m = 100$ and $x_m = y_m = 50$, respectively).

Figure 8.19b shows similar images for the passage of the particle at the distance $l_0 = \frac{0.02}{2\pi}\gamma\lambda$ from the center of the slit. In this case, as expected, the image of the closest edge of the slit is brighter. Such an asymmetric distribution can be used to determine the distance of the accelerated electron beam from the center of the slit. As follows from Fig. 8.19b, when the beam is shifted from the center of the

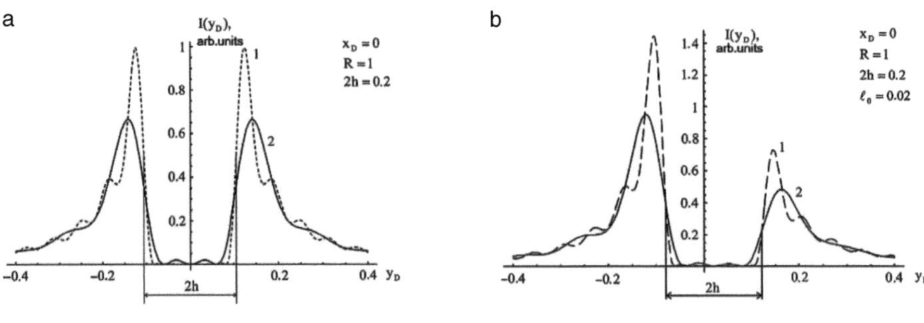

Fig. 8.19 Image of the diffraction-radiation source with the parameters (*curve 1*) $x_m = 100$, $y_m = 100$; and (*curve 2*) $x_m = 50$, $y_m = 50$ in the case of the passage of the charge (**a**) through the center of the slit in the infinite plane and (**b**) at the distance $\ell_0 = 0.02\gamma\lambda/2\pi$ from the center of the slit

slit by 10% of the width of the slit, the asymmetry of the image of the slit can be easily measured even if the resolution of the optical scheme is comparatively low. In particular, the slit width $2H = 0.2\gamma\lambda/2\pi$ for the parameters $\gamma = 2500$ and $\lambda = 0.5\,\mu\text{m}$ corresponds to a value of $40\,\mu\text{m}$ and the beam shift is $L_0 = 0.02\,\gamma\lambda/2\pi = 4\,\mu\text{m}$.

In the preceding sections, the characteristics of transition radiation and backward diffraction radiation were considered for the case of the perpendicular intersection of the target by the particle. This case is generally idealized case. In the general case, when the particle passes near the target inclined at the angle α (see Fig. 8.20), the components of the diffraction-radiation field are calculated by formulas similar to Eq. (8.38), where the phase factors are represented in the form

$$\exp\left[ik\left(r + \frac{\Delta Z_T}{\beta}\right)\right] = \exp\left[i\frac{K}{2a}\left(X_T^2 + Y_T^2\cos^2\alpha\right)\right] \times \\ \times \exp\left[-\frac{K}{a}\left(X_T X_D + Y_T Y_D \cos\alpha\right)\right] \exp\left[ikY_T \sin\alpha\left(1 + \frac{1}{\beta}\right)\right]. \tag{8.50}$$

This expression includes the additional phase shift (see Fig. 8.20)

$$k\frac{\Delta Z_T}{\beta} = k\frac{Y_T \sin\alpha}{\beta}, \tag{8.51}$$

because the almost transverse field of the particle (which is a source of virtual photons) arrives at the point of the target with the coordinates (X_T, Y_T) in a finite time measured from point O, after that virtual photons are transformed into real photons and a spherical wave is emitted. The calculations are more convenient in

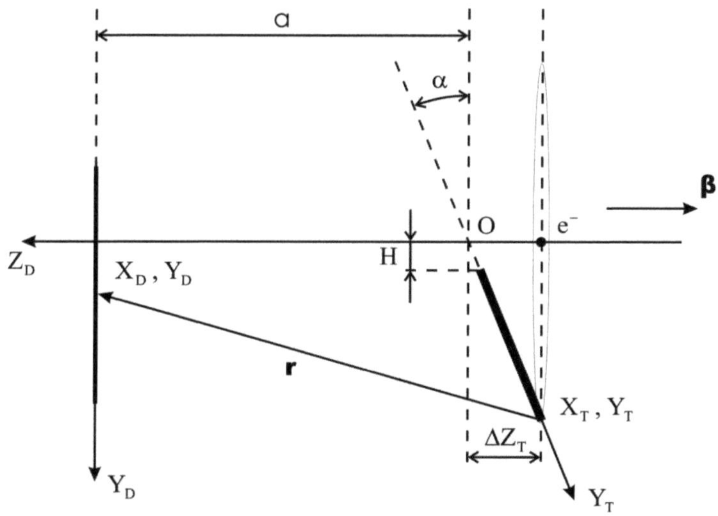

Fig. 8.20 Scheme illustrating the calculation of the phase shift

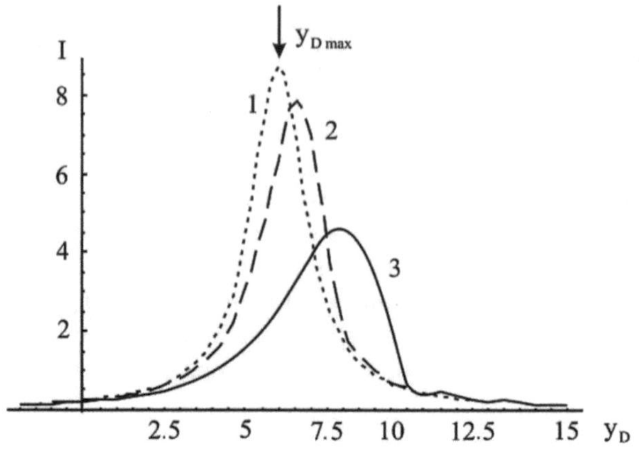

Fig. 8.21 Intensity of diffraction radiation on the surface of the detector from the flat target for various values of the parameter R (zero impact parameter): 1—$R = 10$, 2—$R = 0.5$, 3—$R = 0.1$ ($\gamma = 1000$, $\lambda = 0.5\,\mu$m, $\alpha = 3\gamma^{-1}$)

the dimensionless coordinates (as above). Note that formula (8.50) is derived in the same approximation as formula (8.38).

Figure 8.21 shows the intensity of diffraction radiation from the inclined target $(\alpha = 3\gamma^{-1})$ on the detector plane (at $X_D = 0$) for zero impact parameter and three distances a between the target and detector. As expected, a symmetric intensity distribution (curve 1) is observed in the wave zone and the position of the peak corresponds to the value $y_{D\,\max} = 6 = 2\gamma\alpha$; i.e., the maximum of the distribution coincides with the mirror reflection direction.

In the pre-wave zone ($R < 1$), an asymmetric distribution is observed (curves 2 and 3) and the maximum of the distribution is shifted towards large angles ($y_{D\ \max} > 2\gamma\,\alpha$). When diffraction radiation is generated in the target with the variable inclination angle (e.g., in a paraboloid) and if the detector is placed in the pre-wave zone, such a behavior of "reflected" photons "groups" them near the axis of the paraboloid; i.e., radiation is "focused".

Let us consider the focusing effect for transition radiation from the perfectly reflecting target in the form of the paraboloid along the axis of which the charged particle moves (in the positive direction of the Z_T axis, see Fig. 8.22). In the paper [26] a similar effect was considered for TR in far-field zone. We again consider the geometry of backward transition radiation.

The surface of such a target is described by the equation $2PZ_T = D^2/4 - Y_T^2 - X_T^2$, where D is the maximum diameter of the target, P is the focal length, and L is the distance from the boundary of the target to the detector (see Fig. 8.22).

Since the problem is azimuthally symmetric, we consider the radial components of the transition-radiation field E_r^D. The radial component from the paraboloid target with the diameter D is calculated as

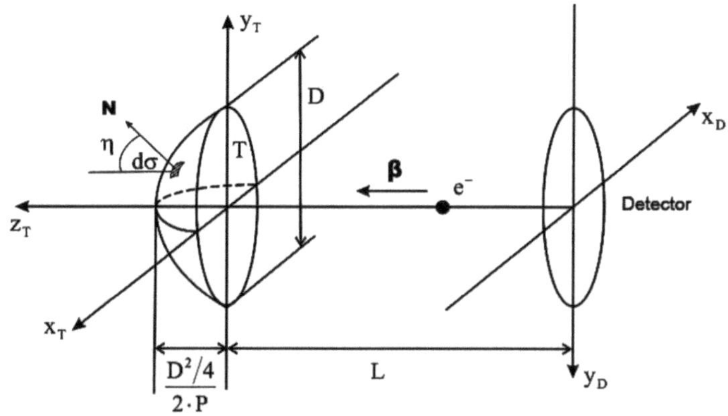

Fig. 8.22 Geometry of the focusing target, where **N** is the vector normal to the surface and $d\sigma$ is the elementary integration area

$$E_r^D(\mathbf{R}_D) = \text{const} \int_0^{d/2} r_T dr_T K_1(r_T) \times$$

$$\times \exp\left[i\gamma\left(\frac{d^2/4 - r_T^2}{2p}\right)\left(1 + \frac{1}{\beta}\right)\right] \exp\left(i\frac{r_T^2}{4\pi R}\right) \exp\left(i\pi R r_D^2\right) J_1(r_T r_D) \tag{8.52}$$

Here, the dimensionless variables are specified by the expressions (cf. expressions (8.44))

$$r_T = \frac{2\pi R_T}{\gamma\lambda}, \quad r_D = \frac{\gamma}{L} R_D, \quad d = \frac{2\pi D}{\gamma\lambda}, \quad p = \frac{2\pi P}{\gamma\lambda} \tag{8.53}$$

where $R_{T,D} = \sqrt{X_{T,D}^2 + Y_{T,D}^2}$. As above, a capital letter and a lowercase letter denote the dimensional and dimensionless variables, respectively.

When calculating the field E_r^D from the parabolic target, the integration is performed over the surface of the paraboloid whose area element is given by the formula

$$d\sigma = \frac{dX_T dY_T}{|\cos\eta|} = dX_T dY_T \frac{\sqrt{X_T^2 + Y_T^2 + P^2}}{P} = \frac{2\pi\sqrt{R_T^2 + P^2}}{P} R_T dR_T, \tag{8.54}$$

where η is the angle between the Z_T axis and the normal **N** to the surface at the point (X_T, Y_T) (see Fig. 8.22). When integrating over the surface of the paraboloid target in integral (8.42), it is necessary to use the area element given by Eq. (8.54) after transition to dimensionless variables (8.53).

First, we compare the intensity of transition radiation from the flat and bent targets with the diameter $D \gg \gamma\lambda$ on the detector placed in the pre-wave zone.

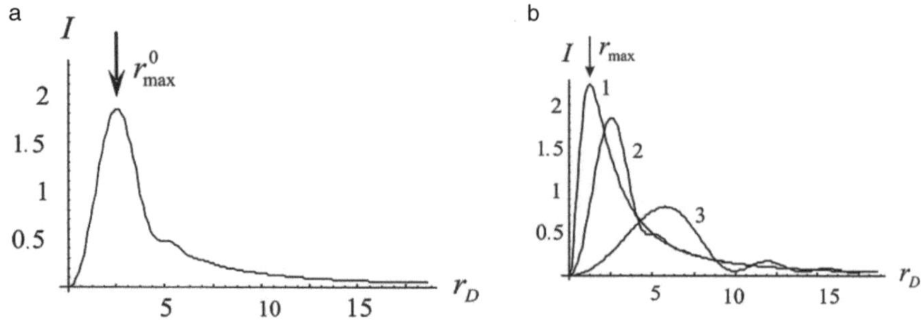

Fig. 8.23 Radial distribution of the transition-radiation intensity for the parameters $\gamma = 2500$, $L = 0.5\,\text{m}$, $\lambda = 0.5\,\mu\text{m}$, $D = 0.1\,\text{m}$, and $R = 0.16$ from the (**a**) circular flat target and (**b**) paraboloid target for various focal lengths (*curve 1*) $P = 5000$, (*curve 2*) $P = 2500$, and (*curve 3*) $P = 800$

Figure 8.23a shows the distribution of the intensity of radiation from the flat target on the detector for $R = 0.16$ (i.e., $L = 0.16\gamma^2\lambda$) and $d = 500$ (or, in the dimensional quantities, $D = \frac{\gamma\lambda}{2\pi}d \approx 80\gamma\lambda$). Figure 8.23b shows the distribution of radiation from the paraboloid target for various focal lengths (curve 1) $P = 2L$, (curve 2) $P = L$, and (curve 3) $P = 0.33L$.

As follows from the results, if the detector is placed at a distance $L < P$, i.e., in front of the mirror focus, the focusing of radiation is observed. If the detector is at the focus, the distributions from the flat and parabolic targets coincide with each other. Finally, when $L > P$, defocusing is observed.

Focusing is the narrowing of the radiation cone from the paraboloid target as compared to radiation from the flat target; i.e., $r_{max} < r^0_{max}$ (where r_{max}, r^0_{max} are the positions of the peak in the radial distributions of the intensity of transition radiation from the bent and flat targets, respectively).

Simple numerical integration indicates that the total intensity of transition radiation generated in the target by the relativistic particle is the same for the flat and bent targets and is determined by the Lorentz factor of the particle and the target diameter.

Figure 8.24 shows the quantity r_{max} for the paraboloid target at $P = 2L\ 0.32\gamma^2\lambda$. Note that the maximum focusing is reached at the distance $L = P/2$. Further, the minimum value of the quantity r_{max} at the point $L = P/2$ reaches one. By analogy with optics, this conclusion is expected to be valid for targets with the large diameter $(D \gg \gamma\lambda)$.

Figure 8.25 shows the dependence of the radius r_{max} corresponding to the maximum of the distribution of transition radiation on the diameter of the paraboloid. This dependence is universal, because it is independent of the parameter R (calculations for R in the range from 0.001 to 1 give the same result with the accuracy better than 10^{-3}).

By further analogy with optics, where the size of the spot of focused radiation, R_f, is determined by the wavelength λ and focusing-mirror aperture $\theta_{ap} = \frac{D}{2L}$:

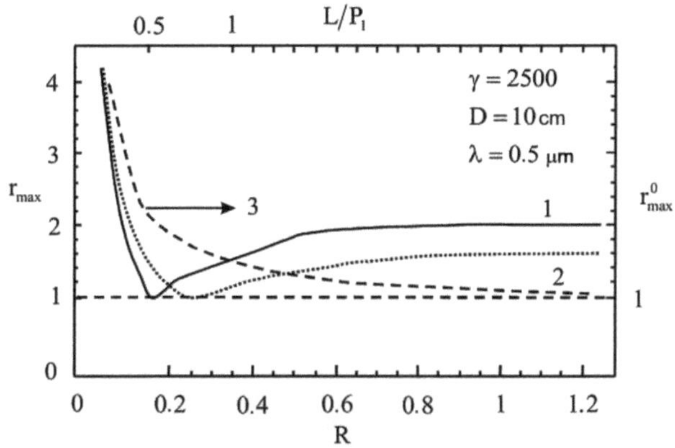

Fig. 8.24 Position of the peak in the intensity of transition radiation from the paraboloid target on the detector for various focal lengths (*curve 1*) $P_1 = 0.32\,\gamma^2\lambda$ and (*curve 2*) $P_2 = 0.5\,\gamma^2\lambda$. For comparison, curve 3 shows the same dependence for the flat target at the parameters $\gamma = 2500$, $\lambda = 0.5\,\mu$m, $D = 10$ cm

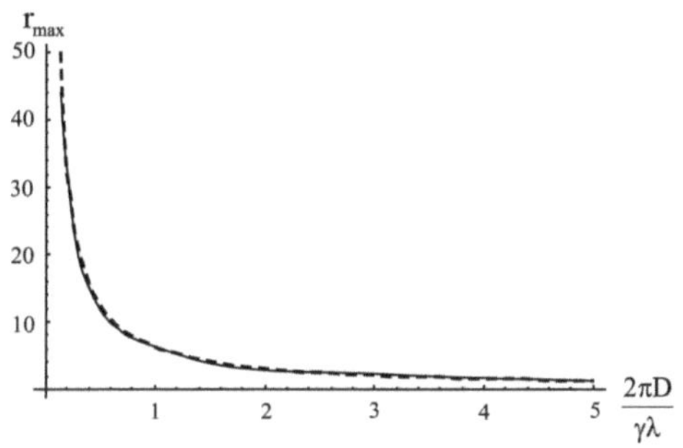

Fig. 8.25 Radius of the focused spot on the detector versus the target size D in (*solid line*) the distribution of transition radiation and (*dashed line*) approximation (8.58)

$$R_f \sim \frac{\lambda}{\theta_{ap}} = \frac{2\pi L\lambda}{D}, \tag{8.55}$$

we obtain a similar estimate for the transition-radiation source. In our case, if the diameter of the paraboloid target is represented in the form

$$D = k\gamma\lambda, \quad k \leq 1, \tag{8.56}$$

then it should be expected that the maximum of focused transition radiation satisfies the relation

$$R_{\max} \approx 2\pi \frac{L}{\gamma} \frac{1}{k} \tag{8.57}$$

or, in the dimensionless variables,

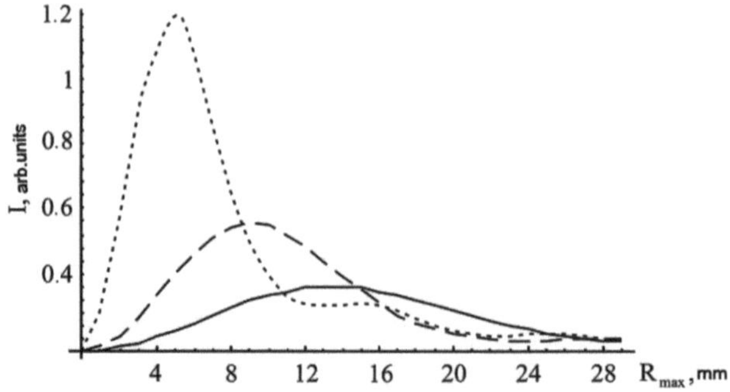

Fig. 8.26 Radial distribution of the intensity of transition radiation from the paraboloid target for the parameters $\gamma = 250$, $d = 0.1$ m, $P = 1$ m, $L = 0.5$ m, and wavelengths (dotted line) $\lambda = 1$ mm, (dashed line) $\lambda = 2$ mm, and (solid line) $\lambda = 3$ mm

$$r_{\text{max}} \approx \frac{2\pi}{k}. \tag{8.58}$$

This dependence is shown by the dashed line in Fig. 8.25.

Let us consider the focusing of transition and diffraction radiations. For particular calculations, we take the parameters corresponding to the experiment reported in [27] ($\gamma \sim 200$ and millimeter radiation). Figure 8.26 shows three curves characterizing the radial distribution of the transition-radiation intensity on the detector placed at the distance $L = 0.5$ m for various wavelengths. As the wavelength increases, the distribution is broadened and the intensity in the maximum decreases, because the parameter R decreases with an increase in the wavelength and, as a consequence, the radius R_{max} increases.

The developed approach allows one to calculate the characteristics of diffraction radiation from the paraboloid target with the hole of the radius ρ for the passage of the electron beam. In this case, the lower limit in the integral (8.52) is ρ_0. The radial distributions of transition and diffraction radiations calculated with the parameters $\gamma = 200$, $L = 0.6$ m, $\lambda = 1$ mm, $P = 1$ m; $D = 0.1$ m, and $\rho_0 = 5$ mm (for the calculation of diffraction radiation) are shown in Fig. 8.27. As follows from this figure, the intensity of millimeter diffraction radiation in the case under consideration is only 20% lower than the maximum intensity of focused transition radiation.

To conclude, we note the following. The field of the virtual photons of the ultra-relativistic particle reflects from the surface of the bent target and is transformed to the radiation of real photons during the process of the generation of transition radiation according to the laws of usual optics, as pointed out in [28]. Thus, it is expected that the use of bent targets (including spherical) for the generation of transition and diffraction radiations will make it possible to obtain a focused radiation spot without optical systems. In particular, for the parameters of the beam used in the experiment [27], owing to the use of the paraboloid target with the focal length $P = 600$ mm, outer diameter 100 mm, and inner diameter 10 mm, the average field of coherent diffraction radiation on a crystal with a diameter of 10 mm placed at a

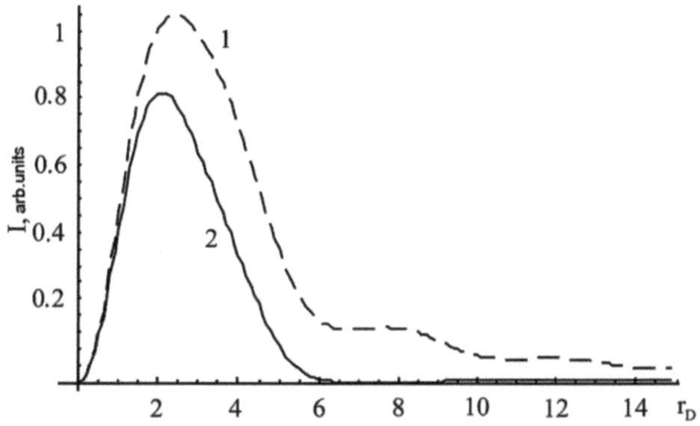

Fig. 8.27 Radial distribution of the intensities of (*curve 1*) transition radiation from the parabolic target without hole and (*curve 2*) diffraction radiation from the parabolic target with the hole of the radius $\rho_0 = 5\,\text{mm}$ on the detector for the parameters $\gamma = 200$, $L = 0.6$, $\lambda = 10^{-3}\,\text{m}$, $P = 1\,\text{m}$, and $D = 0.1\,\text{m}$

distance of 300 mm is approximately an order of magnitude higher than that from the flat target with the same sizes. This characteristic determines the sensitivity of the so-called electro-optical method. The use of diffraction radiation generated in a similar target will make it possible not only to increase the sensitivity of the method, but also to eliminate the undesirable deterioration of the parameters of the beam after its passage through a solid target.

References

1. Ter-Mikaelyan, M.L.: High-Energy Electromagnetic Processes in Condensed Media. Wiley-Interscience, New York, NY (1972)
2. Baier, V.M., Katkov, V.M.: Concept of formation length in radiation theory. Phys. Rep. **409**, 261 (2005)
3. Feinberg, E.L.: Rigid diffraction processes at high energies. Phys.-Uspekhi **58**, 193 (1956)
4. Feinberg, E.L.: Hadron clusters and half-dressed particles in quantum field theory `http://ufn.ru/en/articles/1980/10/a/`. Sov. Phys. Usp. **23**, 629 (1980)
5. Ahmanov, S.A., Nikitin, S.Yu.: Physical Optics. Moscow State University, Moscow (in Russian) (1998)
6. Verzilov, V.A.: Quantum field theory. Phys. Lett. A **273**, 135 (2000)
7. Ginzburg, V.L., Frank, I.M.: Transition radiation in the pre-wave zone. Sov. Phys. JETP **16**, 15 (1946)
8. Castellano, M., Verzilov, V., Catani, L. et al.: Search for the prewave zone effect in transition radiation. Phys. Rev. E **67**, 015501(R) (2003)
9. Karataev, P., Araki, S., Aryshev, A. et al.: Experimental observation and investigation of the prewave zone effect in optical diffraction radiation. Phys. Rev. ST Accel. Beams **11**, 032804 (2008)
10. Ginzburg, V.L., Tsytovich, V.N.: Transition Radiation and Transition Scattering. Higler, Bristol (1990)
11. Landau, L.D., Lifshitz, E.M.: The Classical Theory of Fields. Pergamon Press, (1987)
12. Luke, C.L., Yuan, C.L., Wang, H., Uto, Prunster, S.: Formation-zone effect in transition radiation due to ultrarelativistic particles. Phys. Rev. Lett. **25**(v), 1513 (1970)

13. Jackson, J.D.: The Classical Electrodynamics. New York, NY (1998)
14. Potylitsyn, A.P.: Image of Optical Diffraction Radiation (ODR) Source and Spatial Resolution of ODR Beam Profile Monitor. In: Wiedemann, H. (ed.) Advanced Radiation Sources and Applications, p. 149, Springer Dordrecht (2006)
15. Bosch, R.A.: Focusing of infrared edge and synchrotron radiation. Nucl. Instrum. Methods Phys. Res. A **431**, 320 (1999)
16. Bosch, R.A.: Extraction of edge radiation within a straight. Phys. Rev. ST-AB **5**, 020701 (2002)
17. Kalinin, B.N., Naumenko, G.N., Potylitsyn, A.P. et al.: Measurement of the angular characteristics of transition radiation in near and far zones. JETP Lett. **84**, 136, (2006)
18. Aleinik, A.N., Chefonov, O.V., Kalinin, B.N. et al.: Low-energy electron-beam diagnostics based on the optical transition radiation. Nucl. Instrum. Methods Phys. Res. B **201**, 34 (2003)
19. Potylitsyn, P.: Transition radiation and diffraction radiation. Similarities and differences. Nucl. Instrum. Methods Phys. Res. B **145**, 169 (1998)
20. Karataev, P.V.: Prewave zone effect in transition and diffraction radiation: Problems and solutions. Phys. Lett. A **345**, 428 (2005)
21. Dnestrovskiy, Yu.N., Kostomarov D.P.: Radiation of ultra relativistic charges moving across round aperture in flat screen. Sov. Phys. Dokl. **4**, 158 (1959).
22. Castellano, M., Verzilov, V.: Spatial resolution in optical transition radiation beam diagnostics. Phys. Rev. ST-AB **1**, 062801 (1998)
23. Rule, D.W., Fiorito, R.B.: Imaging micron-sized beams with optical transition radiation. AIP Conf. Proc. **229**, 315 (1991)
24. Potylitsyn, A.P., Rezaev, R.O.: Focusing of transition radiation and diffraction radiation from concave targets. Nucl. Instrum. Methods Phys. Res. B **252**, 44 (2006)
25. Potylitsyn, A.P.: Scattering of coherent diffraction radiation by a short electron bunch. Nucl. Instrum. Methods Phys. Res. A **455**, 213 (2000)
26. Ryazanov, M.I., Tilnin, I.S.: Transition Radiation of ultra relativistic particle from curved surface of media boundary. Sov. Phys. JETP **71**, 2079 (1976)
27. Winter, A., Tonutti, M., Casalbuoni, S., et al.: Bunch length measurement at the SLS Linac using Electro Optical Techniques. Proceedings of the ERAC 2004, Lucerne, Switzerland, p. 253 (2004)
28. Bolotovskiy, B.M., Galst'yan, E.A.: Diffraction and diffraction radiation. Phys.-Uspekhi **43**, 755 (2000)

Chapter 9
Experimental Investigations of Diffraction Radiation Generated by Relativistic Electrons

9.1 Experimental Results on Diffraction Radiation and Comparison with Theoretical Calculations

The theory of diffraction radiation was founded in the 1950s, but the first experimental investigation of the characteristics of diffraction radiation generated by relativistic electrons was carried out only in 1995 [1]. The scheme of the experiment is shown in Fig. 9.1. The measurements were performed with a 150-MeV electron beam consisting of bunches with a length of about 1 mm and a diameter of about 2.5 mm. The population of a bunch was $1.5 \cdot 10^8$ e$^-$. In the experiment, the spectrum of coherent diffraction radiation in a wavelength range of $\lambda = 0.1$–5 mm, as well as the angular distribution of forward and backward diffraction radiations, is measured. Discs with the hole diameters $d = 10, 15, 20$ mm, as well as a transition radiation target (disc without hole), were used to generate coherent radiation.

One of the main results of the cited work is the confirmation of the results of the theoretical work [2] that the intensity and angular distribution are similar for transition radiation and forward and backward diffraction radiations if $\gamma\lambda \gg d$. In addition, the authors of that work demonstrated the possibility of

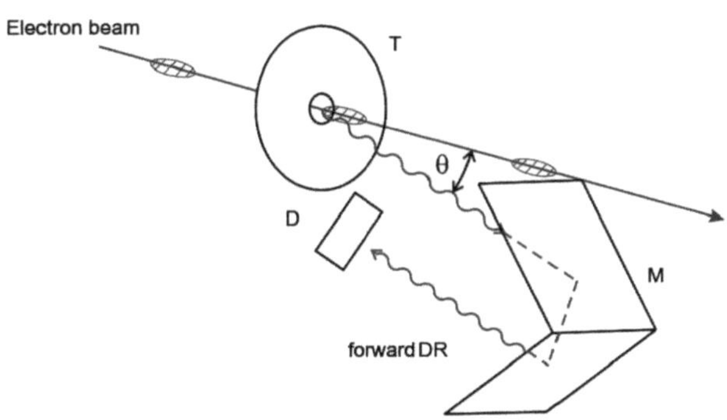

Fig. 9.1 Scheme of the experiment for investigating coherent diffraction radiation from a circular hole in a perpendicular target [1]

A.P. Potylitsyn et al., *Diffraction Radiation from Relativistic Particles*,
STMP 239, 251–277, DOI 10.1007/978-3-642-12513-3_9,

non-invasive diagnostics with the use of diffraction radiation via the determination of the length of the electron bunch from the measured spectrum of coherent diffraction radiation.

Optical (incoherent) diffraction radiation generated by ultrarelativistic electrons at the mirror-reflection angle was observed for the first time in the experiment reported in [3]. The scheme of the experiment that was performed with the internal beam of the Tomsk synchrotron is shown in Fig. 9.2. To reduce the contribution of synchrotron radiation, the measurements were performed for an electron energy of $E = 200\,\mathrm{MeV}$. A polished aluminum target inclined at an angle of 45° to the electron beam was placed in a goniometric holder. The rotation of this target ensured the coincidence of the mirror-reflection direction from the target with the fixed slits of a collimator placed in front of an optical detector (photomultiplier). In addition to optical radiation (which consists of reflected synchrotron radiation, backward transition radiation, and backward diffraction radiation) at an angle of 90° with respect to the electron beam, forward bremsstrahlung was detected in the experiment with the use of scintillation counter. Insert *a* schematically shows the radial distribution $\rho(R)$ of the electron beam in the target region. The shaded area corresponds to the electrons contributing to transition radiation and bremsstrahlung. The beam diameter in the target region was about 2 mm.

In the first approximation, the intensity I_{BTR} of backward transition radiation generated by the electron beam, as well as the bremsstrahlung intensity I_{Br}, is

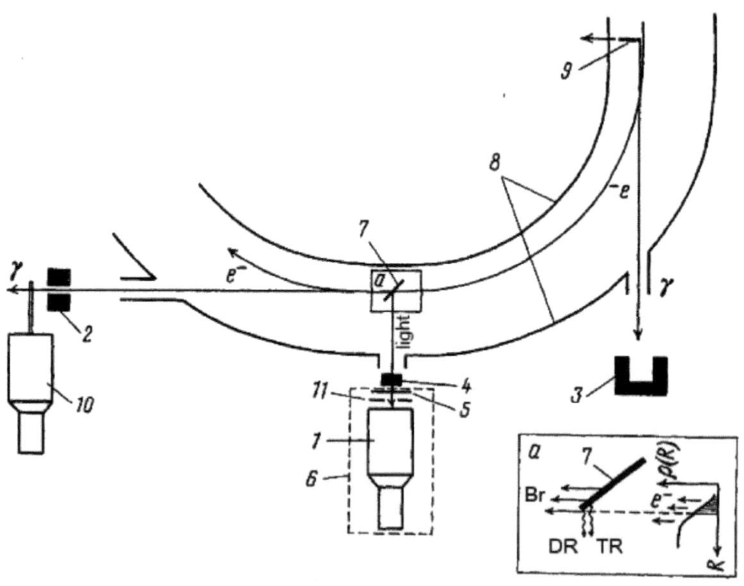

Fig. 9.2 Layout of the experiment setup: (*1*) photomultiplier, (*2*) and (*11*) collimators, (*3*) dump, (*4*) mirror, (*5*) optical filter, (*6*) detecting unit, (*7*) target, (*8*) accelerator chamber, (*9*) scraper, and (*10*) scintillation counter

determined by the number of electrons passed through the target. In this case, these two intensities should be strictly proportional to each other.

In the preceding rectilinear section of the accelerator, thick absorber target (scraper) 9, which could be displaced in the radial direction "cutting" a certain fraction of electrons that could be incident on the main target, was placed. The bremsstrahlung beam from the scraper was absorbed in dump 3.

When the scraper is removed, all electrons are incident on target 7. In this case, the yield of bremsstrahlung and transition radiations is maximal. The number of electrons passing beyond the target near its edge, which determines the yield of diffraction radiation, is also maximal. When the scraper covers a fraction of the electron beam, the yield of bremsstrahlung and transition radiations decreases, whereas the yield of diffraction radiation decreases more slowly.

As mentioned above, backward transition radiation is concentrated in a narrow angular range near the mirror-reflection direction. Figure 9.3 shows the angular dependences of the optical-radiation intensity in the case of the rotation of the target about the vertical axis for various positions of the scraper. The width of the orientation dependences is determined by the divergence of the electron beam, angular aperture of the detector, and self divergence of transition and diffraction radiations. Angle $\theta = 0$ in Fig. 9.3 corresponds to the mirror reflection geometry. Dependence 1 was obtained in the absence of the scraper, when the main contribution to optical radiation comes from transition radiation. The insert shows the same dependence measured with a small scanning step. Although the contribution of synchrotron radiation is quite large, the figure demonstrates the splitting of the peak in the orientation

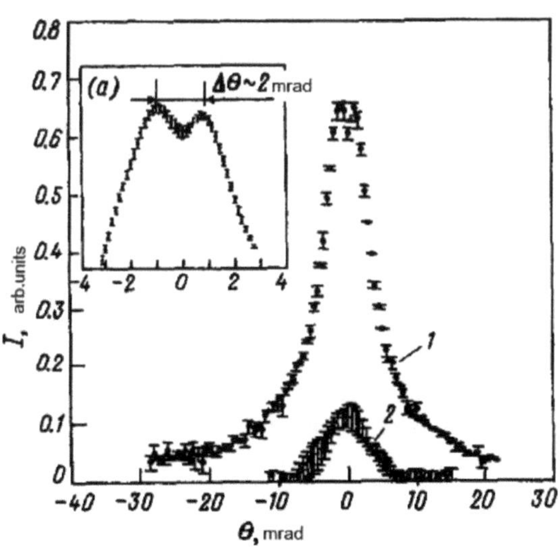

Fig. 9.3 Orientation dependences of the intensity of optical polarization radiation (backward diffraction one and backward transition one) (**1**) in the absence of the scraper and (**2**) in the presence of the scraper at a distance of 1 mm

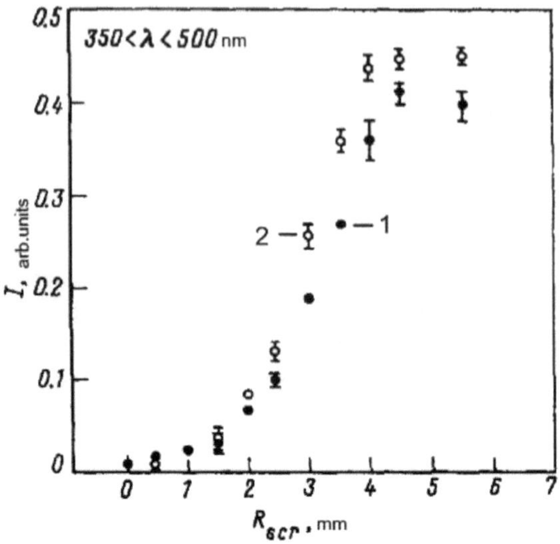

Fig. 9.4 Scraper dependences of the yield of (*line 1*) bremsstrahlung and optical (*line 2*) polarization radiation

dependence, which is characteristic of transition radiation ($\Delta\theta \sim \gamma^{-1} \sim 2$ mrad). Dependence 2 corresponds to the position of the scraper at a distance of 1 mm (see Fig. 9.4). In this position, most fraction of the electron beam is cut by the scraper and is not incident on the target; therefore, significant contribution to the radiation under investigation comes from diffraction radiation. As shown in Chap. 4, the angular distribution of diffraction radiation has a peak in the mirror-reflection direction, which was observed in the experiment. In order to confirm the contribution from diffraction radiation, the scraper dependences (dependences of the yield of radiation from the radial position of the scraper) of bremsstrahlung and optical radiation were measured in the mirror-reflection geometry.

Figure 9.4 shows the scraper dependence for the case where a blue optical filter with the pass bandpass $\lambda \approx 350$–500 nm is placed in front of the photomultiplier. If diffraction radiation were absent, dependences 1 and 2 would coincide up to a constant factor. The detailed analysis of similar dependences performed in [3] for various wavelengths shows that the shift between curves 1 and 2 is a function of the wavelength. In the absence of the contribution from diffraction radiation, causes for the appearance of such shift are absent. Thus, optical diffraction radiation in the mirror-reflection direction from the inclined target was observed for the first time in the discussed experiment in the complete analogue with backward transition radiation.

The quantitative comparison of the characteristics of optical diffraction radiation and optical transition radiation was performed in the experiment [4, 5] in the 1.28-GeV electron beam of the KEK ATF accelerator with extremely low emittance [6], which made it possible to obtain the beam with dimensions $\sim 10 \times 100\,\mu$m in the target region. The target, which is a 300-μm-thick silicon plate with a 0.5-μm

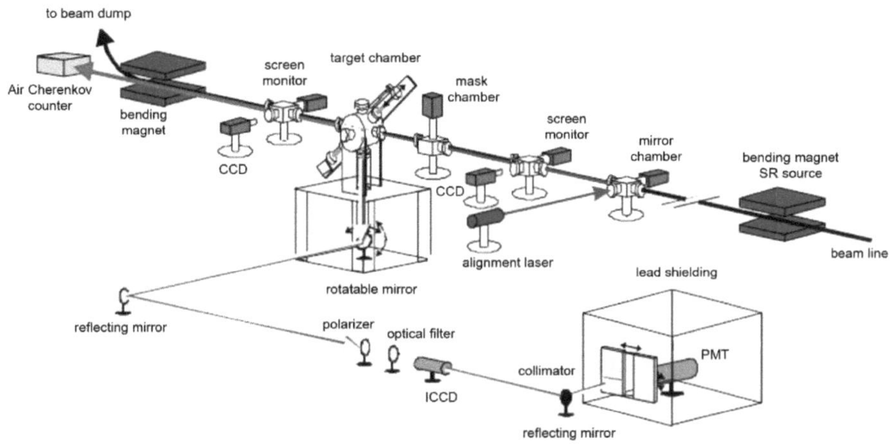

Fig. 9.5 Scheme of the experiment on investigation of the characteristics of optical diffraction radiation with the extracted electron beam of the KEK ATF accelerator

aluminum coating, could be moved at an angle of 45° with respect to the electron beam with a step of 0.5 μm (see Fig. 9.5).

Optical radiation generated through the transition radiation mechanism (when the beam intersects the target) or the diffraction radiation mechanism (when the beam passes near the target) was detected at an angle of 90° with the use of a rotating mirror, stationary reflecting mirrors, and photomultiplier with a collimator in front of it. By rotating the mirror about two perpendicular axes (varying the angles θ_x, θ_y, see Fig. 9.5), the angular distributions of the light flux from the target can be measured.

Figure 9.6 shows the angular distribution of radiation in the wavelength range $\lambda = 0.3$–$0.65\,\mu$m for the case where the target covers the beam. As expected,

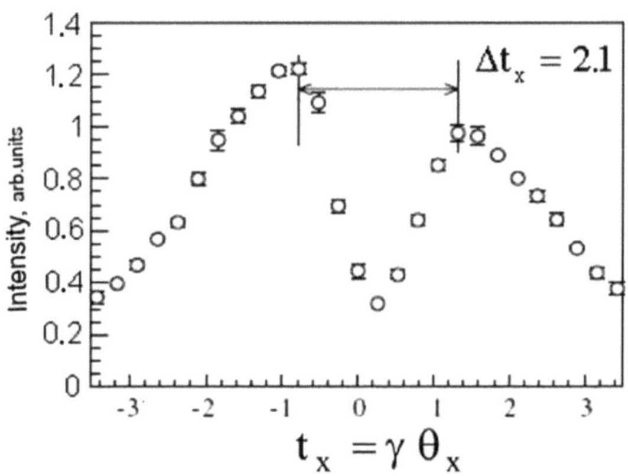

Fig. 9.6 Measured angular distribution of optical transition radiation

optical transition radiation in this case is observed almost in the wave zone (the distance between the detector and target is equal to 2.60 m, $\gamma^2\bar{\lambda} \approx 3.12$ m, and average wavelength is $\bar{\lambda} = 0.42\,\mu$m); as a result, the angular distribution of transition radiation is slightly broadened (the distance between the maxima is $\Delta\theta_x = 2.1\,\gamma^{-1}$ rather than $2.0\gamma^{-1}$). Small asymmetry in the distribution of transition radiation is likely caused by the contribution of synchrotron radiation from a deflecting magnet placed in the path of the electron beam in front of the target.

The measurement of the angular characteristics of transition radiation made it possible to determine the direction of mirror reflection from the target. At the next stage of the measurements, the yield of optical radiation is analyzed as a function of the impact parameter in the mirror-reflection direction ($t_x = t_y = 0$) in comparison with the yield of forward bremsstrahlung, which was measured using a Cherenkov detector with air filling (see Fig. 9.7). As seen in this figure, the yield of optical radiation is maximal when electrons move along the target with zero impact parameter (i.e., at the point, where the bremsstrahlung intensity vanishes) in complete agreement with the theory of diffraction radiation generated by relativistic particles (see [7]). The decrease in the intensity of diffraction radiation with an increase in the impact parameter is described by expression (3.59), but the dependence of the yield of transition radiation generated by a particle near the edge of the target (intersecting the target) is more complicated.

The simplest model describing this process, which can be called edge transition radiation, is illustrated in Fig. 9.8 [8, 9]. The field of edge transition radiation in the wave zone can be obtained from the superposition principle:

$$\mathbf{E}^{\mathrm{TR}}_{\mathrm{ed}} = \mathbf{E}^{\mathrm{TR}} - \mathbf{E}^{\mathrm{DR}} e^{i\phi}, \tag{9.1}$$

where "ed" means edge and $\mathbf{E}^{\mathrm{TR}}$ and $\mathbf{E}^{\mathrm{DR}}$ are the fields of transition and diffraction radiations whose components are given by expressions (3.113) and (3.111), respectively. The phase shift φ is determined by analogy with formulas (3.110). After the

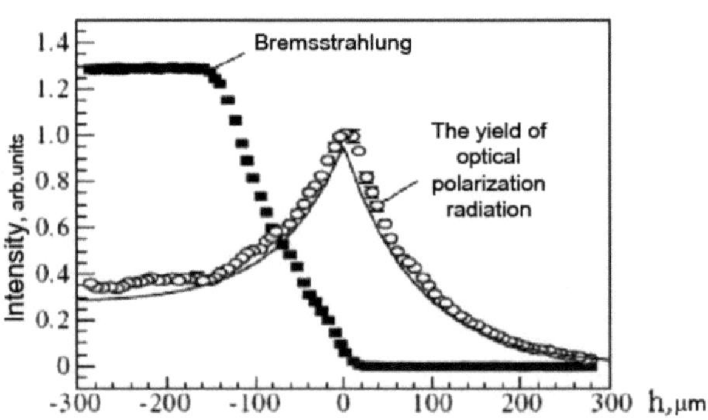

Fig. 9.7 Impact-parameter dependence of the yield of bremsstrahlung and optical polarization radiation. The *solid line* is the calculation with taking into account of a finite spectral resolution and angular aperture

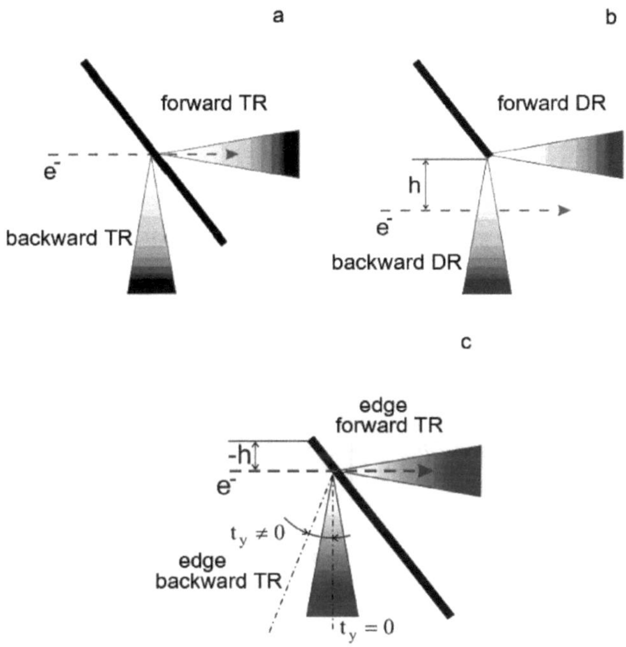

Fig. 9.8 Geometry of (**a**) transition radiation for the infinite interface, (**b**) diffraction radiation, and (**c**) edge transition radiation

calculation of the square of the absolute values of the components of field (9.1) in the coordinate system described by unit vectors (3.50), we obtain

$$\left|E_{1\mathrm{ed}}^{\mathrm{TR}}\right|^2 = \frac{e^2}{(2\pi)^4}\gamma^2 \frac{t_x^2}{\left(1+t_x^2+t_y^2\right)^2}\left\{4+\exp(2\kappa)\left(1+\frac{t_y^2}{1+t_x^2}\right) - \right.$$

$$\left. -\frac{4\exp(\kappa)}{\sqrt{1+t_x^2}}\left[\sqrt{1+t_x^2}\cos\left(\kappa t_y\right)+t_y\sin\left(\kappa t_y\right)\right]\right\},$$

$$\left|E_{2\mathrm{ed}}^{\mathrm{TR}}\right|^2 = \frac{e^2}{(2\pi)^4}\gamma^2 \frac{t_x^2}{\left(1+t_x^2+t_y^2\right)^2}\times$$

$$\left\{4t_y^2+\left(1+t_x^2+t_y^2\right)\exp(2\kappa)-4t_y\exp(+\kappa)\left[t_y\cos(\kappa t_y)-\sqrt{1+t_x^2}\sin\left(\kappa t_y\right)\right]\right\}. \tag{9.2}$$

Here,

$$\kappa = \frac{2\pi h}{\gamma\lambda} = -\frac{2\pi\,|h|}{\gamma\lambda} \tag{9.3}$$

and the impact parameter h is taken negative (see Fig. 9.8c) in contrast to the case where the particle moves near the edge of the target in vacuum (in this case, the impact parameter is positive as in all formulas in Sect. 3.2 that describe diffraction radiation).

The spectral–angular distribution of emitted energy is given by the expression

$$\frac{d^2W_{\text{ed}}^{\text{TR}}}{\hbar d\omega d\Omega} = cr^2\left(\left|E_{1\text{ed}}^{\text{TR}}\right|^2 + \left|E_{2\text{ed}}^{\text{TR}}\right|^2\right) \tag{9.4}$$

and can be defined for any impact parameter:

$$\frac{d^2W}{\hbar d\omega d\Omega} = \begin{cases} \dfrac{d^2W_{\text{ed}}^{\text{TR}}}{\hbar d\omega d\Omega}, & h \leq 0 \\[2ex] \dfrac{d^2W^{\text{DR}}}{\hbar d\omega d\Omega}, & h \geq 0. \end{cases} \tag{9.5}$$

It is easy to see that expression (9.2) for the case of the passage of the particle through the edge of the target ($h = 0$) yields the same result as formula (3.59) determining the intensity of diffraction radiation:

$$\frac{d^2W\,(h=0)}{\hbar d\omega d\Omega} = \frac{\alpha}{4\pi^2}\gamma^2 \frac{1+2t_x^2}{\left(1+t_x^2\right)\left(1+t_x^2+t_y^2\right)}. \tag{9.6}$$

Figure 9.9 shows the calculated yield of optical polarization radiation with the wavelength $\lambda = 0.5\,\mu\text{m}$, which is generated by the electron with the Lorentz factor

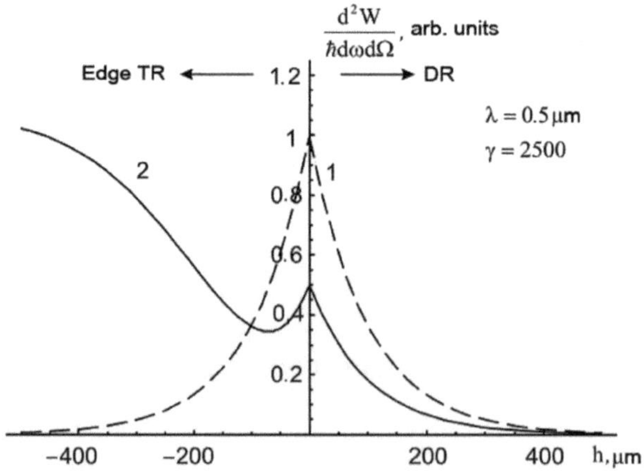

Fig. 9.9 Calculated impact-parameter dependence of the optical-radiation intensity. Line 1 corresponds the radiation at $t_x = t_y = 0$, i.e., at the mirror-reflection angle and line 2 corresponds to the radiation at $t_x = 0$ and $t_y = 1$

$\gamma = 2500$, as a function of the impact parameter in the interval$-500 \leq h \leq 500\,\mu$m (line 1 describes radiation in the exact mirror-reflection direction, $\theta_x = \theta_y = 0$; line 2 corresponds to radiation for the angular variables $\theta_x = 0$ and $\theta_y = \gamma^{-1}$). We use the term optical polarization radiation to describe optical diffraction radiation (for $h > 0$) and optical transition radiation (for $h < 0$). Note that the yield of polarization radiation in the mirror-reflection direction (line 1) is maximal at $h = 0$ and vanishes at $|h| \gg \gamma\lambda$. The suppression of the diffraction radiation yield at $h \gg \gamma\lambda$ is a trivial consequence of formula (3.59), whereas the vanishing of the intensity of edge transition radiation at $t_x = t_y = 0$ when the particle trajectory is displaced inside the target is attributed to the transformation of the angular distribution of transition radiation to the angular distribution of usual transition radiation with zero minimum in the mirror-reflection direction. However, when a detector with a finite angular aperture is used, expression (9.5) should be integrated over this aperture. In the experiment reported in [5], a collimator specified the aperture $-0.4\gamma^{-1} \leq \theta_x, \theta_y \leq 0.4\gamma^{-1}$ (i.e., $-0.4 \leq t_x, t_y \leq 0.4$).

Figure 9.10 shows the calculation results for the indicated solid angle. Note that the ratio of the yield of edge transition radiation at the point $h = -0.3$ mm to the maximum in this case is equal to 0.28 in good agreement with the experimental data (see Fig. 9.7).

The comparison of similar calculations (with finite spectral resolution$0.35 \leq \lambda \leq 0.6\,\mu$m) with the experimental data for angles $t_x = 1$; $t_y = 0$ is given in Fig. 9.11. As above, optical polarization radiation at $h < 0$ and $h > 0$ is edge transition radiation and diffraction radiation, respectively. Thus, the approach used to calculate the characteristics of diffraction radiation is in good agreement with the experimental data. This fact demonstrates the applicability of this theory of diffraction radiation in the wave zone and confirms the possibility of using optical diffraction radiation for the non-invasive diagnostics of low-emittance intense charged beams. This is very important for the creation of new-generation accelerators such as free electron lasers.

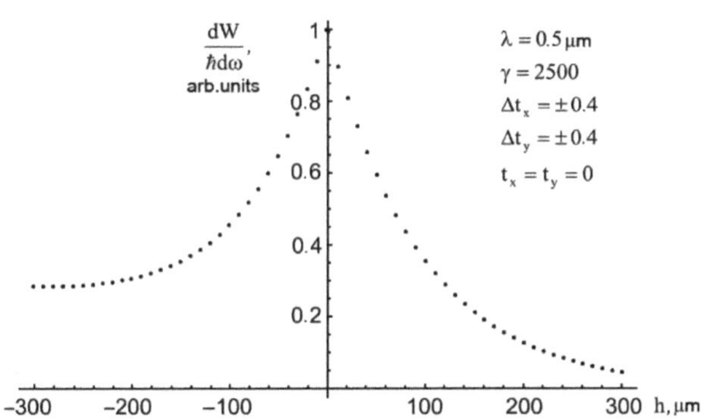

Fig. 9.10 Impact-parameter dependence of the optical-radiation yield for the finite detector aperture $\gamma^2\Delta\Omega = \Delta t_x \Delta t_y$

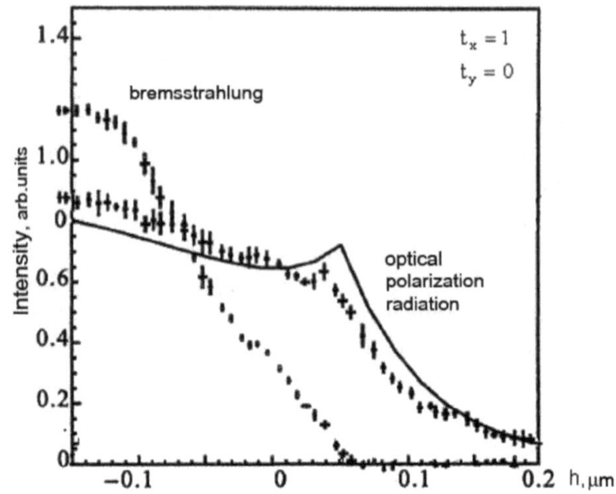

Fig. 9.11 Impact-parameter dependence of the yield of optical polarization radiation at the angles $t_x = 1$ and $t_y = 0$ for finite aperture $-0.4 \leq t_x, t_y \leq 0.4$ for the same parameters as those in Fig. 9.10. Solid curve – calculations based on the developed approach

9.2 Optical Diffraction Radiation from a Slit Target and the Possibility of the Measurement of the Transverse Size of an Electron Beam

The possibility of creating a monitor of the transverse sizes of the beam on the basis of optical diffraction radiation was analyzed with the use of the electron beam of the KEK ATF accelerator in the experiment [10] with the setup described in the preceding section. An aluminized silicon plate with a slit 0.26 mm in width was used as a target (see the lower inset in Fig. 9.12) [11]. In order to vary the impact parameter, the target was displaced at an angle of 45° with respect to the electron beam.

The black points in Fig. 9.12 present the forward bremsstrahlung yield. Owing to the asymmetric displacement of the target with respect to the electron beam, the impact-parameter dependences of the yield of bremsstrahlung generated by the beams intersecting the opposite edges of the target are different. The sharp decrease in the bremsstrahlung yield at the point with the impact parameter $h \approx -0.1$ mm corresponds to the passage of the beam near point A (see the insert), whereas the slow increase at $h = 0.09$ corresponds to the point B. The same figure shows the yield of optical diffraction radiation, which, as above, was measured by the detector with a finite angular aperture. According to the figure, the yield of optical diffraction radiation is minimal when the beam passes through the center of the slit in agreement with the theoretical expectation (see Fig. 3.11). As follows from formula (3.117), the intensity of diffraction radiation generated by the electron beam that has infinitely small transverse sizes and passes through the center of the slit

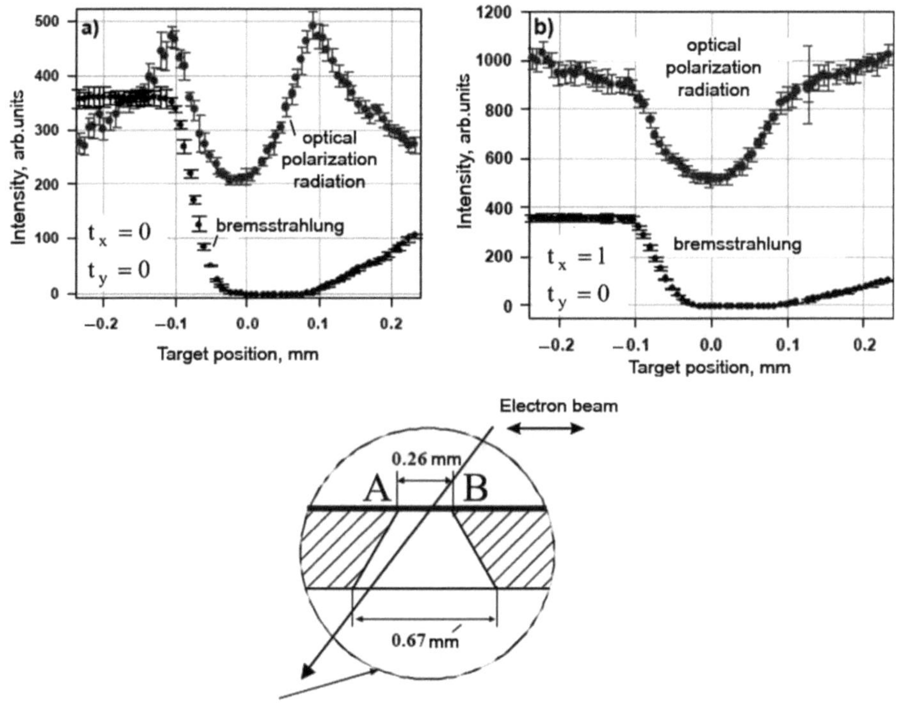

Fig. 9.12 Impact-parameter dependence of the yield of bremsstrahlung and optical polarization radiation generated by electrons passing through a slit at various angles (see *the inset*)

($\ell = 0$) is equal to zero in the mirror-reflection direction ($t_x = t_y = 0$). However, as mentioned above, the intensity of diffraction radiation generated by the electron moving with a nonzero impact parameter is nonzero and this fact makes it possible to use optical diffraction radiation from the slit for the measurement of the transverse size of the beam (in the case of the passage of particles through the slit, the impact parameter is the distance of the trajectory from the center of the slit, rather than from the edge of the target).

Let us consider the simplest case where the distribution of the electron beam in the plane perpendicular to the slit is described by the Gaussian distribution

$$f(y) = \frac{1}{\sigma_0\sqrt{2\pi}} \exp\left\{-\frac{y^2}{2\sigma_0^2}\right\}. \tag{9.7}$$

Changing the dimensional coordinate y to the dimensionless parameter l (see formula (3.115)), one arrives at the distribution

$$f(l) = \frac{1}{\sigma\sqrt{2\pi}} \exp\left\{-\frac{l^2}{2\sigma^2}\right\}, \tag{9.8}$$

where the dimensionless parameter $\sigma = \sigma_0/a\sin\theta_0$ characterizes the size of the beam in dimensionless variables (a is the slit width and θ_0 is the inclination angle).

In the approximation $\sigma \ll 1$, the yield of diffraction radiation generated by the particle beam passing through the center of the slit can be averaged over the transverse beam size for any observation angles:

$$\frac{d^2\bar{W}}{\hbar d\omega d\Omega} = \frac{\alpha\gamma^2}{2\pi^2}\frac{\exp\left(-z\sqrt{1+t_x^2}\right)}{\left(1+t_x^2\right)\left(1+t_x^2+t_y^2\right)}\times$$
$$\times\left\{\left(1+2t_x^2\right)\exp\left[2\sigma^2 z^2\left(1+t_x^2\right)\right]-\cos(zt_y+\chi)\right\}. \tag{9.9}$$

Here, $z = \dfrac{2\pi a\sin\theta_0}{\gamma\lambda}$ and $\chi = \arccos\left(\dfrac{1+t_x^2-t_y^2}{1+t_x^2+t_y^2}\right)$.

Comparing formula (9.9) with expression (3.117) to which the former formula is transformed in the limit $\sigma \to 0$, we note that Eq. (9.9) at $t_x = t_y = 0$, as above, provides a minimum, but it is nonzero:

$$\frac{d^2\bar{W}_{\min}}{\hbar d\omega d\Omega} = \frac{\alpha\gamma^2}{2\pi^2}\left(\exp\left(2\sigma^2 z^2\right)-1\right)\exp\left(-z\right). \tag{9.10}$$

Expression (9.9 is obviously maximal at $t_y^{\max} \approx \pm 1$ [see Fig. (9.13)]; i.e., when the expression in braces in formula (9.9) is maximal $\left(\cos(zt_y+\chi) \to -1\right)$

$$\frac{d^2\bar{W}_{\max}}{\hbar d\omega d\Omega} = \frac{\alpha\gamma^2}{2\pi^2}\frac{\exp\left(-z\right)}{\left[1+\left(t_y^{\max}\right)^2\right]}\left\{\exp\left(2\sigma^2 z^2\right)+1\right\}. \tag{9.11}$$

Measuring the ratio of the minimum and maximum yields of diffraction radiation,

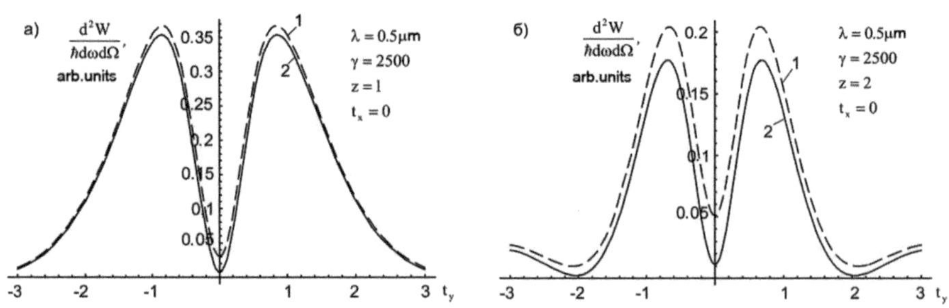

Fig. 9.13 Angular distribution of diffraction radiation in the plan perpendicular to the slit with the width (**a**) $a\sin\theta_0 = 0.2$mm and (**b**) $a\sin\theta_0 = 0.4$mm. Lines 1 and 2 correspond to the parameter $\sigma = 0.1$ and $\sigma = 0.2$

$$R = \frac{d^2 \bar{W}_{\min}/\hbar d\omega d\Omega}{d^2 \bar{W}_{\max}/\hbar d\omega d\Omega}, \tag{9.12}$$

one can determine the parameter σ characterizing the transverse size of the beam:

$$\sigma = \frac{1}{z}\sqrt{\frac{1}{2}\log\frac{1+R_1}{1-R_1}}, \tag{9.13}$$

where $R_1 = R/\left[1+\left(t_y^{\max}\right)^2\right]$. In terms of the dimensional quantities, Eq. (9.13) provides the variance of the transverse distribution of the beam:

$$\sigma_0^2 = \sigma^2 a^2 \sin^2\theta_0. \tag{9.14}$$

Note that this result is reliable under the condition $\sigma \ll 1$. For $\sigma \geq 1$, formulas (9.9), (9.10), (9.11)) are invalid and the theory should be developed implying the averaging of the diffraction radiation yield at arbitrary σ values. At the same time, for a given size of the beam, the slit can be always chosen so that $\sigma \ll 1$.

Figure 9.14 shows the ratio R calculated by Eq. (9.12) as a function of the beam size σ_0 (lower line) disregarding the angular aperture of the detector and (upper line) for the angular aperture $\Delta\theta_y = 0.1/\gamma$. Figure 9.15 shows the experimental and calculated characteristics of the intensity components of diffraction and transition radiations polarized in the vertical plane [12].

The use of a polarizer in the experiment reported in [10] provided the effective cutoff of the background contribution from synchrotron radiation polarized in the horizontal plane (see Fig. 9.15). In addition to the polarizer, an optical filter with the bandpass $\lambda = 0.55 \pm 0.02\,\mu$m was used. Good agreement between the experimental data and theoretical results indicates that background processes (mainly, synchrotron radiation) are almost completely suppressed and the angular resolution of the equipment makes it possible to measure the ratio R given by (9.12) with a sufficient accuracy.

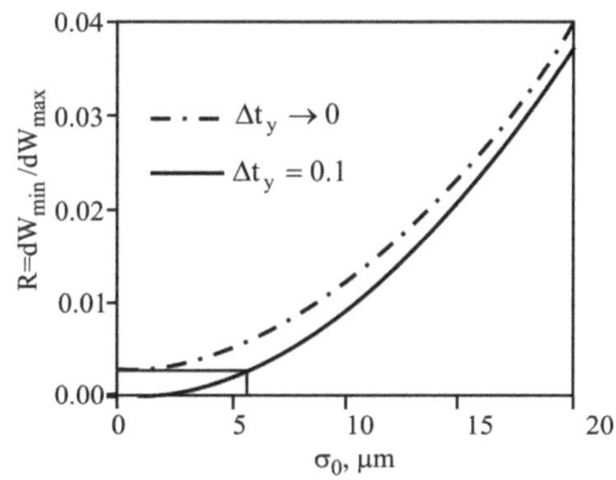

Fig. 9.14 Ratio R versus the transverse beam size (the solid line is the ideal case and the dash—dotted line is averaging over the aperture $\Delta t_y = 0.1/\gamma$)

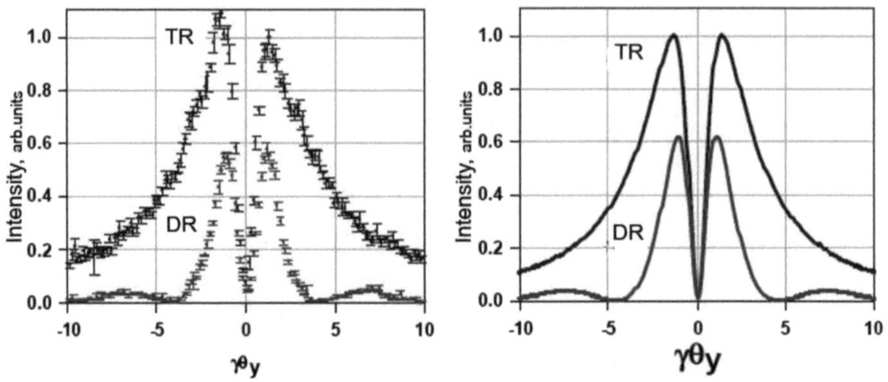

Fig. 9.15 (**a**) Measured angular distributions of transition and diffraction radiations (normalized to the maximum of transition radiation) the (**b**) corresponding theoretical curves

Figure 9.16 shows (solid circles) the transverse size of the electron beam of the KEK ATF accelerator measured using the above non-invasive method based on optical diffraction radiation. This figure also shows the transverse size of the beam measured using two invasive monitors in which a thin tungsten wire intersects the electron beam. The upper panel in Fig. 9.16 shows the arrangement of these monitors, the first of which located in front of the diffraction target measures the size

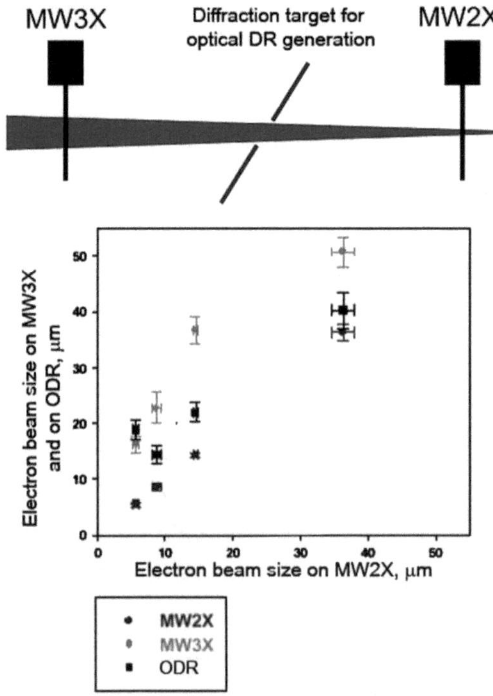

Fig. 9.16 Scheme of the measurement of the vertical size of the extracted beam of the KEK ATF accelerator using wire scanners and optical diffraction radiation monitor and the results of the measurements

of the electron beam with a smaller radius, whereas the opposite situation is realized for the second monitor. The measurement results indicate that the spatial resolution of the proposed non-invasive method based on diffraction radiation is $\sim$ 15 μm.

9.3 Experimental Investigations of the Generation of Smith—Purcell Radiation by Ultrarelativistic Electron Beams

Since the discovery of the Smith—Purcell effect in 1953, there were perfomed about it experimental investigations of the characteristics of this kind of radiation generated by nonrelativistic electron beams from the centimeter range to the optical one. These investigations are stimulated by the possibility of using the Smith—Purcell effect, first, for the creation of intense compact sources of microwave radiation [13] and, second, for the development of new radiation sources with variable wavelength in the millimeter and submillimeter ranges [14].

The first experiments on the generation of Smith—Purcell radiation by relativistic electrons were carried out in the beginning of the 1990s [15, 16]. The experiment reported in [15] was performed in the electron beam with a current of 50—200 mA, an energy of $E = 3.6\,\text{MeV}$, and a size of $\sim$ 3 mm in the direction perpendicular to a grating. The impact parameter h was comparable with the beam size.

The grating with the triangular profile and a period of 769 μm was used in the experiment. The radiation spectrum was measured using a monochromator and an InSb bolometer with helium cooling (see Fig. 9.17). The sensitivity of the bolometer was about 10 nW in a wavelength range of 400–2500 μm. The radiation detection

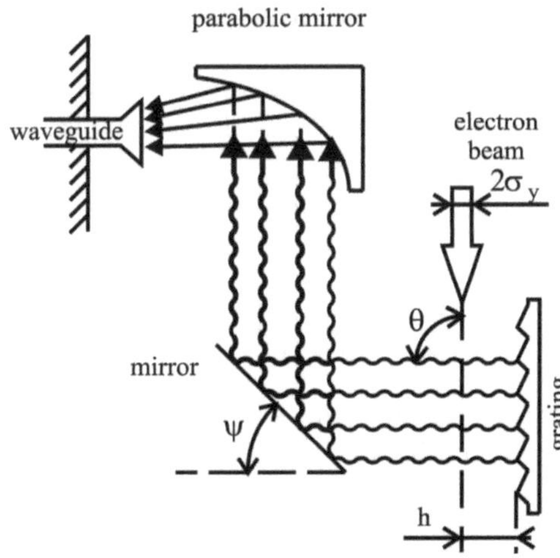

Fig. 9.17 Typical scheme of the experiment on the investigation of Smith—Purcell radiation. Change in the mirror inclination angle ψ gives rise to change in an observation angle of Smith—Purcell radiation θ. The impact parameter is denoted as h and the cross-section of the beam is characterized by the quantity $2\sigma_y$ (in the direction perpendicular to the grating)

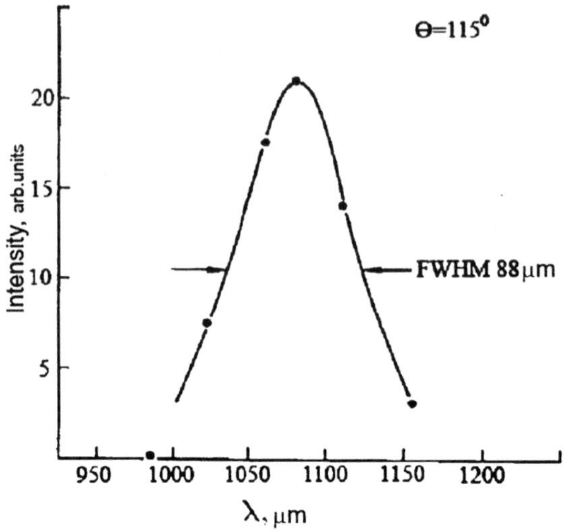

Fig. 9.18 Lineshape of Smith—Purcell radiation measured at the angle $\theta = 115°$ for the grating with a period of $d = 0.77$ mm

angle and, therefore, the wavelength range of radiation incident on the monochromator were varied by varying the inclination angle ψ of the plane reflecting mirror. The Smith—Purcell dispersion law was confirmed in the cited work with a high accuracy in a wide wavelength range. In the measurement of the lineshape of Smith—Purcell radiation for the grating with a period of $d = 0.75$ mm at an angle of $\theta = 115°$ (see Fig. 9.18), the authors of that work obtained a line monochromaticity of $\Delta\lambda/\lambda \approx 8\%$, which was much worse than the energy resolution of the spectrometer and, even more, than the "natural" monochromaticity for the grating with the number of periods $N \approx 90$ ($\Delta\lambda/\lambda \approx \frac{1}{N} \approx 1.1\%$).

One of the possible explanations of this effect is the finiteness of the angular aperture of the spectrometric section. If the angular "acceptance" of the reflecting and paraboloidal mirrors is $\Delta\theta = 10°$, the experimentally measured "broadening" of the Smith—Purcell radiation line coincides with the theoretical estimate

$$\frac{\Delta\lambda}{\lambda} \approx \frac{\sin 115°}{1 - \cos 115°}\Delta\theta. \tag{9.15}$$

In the experiment reported in [16], which was performed with a 2.8-MeV ($\gamma = 5.6$) electron beam according to the scheme similar to that shown in Fig. 9.17, the intensity of spontaneous Smith—Purcell radiation was compared with the predictions of the theory developed by Van den Berg [17]. The electron beam consisted of bunches with a duration of 20 ps (i.e., with a length of 6 mm), a population of about $2 \cdot 10^8$ electrons per bunch, and sizes $2\sigma_x \approx 12$ mm and $2\sigma_y \approx 0.75$ mm. The impact parameter was about 1 mm and most particles of the beam passed over the grating without scattering. To measure the absolute values of the intensity of

Smith—Purcell radiation, the triangular grating with the blaze angle of the "large" surface of the grating tooth $\alpha = 5°$ and a period of $d = 10$ mm was used. For the observation angle $\theta = 17.4°$, the wavelength of the first-order Smith—Purcell radiation was $\lambda = 0.623$ mm, whereas the wavelength for the angle $\theta = 27.8°$ is almost doubled: $\lambda = 1.32$ mm. For both cases, the wavelength is much shorter than the threshold value corresponding to the beginning of the manifestation of coherent effects; for this reason, incoherent Smith—Purcell radiation was observed in the experiment. Figure 9.19 shows the results of observations at $\theta = 30°$. Note that this measurement also provided the linewidth of Smith—Purcell radiation that was larger than the calculated value due possibly to the finite angular aperture of the detector.

Woods et al. [16] performed calculations following the theory developed by Van den Berg [17] for the grating and electron beam parameters described above and revealed that the experimental results for observation angles 17.4° and 27.8° are 200 and 700 times larger than the respective theoretical predictions (ΔW_{exp} ($\theta = 17.4°$) = 0.33 μW and ΔW_{exp} ($\theta = 27.8°$) = 0.11 μW). The authors indicate that one of the possible causes of such a large discrepancy is the possible coherent effect due to the difference of the longitudinal shape of a bunch from Gaussian (see Sect. 7.1). In particular, for a cylindrical bunch, coherent effects can be manifested at wavelengths much shorter than the bunch length.

In one of the latest experiments [18], Smith—Purcell radiation was studied in the optical range with an ultrarelativistic electron beam with an energy of $E =$ 855 MeV. The scheme of the experiment is shown in Fig. 9.20.

Note that the vertical size of the beam (in the direction perpendicular to the grating) in the experiment was no more than 20 μm; hence, the interaction of the beam halo with the grating material is completely excluded. Figure 9.21 shows the measured yields of Smith—Purcell radiation for two wavelength ranges $\lambda \pm \Delta\lambda =$ 546 ± 15 nm (upper panel) and (lower panel) 360 ± 15 nm. The measurements were

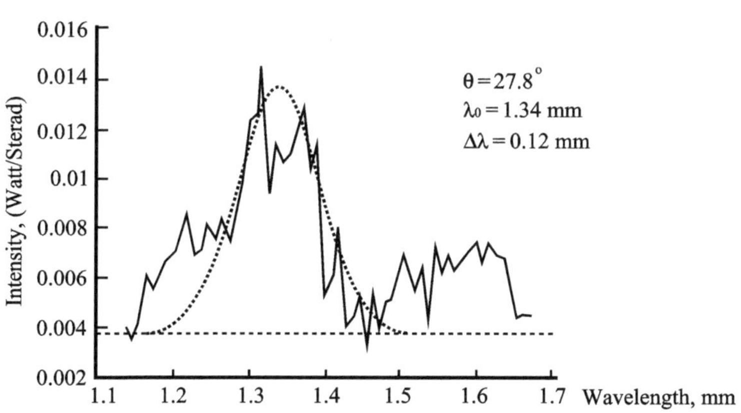

Fig. 9.19 Spectrum of Smith—Purcell radiation (*solid line*) measured for the grating with a period of $d = 10$ mm for the observation angle $\theta = 27.8°$ and (*dashed line*) Gaussian fitting of the experimental data

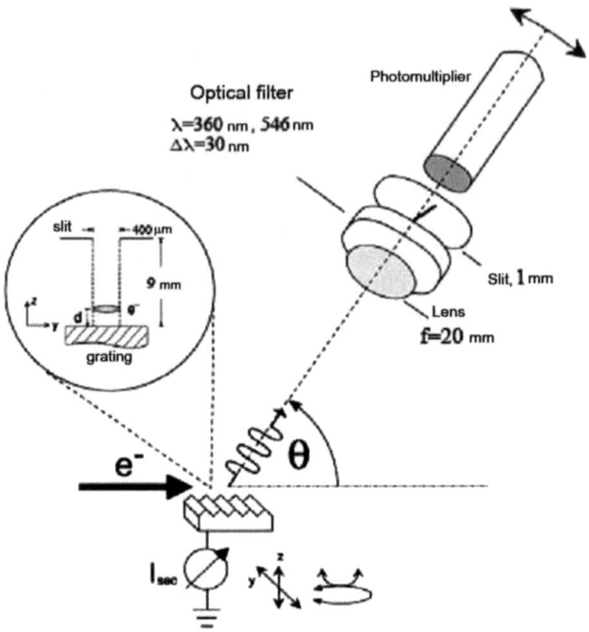

Fig. 9.20 Scheme of the experiment [18] on the investigation of optical Smith—Purcell radiation with the 855-MeV electron beam

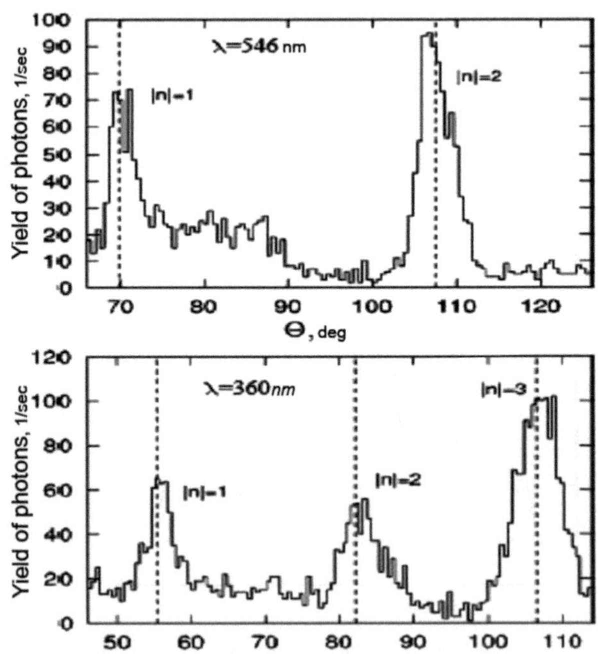

Fig. 9.21 Yield of optical Smith—Purcell radiation with the wavelength λ = (*upper panel*) 546 ± 15 nm and (*lower panel*) 360 ± 15 nm versus the observation angle θ

performed for the beam passing over the grating with a period of 0.833 μm at a distance of 127 μm by varying the observation angle θ (see Fig. 9.20). The expected positions of Smith—Purcell radiation maxima are marked by the dashed lines for various diffraction orders. The experimental value of the photon energy of Smith—Purcell radiation is in very good agreement with the Smith—Purcell formula.

The measured yield of Smith—Purcell radiation with the wavelength $\lambda = 0.36\,\mu\text{m}$ ($\hbar\omega = 3.5\text{eV}$) was $\dfrac{\Delta N}{\Delta\Omega} \approx 10^{-3} \dfrac{\text{phot}}{e^- \cdot \text{sterad}}$ or, in terms of the radiation brightness,

$$\frac{\Delta W}{\Delta\Omega} = 3.5 \times 10^{-3} \frac{\text{eV}}{e^- \cdot \text{sterad}}. \tag{9.16}$$

As mentioned by the authors of the experiment, the value obtained for $\lambda = 0.36\,\mu\text{m}$ is in satisfactory agreement with the theoretical model developed by Van den Berg [17]. Note also that the yield of photons of the (Fig. 9.21, the upper panel) second and (the lower panel) third diffraction orders is higher than the yield of fundamental harmonic radiation ($n = 1$) and this excess is not explained by the existing models.

As mentioned above, Smith—Purcell radiation generated by relativistic particles can be used for diagnostics of accelerated beams. For example, Trotz et al. [19] proposed a monitor for the position of an ultrarelativistic electron beam and Nguyen [20] and Lampel [21] proposed a monitor for the length of an electron bunch using coherent Smith—Purcell radiation. In the latter case, the detector of radiation in the millimeter and submillimeter ranges, which should be comparatively cheap and simple in operation, is of key importance. Bolometers, which are often used for this aim, have a comparatively long response time and require cooling to liquid-helium temperatures. For this reason, they cannot be used as standard equipment for such a monitor.

In the experiment reported in [22], the angular distribution of coherent Smith—Purcell radiation was investigated using a narrowband detector based on an antenna and a diode, which operates at room temperature and ensure the following characteristics: $\lambda_{\text{res}} = 7.2\,\text{mm}$; $\Delta\lambda \approx 2\,\text{mm}$; a sensitivity of 0.18 V/W; response time $\tau = 10^{-8}$ s.

The scheme of the experiment performed with the 5-MeV beam from the linear accelerator at the Institute of Nuclear Physics, Moscow State University is shown in Fig. 9.22. The length of an electron bunch is no more than 2 mm with a population of 10^9 electrons per bunch. The insert in this figure schematically shows the grating with the number of periods $N = 8$ that was used in this experiment. As the detection angle θ varies, the narrowband detector is triggered only for angles at which the wavelength of coherent Smith—Purcell radiation is in the detector sensitivity range.

Figure 9.23 shows the measured yield of coherent Smith—Purcell radiation in comparison with the calculated lineshape of coherent Smith—Purcell radiation. The calculations were performed for the plane grating with vacuum gaps, the number of periods $N = 8$, and the angular aperture of the detector $\Delta\theta = 12\,\text{mrad} = 0.06\gamma^{-1}$.

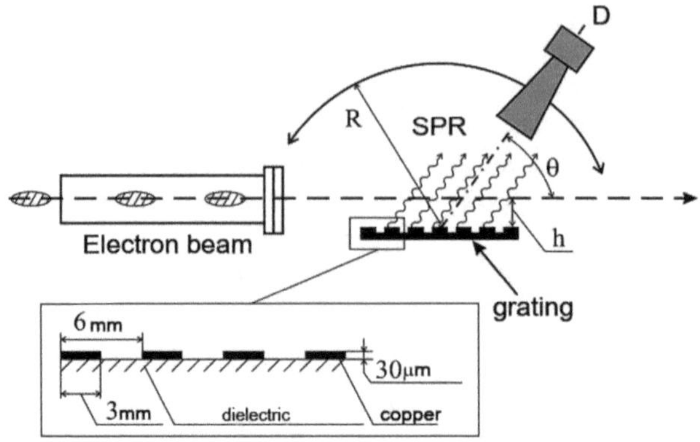

Fig. 9.22 Scheme of the experiment [22] on the generation of coherent Smith—Purcell radiation by the 5-MeV electron beam

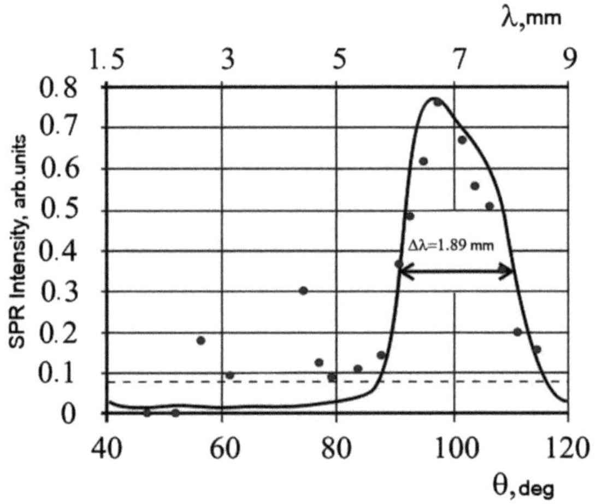

Fig. 9.23 Lineshape of coherent Smith—Purcell radiation (the points are the experimental data and the solid line is the simulation result)

Note satisfactory agreement between the experimental data and calculation. The yield of coherent Smith—Purcell radiation is shown in Fig. 9.24 as a function of the impact parameter h. The minimum at the point $h = 0$ corresponds to the passage of the beam through the target, which in this case is located on the beam axis. Beginning with a distance of $h = 2$ mm, the beam does not touch the target. The measured dependence in the impact parameter range $h = 2$–29 mm (after the subtraction of a constant background) is approximated by the exponential

$$\exp\left[-\frac{4\pi}{\gamma\bar{\lambda}}h\right], \tag{9.17}$$

where $\bar{\lambda} = 7.85$ mm, which is in good agreement with the average value of the sensitivity range of the narrowband detector.

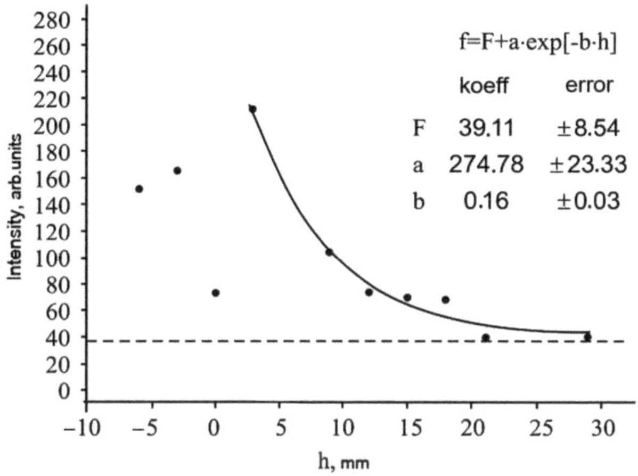

Fig. 9.24 Impact-parameter dependence of the yield of coherent Smith—Purcell radiation

One of the main results of the cited work is the detection of coherent Smith—Purcell radiation from the so-called plane grating for which the ratio μ of the strip height to its period is much smaller than one (in the experiment reported in [22], $\mu = 0.03\,\text{mm}/6\,\text{mm} = 0.005$, see Fig. 9.22). For such gratings, the Van den Berg model provides the radiation yield that is certainly lower than the detector sensitivity threshold [23, 24]. The yield of coherent Smith—Purcell radiation was estimated in [23, 24] for a plane grating consisting of thin strips separated by vacuum slits. In the described experiment, the strips were placed on a dielectric substrate. However, the radiation generation efficiency of such a grating is expected to be lower than that of a similar vacuum grating.

In the paper [25] authors suggested the generalization of the surface current method which allows to simulate Smith—Purcell radiation characteristics without traditional restrictions connected with a high Lorentz–factor, small photon angles, etc.

9.4 Some Prospects of Application of Diffraction Radiation

As mentioned above, optical diffraction radiation can be widely applied in the diagnostics of ultrarelativistic electron beams. At the Advanced Photon Source (APS) storage ring, the diagnostic station was created in 2005 for non-invasive measurements of the parameters of a 7-GeV test electron beam extracted from an injector synchrotron [26, 27]. The size of the beam and its position in the horizontal plane (orbit plane) was measured using not only optical synchrotron radiation traditionally used for diagnostics, but also optical diffraction radiation. The authors of the cited works showed that the coordinates of the center of the beam can be determined with an accuracy of $\sim 10\,\mu\text{m}$ using optical diffraction radiation and an optical scheme similar to that shown in Fig. 8.11.

The geometrical sizes of the electron beams $\sigma_{x,y}$ in moderate-energy accelerators ($E_e \leq 100$ MeV) are usually larger than the characteristic parameter $\gamma\lambda$. For this reason, diagnostics based on the transmission of an accelerated beam through a slit (or near a target with the impact parameter $\sigma_{x,y} < h < \gamma\lambda$) is inapplicable in this case.

Shkvarunets et al. [28] proposed an original diagnostic scheme in which a beam passes through a thin perforated target whose holes are much smaller than the beam diameter (e.g., a target with square holes with a side of 25 μm and a period of 33 μm corresponding to a "transparency" of 55% was considered in [28]).

In addition to optical diffraction radiation generated by the beam particles passing through the holes, transition radiation is generated in such a target by the particles interacting with the target material ($\sim$ 45% of the initial intensity of the beam). The characteristics of radiation generated by a single electron obviously depend on the coordinate of the point of the particle passage through the target. Therefore, the characteristics of radiation generated by the beam are determined by averaging over the transverse sizes of the beam. The authors of the cited work showed that the shape of the forward interference pattern from such a target with backward transition radiation from a mirror placed behind the target carries information on the angular divergence of the initial beam.

Fiorito et al. [29] experimentally demonstrated the capabilities of the proposed diagnostic method (they called it optical diffraction and transition radiation interferometry (ODTRI)) for measurements of the beam parameters of the ATF BNL ($E_e = 50$ MeV) and NPS ($E_e = 95$ MeV) accelerators with an accuracy of no worse than 5% (transverse sizes and divergence).

As was mentioned above the diffraction radiation is generated by charged particles with arbitrary masses, for instance, by protons. In the work [30] author considered the applicability of the diffraction radiation from a round hole for a diagnostics of the FNAl proton beam with energy 980 GeV and showed a possibility to measure a beam offset using detectors for an infrared range.

Blackmore et al. [31] implemented a scheme for determining the length of an electron bunch from the angular distribution of coherent Smith—Purcell radiation. In the cited work, 11 broadband detectors with sensitivity in a range of 0.5–3 mm were used. They were placed at various polar angles with respect to the trajectory of electrons passing over the grating. The results demonstrate the prospects of non-invasive diagnostics based on the use of coherent Smith—Purcell radiation to determine the shape and length of a submillimeter electron bunch.
The technique developed was used to reconstruct the time profile of the electron bunches with 28.5 GeV beam energy [31]. Authors of the cited paper concluded that the Smith-Purcell radiation may be employed to measure bunch lengths not only in the picosecond range, but in the femtosecond range too.

A promising field of the application of diffraction radiation can be the creation of the so-called broadband free electron laser (BFEL). The scheme of the BFEL was proposed in [32, 33] and was based on the use of a preliminarily bunched electron beam (i.e., a periodic sequence of short electron bunches). In traditional free electron lasers, the initial beam is modulated when it passes through an undulator and the period of the bunched beam coincides with the wavelength of generated radiation.

In the BFEL scheme, the spontaneous mechanism of coherent radiation (e.g., transition radiation) provides a continuous spectrum and the enhancement of radiation occurs due to the stimulation of radiation generated by the subsequent bunches during their passage through a cavity. The BFEL based on the coherent transition radiation mechanism was investigated in [33]. Choosing the length of the cavity whose input mirror is a target for the generation of transition radiation, Shibata et al. [33] experimentally obtained the enhancement of radiation in a comparatively wide wavelength range $\Delta\lambda \sim 0.5$ mm near the resonance wavelength $\lambda_{res} = 2.8$ mm for the electron beam with an energy of $E = 37$ MeV and a bunch length of ~ 6 mm. A cylindrical cavity with a diameter of 114 mm and a tunable distance between the mirrors near $L_{res} = 461$ mm (the distance between the bunches of the accelerator was equal to 230.6 mm) was used in the experiment. The experimental estimate of the quality factor of this cavity at a wavelength of $\lambda_{res} = 2.8$ mm was about 50. Nevertheless, the enhancement factor in the experiment at the resonance wavelength reached 14 (see Fig. 9.25).

It is possible to proposed a BFEL scheme based on diffraction radiation with the use of an open confocal cavity, where both mirrors have central holes through which the electron beam passes and generates forward coherent diffraction radiation in the input mirror and backward coherent diffraction radiation in the output mirror. In such mirror targets, first, coherent diffraction radiation is focused and, second, the quality factor of the open confocal cavity can reach $Q \sim 10^3$. The proposed scheme can provide the gain factor noticeably higher than values reached in the cited works.

In the recent experiment reported in [34, 35], the possibility of focusing optical transition radiation, which was generated by the 1.28-GeV ($\gamma = 2500$) electron beam passing through a hemispherical target with a radius of $R = 500$ mm, was demonstrated. The sizes of the target were 15×7.5 mm, which are much smaller than $\gamma^2\lambda$ (for $\lambda = 550$ nm), but are noticeably larger than $\gamma\lambda$.

Since the transverse sizes of the target are much smaller than the focal length, a spherical target can be used instead of a paraboloidal target.

Figures 9.26 and 9.27 show the angular distributions of optical transition radiation from the plane and spherical targets, respectively, which clearly demonstrate the focusing effect.

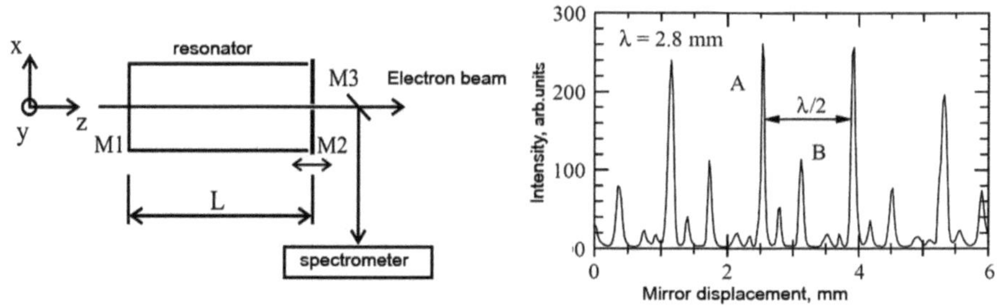

Fig. 9.25 (**a**) Scheme of a broadband free electron laser based on transition radiation and (**b**) cavity-length dependence of the yield of stimulated transition radiation at a wavelength of $\lambda = 2.8$ mm

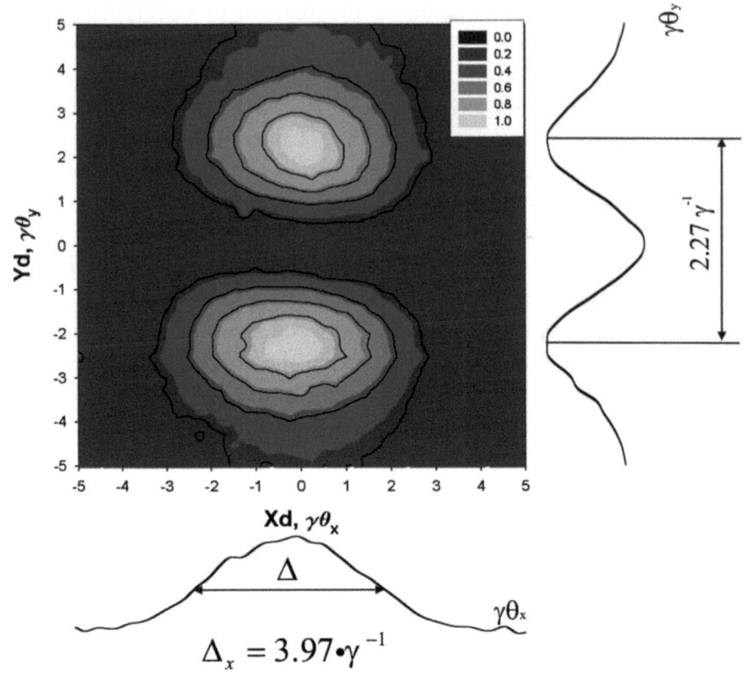

Fig. 9.26 Angular distribution of optical transition radiation from the plane target

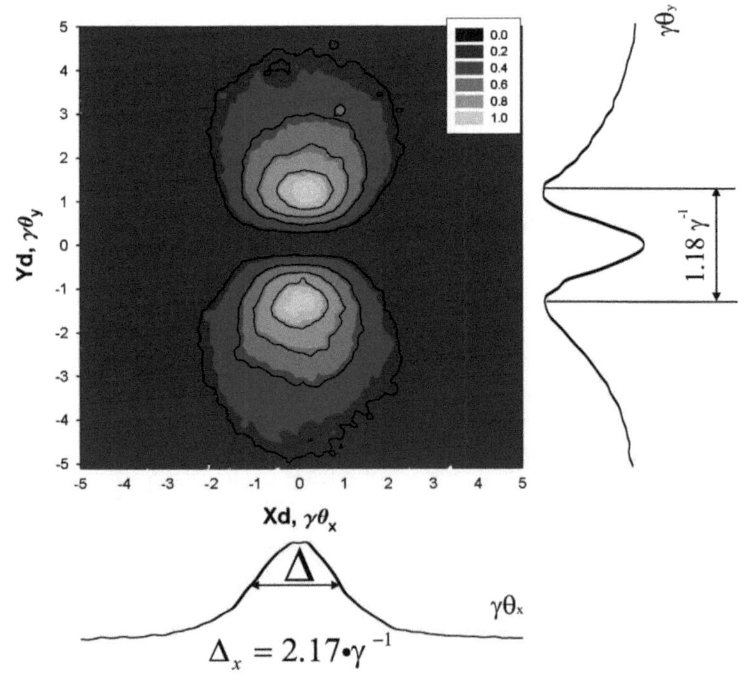

Fig. 9.27 Angular distribution of optical transition radiation from the focusing target

An alternative scheme of the free electron laser was proposed in [36–38]. In this scheme, spontaneous radiation is Smith—Purcell radiation, which is generated by a nonrelativistic electron beam passing over a grating with a period of several millimeters. Feedback is ensured by using a parallel optical grating located on the opposite side of the beam, of parameters which is determined by the beam energy. In other words, two diffraction gratings form a transverse cavity and ensure distributed feedback. Generation of radiation at a frequency of $\nu_0 = 54\,\text{GHz}$ ($\lambda_0 = 5.56\,\text{mm}$) was observed in the experiment reported in [36]; in this case, the frequency ν_0 was governed by the rotation of the grating that ensures feedback. In that experiment, the rotation of the grating by 5° about the axis perpendicular to the main grating plane gave rise to a radiation frequency shift of 0.7 GHz.

The above examples illustrate only some of the possible application fields of diffraction radiation.

As indicated in the first chapter, diffraction radiation can be considered as one of the types of polarization radiation. Types of polarization radiation such as Cherenkov radiation and transition radiation are widely used in modern physics. Cherenkov radiation is used in detectors of elementary particles and for creation of free electron lasers, whereas transition radiation is used in detectors of particles and for the diagnostics of beams in accelerators. It is expected that diffraction radiation will also be widely applied in modern physics.

References

1. Shibata, Y., Hasebe, S., Ishiki, K. et al. Observation of coherent diffraction radiation from bunched electrons passing through a circular aperture in the millimeter- and submillimeter-wave length regions. Phys. Rev. E **52**, 6737 (1995)
2. Dnestrovskiy, Yu.N., Kostomarov D.P.: Radiation of ultra relativistic charges moving across round aperture in flat screen. Sov. Phys. Dokl. **4**, 158 (1959)
3. Vnukov, I.E., Kalinin, B.N., Naumenko, G.A., Padalko, D.V., Chefonov O.V.: Experimental observation of optical diffraction radiation. JETP Lett. **67**, 760 (1998)
4. Urakawa, J., Hayano, H., Kubo, K. et al.: Feasibility of optical diffraction radiation for a non-invasive low-emittance beam diagnostics. Nucl. Instrum. Methods Phys. Res. A **472**, 309 (2001)
5. Muto, M., Araki, S., Hamatsu, R. et al.: Observation of incoherent diffraction radiation from a single-edge target in the light region. Phys. Rev. Lett. **90**, 104801 (2003)
6. Kubo, K, Akemoto, M., Anderson, S. et al.: Extremely low vertical emittance beam in accelerator test facility at KEK. Phys. Rev. Lett. **88**, 194801 (2002)
7. Potylitsyn, P.: Transition radiation and diffraction radiation. Similarities and differences. Nucl. Instrum. Methods Phys. Res. B **145**, 169 (1998)
8. Karataev, P., Araki, S., Hamatsu R. et al.: Status of optical diffraction radiation experiment at KEK ATF extraction line. Nucl. Instrum. Methods Phys. Res. B,**201**, 140 (2003)
9. Karataev, P.V.: Investigation of Optical Diffraction Radiation for Non-Invasive Low-Emittance Beam Size Diagnostics. PhD Thesis, Tokyo Metropolitan University (2004)
10. Karataev, P.V., Araki, S., Hamatsu, R. et al.: Beam-size measurement with optical diffraction radiation at KEK accelerator test facility. Phys. Rev. Lett. **93**, 244802 (2004)
11. Karataev, P., Araki, S., Hamatsu, R. et al.: Observation of optical diffraction radiation from a slit target at KEK accelerator test facility. Nucl. Instrum. Methods Phys. Res. B **227**, 158 (2005)

12. Karataev, P.V., Hayano, H., Muto, T. et al.: Development of optical diffraction radiation beam size diagnostics at KEK accelerator test facility. Proceeding of EPAC 2004, Lucerne, Switzerland, p. 2643 (2004)
13. Shestopalov, V.P.: The Smith-Purcell effect. Nova Science Publishers, Commack, NY (1998)
14. Urata, J., Goldstein, M., Kimmitt, M.F., Naumov, A., Platt, C., Walsh, J.E.: Superradiant Smith-Purcell emission. Phys. Rev. Lett. **80**, 516 (1998)
15. Doucas, G., Mulvey, J.H., Omori, M. et al.: First observation of Smith-Purcell radiation from relativistic electrons. Phys. Rev. Lett. **69**, 1761 (1992)
16. Woods, K.J., Walsh, J.E., Kirk, H.G., Fernow, R.C.: Forward Directed Smith-Purcell Radiation from Relativistic Electrons. Phys. Rev. Lett. **74**, 3808, (1995)
17. Van den Berg, P.M.: Smith-Purcell radiation from a point charge moving parallel to a reflection grating. J. Opt. Soc. Am. **63**, 1588 (1973)
18. Kube, G., Backe, H., Euteneuer, H. et al.: Observation of optical Smith- Purcell radiation at an electron beam energy of 855 MeV. Phys. Rev. E **65**, 056501 (2002)
19. Trotz, S.R., Brownell, J.H., Walsh, J.E., Doucas, G.: Optimization of Smith-Purcell radiation at very high energies. Phys. Rev. E **61**, 7057 (2000)
20. Nguyen, D.C.: Coherent Smith-Purcell radiation as a diagnostic for subpicosecond electron bunch length. Nucl. Instrum. Methods Phys. Res. A **393**, 514 (1997)
21. Lampel, M.C.: Coherent Smith-Purcell radiation for use in electron beam diagnostics. Nucl. Instrum. Methods Phys. Res. A **395**, 19 (1997)
22. Aryshev, A., Kalinin, B., Naumenko, G. et al.: Experimental investigation of coherent Smith-Purcell radiation from a "flat" grating. Nucl. Instrum. Methods Phys. Res. B **227**, 175 (2005)
23. Kube, G.: Comparison of Smith-Purcell radiation models and criteria for their vertification. Nucl. Instrum. Methods Phys. Res. B **227**, 180 (2005)
24. Karlovets, D.V., Potylitsyn, A.P.: Calculation of Smith-Purcell radiation from a volume strip grating. Phys. Rev. ST-AB **9**, 080701 (2006)
25. Karlovets, D.V., Potylitsyn, A.P.: Generalized surface current method in the macroscopic theory of diffraction radiation. Phys. Lett. A **373**(22), 1988 (2009)
26. Lumpkin, A.H.: Nonintercepting diagnostics for transverse beam parameters: From rings to ERLs. Nucl. Instrum. Methods Phys. Res. A **557**, 318 (2006)
27. Lumpkin, A.H., Berg, W.J., Sereno, N.S. et al.: Near-field imaging of optical diffraction radiation generated by a 7-GeV electron beam. Phys. Rev. ST-AB **10**, 022802 (2007)
28. Shkvarunets, A.G., Fiorito, R.B., O'Shea, P.G.: Optical diffraction-transition interferometry and its application to the measurement of beam divergence. Nucl. Instrum. Methods Phys. Res. B **201**, 153 (2003)
29. Fiorito, R.B., Shkvarunets, A.G., Watanabe, T. et al.: Interference of diffraction and transition radiation and its application as a beam divergence diagnostic. Phys. Rev. ST-AB **9**, 052802, (2006)
30. Sen, T.: Optical diffraction radiation from a beam off a circular target. arXiV:0808.3962 (2008)
31. Blackmore, V., Doucas, G., Perry, C. et al.: First measurements of the longitudinal bunch profile of a 28.5 GeV beam using coherent Smith-Purcell radiation. Phys. Rev. ST-AB **12**, 032803 (2009)
32. Shibata, Y., Ishi, K., Takahashi, T. et al.: A prebunched FEL using coherent transition radiation in the millimeter wave region. Nucl. Instrum. Meth. Phys. Res. B **145**, 49 (1998)
33. Shibata, Y., Ishi, K., Ono, S. et al.: Prebunched free electron laser with a broadband spectrum. Nucl. Instrum. Methods Phys. Res. A **528**, 162 (2004)
34. Sukhikh, L.G., Naumenko, G.A., Potylitsyn, A.P. et al.: Focusing of optical transition and diffraction radiation by a spherical target. Proceedings DIPAC07, Italy (2007)
35. Sukhikh, L.G., Aryshev, A.S., Karataev, P.V., Naumenko, G.A., Potylitsyn, A.P., Terunuma, N., Urakawa, J.: Observation of focusing effect in optical transition and diffraction radiation generated from a spherical target. Phys. Rev. ST Accel. Beams **12**(7), 071001 (2009)
36. Baryshevsky, V., Batrakov, K., Gurinovich, A. et al.: First lasing of a volume FEL (VFEL) at a wavelength range $\lambda \sim 4$–6 mm. Nucl. Instrum. Methods Phys. Res. A **483**, 21 (2002)

37. Baryshevsky, V.G., Batrakov, K.G., Gurinovich, A.A. et al.: Progress of the volume FEL (VFEL) experiments in millimeter range. Nucl. Instrum. Methods Phys. Res. A **507**, 137 (2003)
38. Aryshev, A., Araki, A., Fukuda, M. et al.: Development of microwave and soft X-ray sources based on coherent radiation and Thomson scattering. J. Phys. Conf. Ser. **236**, 012009 (2010)

MIX
Papier aus verantwortungsvollen Quellen
Paper from responsible sources
FSC® C105338
FSC
www.fsc.org

If you have any concerns about our products,
you can contact us on
ProductSafety@springernature.com

In case Publisher is established outside the EU,
the EU authorized representative is:
Springer Nature Customer Service Center GmbH
Europaplatz 3, 69115 Heidelberg, Germany

Printed by Libri Plureos GmbH
in Hamburg, Germany